DUALITIES IN HEAVEN AND EARTH

A NEW LOOK AT GOD AND CREATION

By

ZOLTAN KISS

Copyright © 2019 Zoltan Kiss
All rights reserved
First Edition

PAGE PUBLISHING, INC.
New York, NY

First originally published by Page Publishing, Inc. 2019

ISBN 978-1-64350-652-4 (Paperback)
ISBN 978-1-64350-653-1 (Digital)

Printed in the United States of America

CONTENTS

Introduction 5

1. One Friend Missing from the Poker Party 9
2. Doug Recounts His Discussion with Peter about Religion and Science 49
3. More Discussion of Religion, Science, and Evolution at the Poker Meeting 67
4. Doug Couldn't Stop Reliving His Conversation with the German Guests 114
5. Helga and Ulrich Make Big Decisions 120
6. A Tragic Event Occurred 146
7. Suspected Causes of Peter's Death 169
8. Zinc: A New Suspected Killer Substance 188
9. Zinc as Double-Edged Sword and the Possible Case for an Accomplice Substance 194
10. A Busy Saturday for Doug 205
11. The Funeral and a New Twist in the Mystery of Peter's Death 208
12. Steve Might Have Found the Accomplice Substance 219
13. Hunting for the Accomplice 226
14. Were Helga and Ulrich Meant for Each Other After All? 248
15. Events Leading to Peter's Death 279
16. More Findings, More Analysis 325
17. Agnes and Helga May Get Off the List of Suspects 336
18. Doug and Atheist-Turned George Discuss God and Multiple Universes 350

19. Carl's Report 361
20. A Sudden Turn in Doug's Life 365
21. Agnes and Doug Discover Each Other 373
22. Ulrich Takes Doug's Job 450
23. Doug's Last Struggle, a New Beginning for Agnes 454
24. Agnes Got New Friends..457
25. Agnes's Big Decision..464

Main Literature 477
Acknowledgments 485

PART 1

INTRODUCTION

This book is a mixture of fictional and real elements. It is pure fiction in the sense that neither the characters nor their interactions are real. It can also be regarded as mostly fiction, when the main characters talk about, for example, God's role in Creation and our lives. On the other hand, all the science presented is based on published data in peer-reviewed journals or patents. There are many other real issues also included that people in the medical profession, philosophers, astrophysicists, people of faith, and people from many other segments of society grapple with day after day.

However, because each character in this book is entirely fictional, the often outlandish, and on the surface even childish, views and theories presented by some of them will hopefully not offend anyone. At the same time, even though these ideas are presented under the cover of science fiction, I hope that some of them will nevertheless serve to stimulate some out-of-the-box thinking about Creation and the nature of a possible supernatural being. I hope that some of the readers will pause for a minute or two and consider, What if there is a grain of truth in what these characters say about, for example, how certain precreation events might have led to the big bang and, if a god had been involved, why He chose evolution instead of frequent interventions as the driving force to create the universe and eventually humankind? No doubt many readers, both in the atheist and religious camps, will consider the main characters' ideas as vicious, skewering of their world view only befuddled minds can come up with; I hope that they will at least enjoy the ideology-free chapters of the story.

A centrally important question that the main characters grapple with is whether faith, the belief in a supernatural being that is much greater than we are, can ever be reconciled with science. One of the main characters, Doug Lowry, argues that this is possible, but only if the faithful start thinking about that supernatural being, or God, very differently. He introduces the concept of a "material god," who can interact with humankind via material, not spiritual, means and who, after setting up the initial rules and conditions, allowed evolution to completely take charge in directing the development of the universe, including life on Earth. Once Doug made this conceptual leap, he declares that "science is God's science, and therefore science can never contradict Him or produce anything that could be held against Him or His followers." Doug argues that such a conceptual leap should be made not only because it makes sense but also because otherwise the faithful will eventually lose ground to nonbelievers. As he sees it, all religions, including the Catholic Church, are walking on thinning ice, and the faithful have no choice other than to accept science and use its methods to find an answer to the mystery of God's nature and His relationship with humankind. Essentially, he urges the faithful not to be afraid of fully embracing good science. Be intransigent when the essentials of your faith are attacked but also be very flexible to accept well established facts of science, Doug urges the Christian churches, which may apply to other religions as well. Doug also argues that this will take some "cleaning of the clutter" in the church's teaching, something easier said than done.

As things stand now, it may appear that atheists can claim that *reason* is on their side. However, according to Doug, this is only how it appears on the surface; many of the atheists' arguments hold no water for him. The deeper he digs, the more convinced he becomes that atheism, which in his high school years he embraced so enthusiastically, cannot be the full explanation for the world. He concludes that for the big questions, like the apparent creation of the universe from nothing, how living organisms developed from the inorganic material, how it is possible that our mind (as an observer of the quantum world) can influence quantum events, and what is consciousness, atheism does not and cannot provide answers; however, materialism might. Here Doug, who clamors for answers, makes another conceptual leap; he declares that atheism and materialism must be viewed as two different concepts. In contrast to what many people believe, materialism does not a priori exclude the existence of God; only atheism does. Sticking to this idea, Doug attempts to explain God's nature through purely materialistic means and endows God's apparent "spiritual" existence with a new materialistic meaning. He and his wife, Agnes, try to strike a balance between the dualistic, material and spiritual, sides of God.

To fill the gap between God's materialistic nature and the energy and matter forms of our universe, Doug suggests the existence of a new entity, *information*. In his view, "spiritual God" is the totality of information, while "material God" is a very highly organized information, the two forms of God existing in unity, thus forming the greatest duality of all. Following this line of reasoning, Doug further suggests that material God was the one who created the "speck" from units of information that contained the instructions which types of elementary particles should be formed as well as which kind of forces and natural laws should be established right after the big bang. Doug assumes that information, which comes in variable units like the three-letter codes in the DNA, is an entirely new entity that is the "mother" of all known energy and matter forms present in the universe. Doug thinks that information suffuses space and is still involved in new particle formation, even if temporarily. As an example, new particles, popping out of vacuum and then disappearing within a fraction of a second, borrow energy from information for their existence, and thus the law of energy conservation is not broken even for a short period of time, which is contrary to what physicists think today.

On another level, this book also provides examples of how lives can be devastated, or redeemed, when the unknown and unknowable emotions residing in all of us come to the surface. Are these unknowable emotions still part of our consciousness? What triggers them? Doug, like all of us, struggles with these questions, and yet they remain a mystery for him forever, perhaps for the better.

The reader might wonder if I fully identify with what the characters say, and my short answer is, not necessarily. However, as an observer who believes in the existence of a higher authority, I attempt to remain rational with keeping my mind open to all the possible options that the main characters offer. I would not like to close my mind and miss out on new ideas about God, Creation, the soul, and consciousness, however impossible they might sound at first. I believe the main characters raise intriguing questions that someday scientists, philosophers, and theologians might be able to answer, with a combined effort, one way or the other, without diminishing the roles of religions and science in explaining how our world and the universe work from the very beginning.

CHAPTER 1

ONE FRIEND MISSING FROM THE POKER PARTY

The year 2011 was an exciting political year leading up to the presidential election the following year, but on that late summer Monday evening, close to seven o'clock, three friends only wanted to play poker and chat a little bit. They had been doing this every Monday evening for years, not because they were so much in thrall to the poker, but it proved to be a good stress reliever. Now they were agog over waiting for the fourth friend to show up.

They all lived in Rochester, a rapidly growing southeastern Minnesota town that is also known as Mayo town due to the presence of the world-famous Mayo Clinic. In the late eighties and early nineties, they all played football at the same local high school and had kept their friendship going ever since. They attended different universities spread across the country, but by 2006, one by one they each returned to Rochester. There were many reasons they all wished to return, but one of them stood out from the rest; despite the recent growth of the city that had changed the earlier somewhat cloistered enclave-like atmosphere, it still retained the charm and safety of a small town. Their parents and relatives used to say that Rochester was a wonderful city in which to raise children, and it was true. The city had all the needed amenities in abundance, the schools and health care were of exceptionally high quality, well-paying jobs were widely available, and crime rates were low. Even the recession starting around 2007 had relatively little impact on the lives of most people. Sure, predatory lending by certain big banks selling mortgages on false pretenses had caused financial troubles to some, and the overall economic situation was more brittle and fraught with more dangers than before. But by 2011, things were beginning to look brighter again.

    The first person to return to Rochester, back in 2002, was Mike Collins, a music teacher. His friends thought of him as a man of poise. He was an accomplished pianist playing concerts all over Minnesota and beyond. He also had a college degree in music and was fortunate enough to have found a position as a music teacher at his old high school. Unlike his friends who were born in Rochester, he had moved there when he was eight years old from an area in Oklahoma, called Tornado Alley. He still had a hint of the characteristic Oklahoman accent that after all those years was still recognizable. He used to tell the story of a massive F5 tornado that hit the edge of his town. His house wasn't in the direct path of the tornado, yet it was seriously damaged by a roof girder that the tempest tore from a pulverized nearby house and flung it like an artillery shell. During the minute-long rambling and ensuing chaos, they cowered in the basement and remained unscathed, but six people died on that occasion, including an infant. Mike never figured out whether it was because of the frequency of these tornadoes or something else, but soon after this incident, his father quit his job and moved his family to Rochester. Ironically, shortly after their move, a tornado ripped through the area just one mile away from their new home, splintering trees and snapping utility poles like small twigs in its sway. This time it did not damage their house but riddled the nearby gridiron, where he used to play with friends, with timber and goalposts torn asunder. Realizing that tornados and bad weather can occur at almost any part of the country, they shrugged off the incident and stayed put.

    In his spare time, Mike had composed a three-hour opera, but no theater had showed any interest in it so far. He had been scrupulous in his musical creation, and some directors who turned down his offer to stage the opera acknowledged that this came through in his work; however, exactly because of that, it missed a certain level of freedom and experimentation. Some others suggested that the internal harmony of the piece was ragged, perhaps because it was too long, and he had to work on it more to make it more flexible. One reviewer was especially hard on him; he suggested to Mike without prevarication that he should start entirely from scratch and stop fiddling around with the version he had. But on the other side of the coin, as one of his friends knowledgeable in art matters explained to him, was the fact that major established opera houses seemingly gave preference to more classical pieces, in part for box office reasons.

Mike was initially frustrated and hamstrung by the lack of interest and understanding, but in recent months, cognizant of the criticism and the fact that there are always some theaters that like to show new pieces, he resumed working on an abridged version. He didn't feel at all that his previous effort was wasted and that he should give it up; in contrast, he was elated when he found some time to work on the piece, even if it was just scrawling a few notes on a piece of paper. As he gradually made progress with the revision, he began to regain confidence that his opera would assuage previous objections and finally find a taker. But unlike before, he was now very secretive about his progress with his friends and even his wife, and they really had no idea where he was with the revision. All they saw was that his ambition once again gnawed at him like a wolf; his previously sullen face had changed to one glowing with determination.

Mike and his wife, Nancy, had had a decorous relationship for several years before they finally decided to tie the knot. She had formal training in visual arts and had mastered oil painting during the past few years, but now she stayed at home with their two small children. However, she had remained part of Rochester's art scene, which just recently received a boost in the form of a new art center. Although she consciously made the choice to stay home, from time to time she daydreamed about renting a studio and starting to paint again. She sometimes felt a sort of guilt at the thought that she was losing the desire, or rather the necessary will, to further experiment with painting. On those occasions, she asked herself if the children were only subterfuges to cover her mental laziness. Realistically, since she worked with an oil medium that had a characteristic and unpleasant smell, she could not possibly paint at home with the small children around. The children already generated enough headaches for Nancy and Mike without the paint's odor. So one day she woke up begrudging those who had the opportunity to paint; the next day she was happy that God blessed her with children.

It is also a fact of life that creating quality art and developing an idiosyncratic style require full concentration and being able to work unfettered. Having two noisy and squirmy small children requiring constant attention is such a drain on one's mental energy that not much of it is left for anything else that requires creativity or critical thinking. Perhaps a little goading from Mike would have helped, but he was not the right person to do that. Thus, Nancy's dream of owning a studio would have to wait for at least a couple of years, until her children started school, which would free up a good portion of the day. After thinking through her options for the next couple of years, her twinge of conscience usually came to a rest; she was after all making the right choice for the time being. But she was careful not to lose her artistic dexterity, and she sketched little drawings whenever she had some time for it.

Steven Milken, the quarterback in his high school team, was raised by parents who were financially not well-off, but they did everything in their power to give him a good education so he could move up the ladder of life. It worked; he succeeded beyond the wildest dreams of his parents. Six months back he defended his PhD in a particularly interesting area of medicinal chemistry. His thesis focused on patients who were suspected of being treated inadvertently or on purpose with excessive drug doses or who had received multiple medications resulting in unforeseen drug interactions that wreaked havoc on their bodies. These relatively understudied and often underreported problems in fact occur much more frequently than the public is aware of. These "mistreatments" are due to a combination of human errors, the unpredictability of the human body's response to drugs, or sometimes even foul play.

Taking multiple medications is always fraught with risk, and it is often difficult to find the reason why. As he put it in his dissertation, Americans use about six thousand medicinal drugs, and many patients use multiple ones at once in various combinations that can yield many, often unknown, cross reactions showing up as side effects. Due to these cross reactions, the efforts to effectively treat the primary disease of major concern may come to naught.

It is difficult to keep track of all these drugs, particularly if the patient also takes some over-the-counter drugs that he or she fails to inform the doctor about. It is perhaps not surprising, as Steve learned after completing his dissertation, that in Minnesota's nursing homes, in 2007 alone, death from suspected medication errors occurred in 199 cases. When misuse of medication could be proven, it turned out that often the nurses were nonchalant, or in a few cases, even deliberately poisoned the patients who were in their care. However, in most cases when serious side effects occurred, Steven would argue that the medication was given as prescribed by the doctors, but the patients just didn't respond in the way they were expected to. The patients may have simultaneously taken a different medication as well or could have some unresolved inflammation or were dehydrated at the time when the pills were taken, which would alter the drug's distribution in the body. Also, many of these situations arise when the patient has a serious but unrecognized underlying disease that makes the drugs' actions and interactions with other drugs unpredictable. Age is a special factor too, because older people do not absorb and/or tolerate drugs as well as young people. And then some drugs are simply incongruous with others. In his thesis, Steven discussed all these situations, backing them up with a large amount of observations and data compiled from the records of several hospitals and nursing homes. He advanced the idea, among others, that many older people are simply unnecessarily "overly drugged." No wonder, many doctors drew from his work, helping them to navigate through the jungle of drugs. When doctors sought his advice, he had a simple message: less is often more, meaning that prescribing fewer medications may provide more benefit to the patient particularly if the patient is old.

From the summer of 2003, Steve worked primarily as a pharmacist in the St. Mary's hospital in Rochester, the flagship hospital of the Mayo's health system. He was very much liked and highly regarded. But even if doctors often turned to him for advice in complicated medical cases, which gave him some satisfaction, he felt that in his present job, he was stuck in a minor role with no room to grow, and that made him a little nervous about his future. He wasn't craven to look for a change, and for years he had constantly been planning to develop his own business in the health-care area without making much progress; this was not, however, for lack of trying. Sometimes he found himself wondering if his attempts were just too quixotic, or if he was simply unlucky in trying to find the right partners. One problem that he had, common to scientists trying to develop businesses, was that without previous business experience, it had been very difficult to prove his mettle, which is essential in the business world. And for potential investors, this was always a valid concern that he could not shake off. In his spare time, he often thought of how to solve this problem, but so far he had found no solution except to go to a business school for which he had no time. Unfortunately, another option, to find investors through mere serendipity, is virtually impossible for start-up companies. So the idea of starting a business was beckoning and forbidding at the same time, at least for the time being. On the bright side, his primary job was stable, and although he was stuck in the quandary of how to go about building a business, he at least had a steady flow of income. Thus, after each unsuccessful attempt to raise capital, he could afford to hold his breath and keep the idea of developing a company in limbo. Now he was again in a period of knuckling down to search for a wealthy partner with a better record in the business, hoping that the propitious moment will come soon.

Steve's wife, Elisabeth, worked part-time in an accounting office, and she was halfway through her first pregnancy. She was at the point when the everyday grind had begun to take a toll on her, and she would have been ready to negotiate for a leave of absence with her employer. Unfortunately, her immediate boss was quite an arrogant person who apparently did not like expectant women in the office. When an opportunity arose, he never failed to drop a hint to Elisabeth that she was dispensable; and this did not bode well for her future at that firm. It seemed that Elisabeth had two choices:

either stay on the job and somehow manage to ignore her boss without grousing about the situation, or quit. For the time being, she just hunkered down in her office without griping and opted not to relinquish her position. But, as her boss already implied, she couldn't expect to return to her job after her maternity leave was over, and she also decided that she would not try to contest the matter. Sooner or later, her boss would have found a subterfuge to fire her.

Doug Lowry had graduated from Harvard Medical School and started to work there while specializing in pathology and toxicology. He had had a long-standing interest in treating patients who have more than one medical condition. It is a real nightmare for a clinician to treat, for example, a diabetic cancer patient who may even have additional problems like cardiac disease. All these conditions would require separate treatments with drugs that may interact with each other, making the outcome often quite unpredictable. Not infrequently and often unbeknownst to doctors, patients may also deaden their senses and immune systems with drugs, unauthorized alternative medicine, or alcohol, further complicating the treatment plan. The common interest of Doug and Steve in these vexing problems of toxicology, leading to day-to-day contact between them, was an additional strong force that kept their friendship alive. They always had time for each other. One day Doug would call Steve and ask if he knew how to prevent the cancer drug cisplatin from further deteriorating diabetic neuropathy. Then some other day, Steve would seek Doug's advice on what he thought about recent concerns that the antidiabetic drug insulin may promote tumor growth, and so on. In a way, both had highly strung jobs, so it was no wonder that that their weekly meetings helped to ease the tensions.

In the spring of 2005, Doug accepted a pathologist position at the Mayo Clinic and was placed in an office with a spacious laboratory in the building of St. Mary Hospital. Part of his deal with the Mayo Clinic was that he had to provide laboratory services to the various departments. In exchange, he received a start-up fund and a decent yearly budget that was sufficient to pay part of the salaries of his two assistants and purchase diagnostic kits. Soon after his arrival, Doug also started taking special courses in nutrition to keep up with new developments in that field. Toxicology, pathology, and nutrition are all intertwined on several levels. For example, as Doug learned, antioxidants and rare elements like zinc and selenium are present in many fruits and vegetables, and they may affect, for better or for worse, the metabolism and effectiveness of certain drugs. Some food components and the acidity of fruits and juices can also modulate the amount of drug absorbed by the diseased tissue. For example, grapefruit juice interferes with the action of certain types of blood pressure drugs. There are suddenly so many issues to deal with when a person starts taking multiple medications at once that almost every day doctors have a new battle to fight. Still, from any viewpoint, Doug had a sublime job that he loved and would not forsake for anything.

Doug was much revered by the people around him. He had a benevolent, unflappable, and always reassuring nature that was widely respected by everyone who knew him. Whenever there were little squabbles among his assistants and colleagues, almost unavoidable at any workplace, Doug was famously able to keep his calm and smooth out the situation. His ability to provide calm, stemming from his natural altruism, was accompanied by an unflagging work ethic, which was occasionally abused by others at the clinic because everybody knew that he wouldn't say no if something was requested of him and that he wouldn't renege on his promise to help out. He usually worked with two technicians, who had to be experts at handling the sophisticated equipment that they used for the analysis of pathological samples as well as drugs, natural substances, potential poisons, and the thousands of components of a human body, such as proteins, lipids, carbohydrates, lipids, DNA, and minerals. To keep up with the demand, he sometimes had to prod his assistants to work a little harder and for longer hours, which they did without any complaint. But even during high-pressure times, he remained even-keeled in front of his associates and kept his occasional struggles internalized.

Five years ago, Doug had received a grant from the National Institutes of Health, widely known as NIH. The grant enabled him to hire a postdoctoral fellow whose duty was to determine how a certain blood component might affect the distribution and effectiveness of a series of anticancer drugs. But not long ago, he had to fire that person because his grant was not renewed after the first round of the review process. It was an anguished moment in his life, which tested his tenacity to remain in science. Admittedly, part of the problem was that his postdoc's initial enthusiasm lately gradually ebbed, and he wasn't very dedicated to his work, although he had incredible talent. Doug explained to him many times that in a laboratory, a brilliant brain isn't sufficient; it must be paired with hard work at the bench. Such warnings made no difference, and now there was a real danger that some other laboratories will upstage him.

Over the years he spent a lot of time and much of the grant money that he received from various companies as well as the NIH and the Mayo Clinic to equip his laboratory with the most essential and at the same time most sophisticated analytical tools. His armory of equipment was equally suitable to analyze both large and complex molecules as well as smaller ones; thus, he never had to ask others for help. It is no wonder that Doug was very proud of the quite impressive analytical power of his laboratory. "If only I had more people in the lab," he often lamented, "I could take on more projects and thus use this equipment a lot more efficiently." Equally important is the fact that by producing high-quality research, for which he had the equipment but not the dedicated manpower in the lab, he could publish his papers in first-rate scientific journals, which was the gateway for gaining more research support. All in all, at least for the time being, his high-value analytical tools were of little avail to him in chasing grants. But he wasn't bereft of hope.

These days, it is almost a platitude for most scientists doing research full-time that merely staying in the business is an endless battle, involving much more practicality than passion. A scientist's life can easily get into a groove circling around grant writing, rejection, revision, another rejection, and perhaps, only perhaps, acceptance of the grant with reductions of the requested years and budget. However, this is not true for all scientists. He had observed that for some of them, NIH acts like a big smorgasbord; they just go to the table and easily gather what they need and often more than that.

Unfortunately, Doug was not lucky with grant agencies in fulfilling his dream to populate his lab with scientists, and this, with time, made him feel a sort of depressing insufficiency. Despite his tenacity both in his research and grant writing, he had received only one major grant but had not been able to obtain any others. Evidently, if he wished to get closer to the smorgasbord, he had to do more than just working and submitting grant proposals. When he analyzed himself for the reasons of his many failures, he became aware that he lacked some personal characteristics necessary to fare better in shepherding through the grant award process. Above all, in addition to doing frontline research, some level of audacity and boldness as well as seeking supporting friends and providing more powerful arguments for his case had to be added to his "toolbox" and more.

Doug was an introvert by nature, and making social contacts with people at scientific meetings was quite taxing for him. He avoided the swanky parties often held for scientists before the start of the main event. That was unfortunate because those gatherings are a means for scientists to network with grant providers and potential reviewers and thus are an integral part of a scientist's career. Not participating in such events could put a scientist's future in mortal peril, leaving the grant smorgasbord out of reach. He had also made another major mistake; he thought that the quality of his work was far more important than the volume of his voice or trying to insinuate himself into the favor of somebody close to the decision-making about grants. The harsh reality is that the merit of a proposal alone is rarely sufficient to convince the granting agencies. Successful scientists also need to have the mastery of politics and know how to manipulate people to their advantage. Most grant reviewers are not impervious to the applicant, but knowing each other helps to give some assurance that the preliminary work presented and the proposal built on it can be trusted. To his credit, Doug was beginning to confront all these facts, and with his intellectual acuity, he was fast learning how to address them. So, after his last failure, which left him relatively unruffled, he once again prepared himself to start anew his perennial struggle for grants, but this time trying to use his new toolbox to full effect.

On the positive side, despite being an introvert, his well-equipped laboratory served as a magnet to draw collaborators, although this was usually not accompanied with significant money or long-lasting collaborations. His collaborators would ask him to do some analytical work within the realm of toxicology, and he did not need a lot of suasion to agree. For example, a recent collaboration that was related to his own research involved the determination of the binding of a prospective anticancer agent to a certain blood protein. This protein was notorious for its promiscuous behavior. Others showed that it interacts with many drugs and drug candidates, diminishing their actions, because drugs act efficiently only if they can freely access their target tissue, which can be a tumor or any other diseased tissue. How much free drug is left depends on how much of this ravenous protein is present in the circulation and how strongly the drug binds to it. As it turned out, the protein indeed sucked up most of the drug, making it far less effective than it should have been. Then Doug came up with an idea for modifying the drug to make it less attractive for the protein and thus more effective on the tumor. A drug company could carry out this procedure, but first he wanted to patent the idea and the actual structure of the new compound.

When Doug raised the issue of filing a patent application with his collaborators from Columbia University, he received enthusiastic support and money from the Mayo Clinic. He was ready with the application and planned to file it with the US patent office as soon as the Mayo and Columbia agreed how to share the costs and potential profits. Considering a 5 percent royalty for at least ten years and a potentially large market, obviously a lot of money was at stake if the method worked well. As big universities and health centers began to feel the recession, forcing them to squeeze their budgets, they couldn't afford to risk losing any kind of potential income anymore. A settlement was eventually made, and the patent had been filed more than a year ago. An answer from the US patent office was expected within two years, although they had a serious backlog of applications under review. Of course, the application had to pass muster of the patent office, which is always a big *if*.

Doug considered himself a happily married man; he really relished his marriage. But he held the somewhat naive belief that all that is needed to maintain a stable marriage is full commitment. When he and his wife, Sarah, made the decision to marry, for him this meant that the two of them would be together forever, and he never imagined anything could ever come between them. Because of this belief, which he believed was an incontrovertible truth, from that point on, he had perhaps devoted less energy than he should have into keeping his marriage fresh and exciting; he simply didn't discern the importance of it. In that respect, he was certainly indolent. He had heard eerie stories about seemingly happily married couples parting ways after twenty, thirty, or even forty years of marriage, but he never considered that such a thing that had brought so much misery to so many couples could ever happen to them. Their two beautiful boys, Roy and Brian, six and four years old, were also very strong linchpins between him and Sarah, he thought.

Sarah had attended the same high school that Doug and his friends did. It was an interesting and fortuitous event that eventually led to Sarah and Doug becoming friends and eventually lovers. At the beginning of her senior year, Sarah was still dating the fourth friend at the poker party, Peter Cartwright, a slender boy who was rather shy with girls. At first Sarah was attracted to Peter very much not only because of his physical appearance but also because he acted so innocent and candid. But as time went on, Sarah decided that Peter wasn't sufficiently mature for her, although for a while she couldn't find an appropriate occasion to explain the situation to him. At that time Peter was uncertain about what he wanted to do with his life, and that sent an alarming signal to Sarah; she became apprehensive of the future she imagined with him. Perhaps Peter also lacked a certain level of sophistication that would have been needed to keep her interested. But Peter was very much in love and didn't understand why Sarah often behaved queer and turned up her nose without apparent reason. With more experience, he should have known that Sarah was just simply looking for a least hurtful subterfuge to end their relationship.

Then, two months into the school year, Sarah and Doug participated in an educational bus trip to Washington, DC. Peter was not there, so Sarah sat beside Doug. In retrospective, they could not figure out if sitting together happened by chance or by design due to an early, perhaps even unconscious, attraction between them. During the long trip, Doug and Sarah were completely relaxed and they got to know each other so well and so favorably that on their return, they made sure they again got to sit beside each other. By the time they arrived back to Rochester, Sarah and Doug had become very close friends, and their relationship soon developed into a romantic one. Sarah felt that she found in Doug a man of grit who possessed the level of sophistication she was looking for and would never leave her in the lurch. For her, sophistication and articulation of thoughts in a man were sure signs that he knew his goals and would find the means to achieve them. That is, he would provide the security that is so important for a family with children. With a man like Doug, life would be a cakewalk, she thought. Sometimes, however, Sarah found herself vaguely wondering if she indeed found the same romanticism within Doug that she experienced with Peter. But at this stage of her life, other considerations swept aside her worries. Doug was worried that the news of him and Sarah becoming romantic partners would lose him Peter's friendship, but Sarah said they should not have felt any remorse and that she would talk to Peter. After their return from Washington, she told Peter deftly but without any prevarication what had happened between her and Doug during the trip. Although at first Peter was understandably hurt and eyed the two of them jealously, he never accused Doug of "stealing" Sarah from him. Admittedly, he felt slack for a couple of days, but eventually he accepted the situation; the ties between him and Doug were strong enough to keep their friendship afloat. However, ending romantic relationships at a young age always has the potential to lead to unpredictable developments much, much later. Doug used to muse that these and other human relationships were a demonstrable everyday examples of chaos theory.

Even though Doug went to Harvard and Sarah to the University of Minnesota, Doug was pleased to see that they had managed to keep their intimate relationship going without any major hiccups; at least, that is what he thought. The reality during their first few months of separation was, however, a different matter. At the University of Minnesota, like at other universities in the US, freshmen were obligated to occupy their rooms and participate in a weeklong orientation event before starting their first semester. In the group Sarah belonged to, there was a handsome and muscular boy, Roy, who almost immediately attempted to catch her eye, and at the earliest convenient time, they struck up a conversation. That first evening they were already eating dinner together, and during the following days, they spent as much time together as they could. Roy was obviously in love with her, a feeling that for several weeks Sarah didn't reciprocate. She regarded Roy only as a friend who helped her overcome loneliness. Besides, she considered Doug as her boyfriend, something that was always in the back of her mind, although not for too long. One night, Roy broke Sarah's resistance, and after that they were lovers for a couple of months. From time to time she thought of Doug and felt guilty for cheating, not knowing if she can ever exculpate herself for what she did; but Roy was always around, and in his presence, she brushed aside her concerns about Doug. Roy admired Sarah's beauty and intelligence, and he promised that his love was eternal. Sarah's emotional side liked to hear that. Doug didn't say such nice things to her very often, and now that she got it from Roy every day, she realized that she missed it.

Back then Doug knew nothing about Sarah's infidelity. He did find it a little strange that Sarah couldn't find time for longer telephone conversations anymore, and when he could talk to her, she sounded jaded and less enthusiastic than she used to. He himself was very busy with his studies, and he ascribed Sarah's reticence to her being busy too. Doug had an exceptional aptitude to not suspect anything; for him, trusting someone's word overrode everything else. He was the kind of person who was always the last to recognize changes in a person's behavior, if ever. He would have never considered Sarah capable of betraying their relationship like that. How could he ever question Sarah's faithfulness? The very idea was so distasteful to him!

However, as much as Sarah liked Roy's uncluttered mind and his constant dalliance and devotion to her, with time she figured that she wouldn't be able to get much substance out of him. He was talkative and seemed to never run out of topics to discuss, but Sarah found these topics more and more superficial; in short, she was beginning to find Roy boring, a sort of all-platter, no-substance guy. For many girls Roy would have been an ideal boyfriend, but Sarah once again desired a higher degree of sophistication. This is what she missed in Peter, and now in Roy, but found in Doug. At first, she tried to delay for as long as possible a direct confrontation with Roy. But a few weeks before Christmas, Roy could not contain himself and asked if something was wrong. In that moment, Sarah decided finally just to be direct with him. She explained to Roy that they were not going to have a future together, although she would like to remain his friend. Roy took it rather hard and sobbed for a while, but then he composed himself and told her that he understood and accepted the situation without any indignation. "Sarah, I tell you in the earnest that every minute you spent with me was a treasure, and that is how I will remember you. Even if it can only be from a distance, I know I will always love you. I ask only one thing from you; if I write to you, please answer my letter. It would mean a lot to me." Despite her firm decision, seeing Roy's sentient side, Sarah was close to tears. She promised that she would always answer his letters and return his calls.

After they separated, Sarah kept wondering if she had made the right decision. Who is worth more: a rather simple man who loves you completely and without recompense, or a man who may not have burning love for you but who challenges you intellectually? But for the second time, she chose Doug, and when they met during Christmas break in Rochester, they soon were back to the same level of intimacy that they had enjoyed before.

Doug radiated intellectualism and security, and for that, Sarah respected him at the time to an extent that was perhaps beyond rational. This often happens early on in relationships, and sometimes it may take decades until the "loving wife" or "loving husband" realizes that it was not true love but a high level of respect, perhaps combined with low self-esteem. In these situations, even the smallest change in the family's life that somehow erodes that respect and highlights the faults of the partner can lead to divorce. Sarah was fully aware of this common conundrum, but she convinced herself that she felt both respect and love for Doug; she thought it was not either or. Since her affair with Roy, however morally repugnant it was, had been over before Christmas, this gave her some justification not to talk to Doug about it.

The Christmas break came at the best possible time for Sarah and Doug. With her infidelities behind her and almost forgotten, Sarah complied with her brain's behest and smoothly transitioned back to Doug, who never really realized that there had ever been a problem at all. Their commitment to each other had remained stable during the remaining school years, without any other incidents, and they married within one year of Sarah's graduation. Sarah became an English teacher and moved to Boston to teach until Doug finished his remaining years in school. Soon after their return to Rochester, Sarah again worked as a teacher in a middle school, but after her first and then second son were born, she decided to stay at home full-time until the children started school. She had the same dilemma that Nancy Collins and millions of other American women faced every year; should she continue to work and spend a large chunk of her salary on child care and taxes, or instead should she provide the warmth and attention to her children during their most formative years at home that no day care could provide? She, like Nancy, chose the latter option and in a way, at least temporarily, squandered her professional life not least because Doug's salary alone was sufficient to provide an upper middle-class lifestyle for them both. This was exactly the kind of security she expected and appreciated so much.

Sarah also balked at the idea of entrusting her children to a day care service, since she thought no day care could provide the love that her children needed. Doug agreed with Sarah staying at home, although as a scientist, he didn't agree with the reasoning. One evening when Sarah brought the issue up again, Doug cited, just as a matter for discussion, a large study he read recently. That study was based on the analysis of more than forty-five years of research at seventy different locations. It stated, as Doug quoted it to Sarah word for word, that "when it came to academic achievement and behavioral issues, children of working moms on average didn't turn out to be worse than kids of stay-at-home moms." But after this short lecture, Doug repeated that he had no intention to reverse Sarah's decision. As a scientist, he just liked the facts, and in conversations, he often displayed this quality of his.

Doug never consciously doubted that Sarah was happy with their marriage, but on occasions, he noted that she seemed to be a little blasé and her actions appeared somewhat unfathomable, which to an objective outside observer might have indicated her dissatisfaction with something. For example, he recently noticed Sarah's forbearance from expressing her opinion about their friends, particularly when they talked about Peter Cartwright. The fact that he could not fully explain the occasional muteness of his wife made him feel somewhat uncomfortable, but he always managed to assign this feeling to the category of the unassailable platitude that "women can never be fully understood" and that it would be of little avail to him to deal with this issue any further. Still, from time to time it crossed his mind that these were perhaps the incipient signs that their marriage had slid into a phase that experts characterize as more mature but less romantic. It had never occurred to him, however, that what he had thought of as a previously wonderful romantic relationship with Sarah might not have had the same meaning to her.

Perhaps stemming from his job, Doug was an analytical kind of person and always had the propensity for getting to the bottom of any subject that happened to be in the center of a discussion. This meant that Sarah had full access to his innermost feelings about anything. He would have liked to see the full reciprocity that he thought he had received during the first years of their marriage, not realizing that even then he did not. The marriage of Doug's parents had had a very rocky start, but later they figured out each other's needs, which slowly transformed their lives for the better. Based on that experience, Doug's father once advised him that in marriage, the lack of reciprocity in sharing deep feelings, even if only suspected, is often a ticking time bomb. Usually it becomes clear when a breakup is near that there had been signs to foretell such event, signs that earlier the spouses had not taken seriously. "Remember this," Doug's father had told him, "in a good marriage, silence is definitely not golden. Proactive discussions, even if they appear importunate at the time, often clear the air. I always started a conversation with your mom when I found her face even slightly inscrutable. This made the air around us stifling, which I had to clear up. The other important thing is that you should try to find things that are new and exciting for your wife. Do not get stuck with old routines that can become boring after a while. Early in my marriage, I didn't follow these rules, and it almost led to our divorce. In many cases, the problem starts with us, men."

Although Doug often remembered his father's advice, he didn't feel that there was any need to discuss his perception with Sarah or change the daily routines. In his mind this would have meant acknowledging that there was in fact a problem in their marriage, but he was certain that there wasn't. Despite his sophistication in other matters, it didn't occur to him that a little more flattery and perhaps going home with flowers more often and earlier, as he did in past years, could perhaps postpone that less exciting phase of their marriage. Many husbands have knee-jerk reactions to the challenge of remaining attentive to their wives' needs and continue to do these things just as they had done earlier in their relationships. Doug was not one of them; for him their marriage was a done deal inextricably binding him and Sarah together with no reinforcement needed.

On this late August evening, the friends met in Mike's house. They were in a boisterous mood, ready for poker and chatting. Steve arrived first, exactly at 7:00 p.m. He had made great progress in setting up his business with a wealthy partner, and he hoped to sign a final agreement within a couple of months, pending finalization of a strong business plan that would allow securing the bank loan that he still needed. Doug was about ten minutes late because Sarah had come home only shortly before 7:00. She had said that she would be home by 5:30, but around 6:30, she had called Doug, asking him to stay with the kids a little while longer. Doug was in such a hurry that when Sarah came home, he forgot to even ask her where she had been. On his way to Mike's house, he vaguely thought about Sarah's late arrival, but then he pushed it out of his mind by assuming that Sarah had probably done some shopping that took her longer than expected. This wasn't unusual and had happened before, and by the next day, Doug had completely forgotten about it; he had more important things to worry about. Besides, they allowed each other substantial freedom of movement, and they would never subject each other to 'cross examination' as other couples often did; this just seemed so untoward and unnecessary. They knew a man working in a department close to Doug, who required almost constant updates of his wife's activities; they both despised the man. They would voluntarily talk about their daily activities over dinner, when they managed to have one together, and if they did not mention something, then it was simply not important.

They were so deeply involved in discussing Steve's progress, local gossip, sport news, and possible election outcomes that they didn't notice until about 7:30 that Peter had still not showed up. Peter, respected by his friends for his trenchant wit and abundant good qualities, had become a business professional and worked at the local IBM facility. He was a Yale graduate, and shortly before finishing school, he had married his college sweetheart Anna. Later Peter often wondered what would have happened if school had lasted just one year longer. He wasn't sure that in that case their relationship would have ended in marriage. They would have had more time to get to know each other without the pressure the ending of school had brought upon them. They both found jobs in New York, Peter at an IBM affiliate and his wife at a famous accounting office. Within three years of their marriage, they had two boys, but after that their relationship took a nosedive and ended in bitter divorce in early 2005. Even after all these years, his emotional wounds hadn't completely healed; after a while it was not so much about Anna but more about his children. He tried to see them as often as he could, at the expense of building a new serious relationship, but the children went through their most formative years without his presence. He often reproached himself if he did the right thing by divorcing Anna.

To make matters worse, Peter's divorce coincided with losing his job due to one of the long-lasting aftershocks of the 9/11 attack in 2001, which sent so many businesses and financial institutions into a downward spiral. After several months of unemployment and exasperation, he finally found a new job at the IBM facility in Rochester, which he occupied just one month before Christmas in 2005. His friends helped him in many ways to start a new life. All his friends had stable marriages, and they were happy to see Peter's happiness two years later when he married a girl named Helga. When Peter introduced Helga to them, all were in raptures. At that point in his life, everything seemed to be given magically to Peter to revamp his life. Several months ago, he had hired a young single woman as his secretary, who helped him a lot; in fact, she helped Peter in more ways than the friends would have suspected. There was only one disconcerting small cloud on the sky for him; IBM had begun to allow its workforce to dwindle, first through attrition and then through layoffs. However, in the foreseeable future, these measures were unlikely to affect Peter's position because he had attained a key role in representing IBM in a new effort to spread the company, both domestically and internationally. He was highly regarded at the firm for his skills in regaining and maintaining the allure of IBM in a very competitive marketplace.

Unfortunately, less than two years after tying the knot, it became obvious to the friends that Peter's marriage was in danger once again and that his divorce from Helga was inevitable. For a long time, Peter had not divulged anything to them, but whenever Helga accompanied him, they just seemed too incongruous for each other. Despite their best intentions, all that the friends could do beyond their commiseration with Peter was to hope that he would somehow get through this difficult period and patch things up with Helga; but it didn't work out that way. Not long ago, after a lacerating few months, Helga had moved out to her own house near the airport, and as soon as she had settled down, she filed for divorce, which Peter's friends again learned indirectly. One neighbor across the street who spent most of his time outside either painting an urban landscape or tending the lawn noticed that occasionally Helga still visited Peter.

The neighbor met Nancy at the art club; this is how the rumor was spread to the friends and beyond. Nobody was sure, although everybody who knew Peter were dying to know, if these visits were about resolving disputes before adjudication or if their relationship still had some chance, perhaps realizing how improvident their decision was. Then, at one of their previous poker parties, Peter informed them that finally his marriage with Helga was over. With this announcement, he wanted to make sure that if he might be seen with another woman, he should not be judged as a sleazy, unfaithful husband cheating on his wife, which, considering his character, would have been a plumb nonsense. The way it was presented to them, this news should have given a clue to the friends that there already was another woman in Peter's life, but somehow this subtle information failed to sink into their minds.

Around eight, the friends began to realize that Peter might not show up at all. As almost always happens when someone is missing from a gathering, they started to talk about Peter and his terrible luck with marriages, particularly this second one with Helga. Although at the beginning of their marriage Peter had talked a little bit about her during their poker games and even introduced her to his friends, she was still a mystery to them. Apparently, she remained a mystery to Peter as well, even though factually he knew almost everything about her and her earlier relationship with Ulrich, her German ex-boyfriend. At their recent poker meetings, Peter was particularly obsessed with Ulrich as if he had had the presentiment that their lives would intersect in a crucial way.

Helga Schmidt was an immigrant from Stuttgart, Germany. She majored in biology and chemistry at the University of Stuttgart. She was nineteen when she met Ulrich Siebert, an aspiring medical student, at the same university and fell in love with him. For Helga, Ulrich was her first boyfriend, while Ulrich had had several dead-end relationships before Helga finally captured his heart. They looked fabulous together. Helga had beautiful long blond hair, blue eyes, and a very attractive body. Ulrich was a tall slender man, also with long hair at the time and a slightly jut-jawed face. Both liked track and field, and for a while they belonged to their local athletic club that allowed their participation in competitions. Ulrich experimented with decathlons, but he competed only in the javelin-throwing event at which he was very accomplished. Helga was good at running both the 800-meter and the 1,500-meter distances, but she competed only in the former event. In their free time they often ran on the cinder track and talked under their breath while completing the laps they had planned for the day. They both liked to talk, and it wasn't unusual that after a good training, they would converse until late in the morning. Occasionally, Ulrich would grab a beer or two, but Helga at that time was a teetotaler and she also hated sugary drinks, so she usually stayed with water or milk.

Helga and Ulrich were in many respects alike, and at this stage of their lives, they had no qualms about each other except one major difference. As their relationship evolved, Helga occasionally hinted at the idea of marriage, but Ulrich, to say the least, was dismissive of the thought of becoming a married man. He wouldn't say it openly, but whenever Helga brought up the issue, in however subtle a manner, it sent shudders through his body; he simply wasn't ready for it. Other than that, their loving relationship continued, with occasional breakups and reunions, over the entire period of their university years. The breakups always centered around the marriage issue, which were quickly followed by their reunions because they loved each other so much. Overall, they had a great time together, particularly during the periods when Helga refrained from mentioning the frightening word *marriage*.

One important tie between them was that in their youth, both had had bad experiences with the Catholic Church and the people in it. When in their elementary and middle-school years they were given no other option than to go to church service with their parents, they were dumbfounded when told about the miracles that clearly flew in the face of their everyday experiences governed by physical laws. They also abhorred the overly fervent people who participated in the services. Both agreed with the usual confidence of young people that many of them were just smug and their fealty to God was in many cases faked. Already at that time in Germany, most people, including particularly the younger generation, were nonbelievers, and for them religion was superfluous; so they weren't alone in their dismissive opinion of Christianity. It didn't help to attract young people to the church when reports of child abuse by priests began to surface in the press. Many journalists were all too happy to report, and quite often inflate, these cases, which usually appeared on the front page. Even many of those trying to stay the course in their religious beliefs had been caught in the backwash of this negative publicity and were shaken. On the other side of the coin, and contrary to the church's teaching of "love your neighbor as you love yourself," in many cases those priests received little or no empathy or help.

Instead of going to church, their favorite pastime besides sports was to sit in a jazz bar and listen to good music, Ulrich sipping beer brought to him in a large traditional mug, and Helga drinking orange juice. Ulrich so obviously enjoyed his beer that one day Helga could not withstand the desire to taste it. It was a perilous act, since at that moment, she instantly became hooked on beer. Gay men also frequented the bar they liked, but that did not disturb them at all. This just gave them more opportunity to discuss socially sensitive matters that often divided even close-knit families. Both were blessed with verbal dexterity when defending their views on issues like gay marriage and others. They agreed that gay couples should generally have the same rights as heterosexual couples, but when it came to the question of marriage, they differed. In Helga's opinion, marriage between gay people should be acknowledged but under a different name. "Why not distinguish marriages between heterosexual and homosexual couples by giving them the distinct names of, for example, hetero-marriage and homo-marriage?" Helga insisted. "After all, these marriages are different in the key aspect that only marriage between heterosexual couples has the potential to procreate and maintain humanity." Helga failed to see why no such distinction between the two types of marriages could be made, but Ulrich thought that such separation wasn't important. After all, many heterosexual couples cannot have or do not want children either, he used to say, and for the time being, overpopulation, not underpopulation, was the bigger problem in most parts of the world, except Europe.

Then, similarly to many other unmarried student couples completing their final years at the university, just as it happened to Anna and Peter several thousand miles away, Helga and Ulrich were confronted with the question of how to continue their relationship after school was over. They chose to stay together, although Ulrich still was not ready to get married, and after thinking over her options, Helga grudgingly seconded this for the time being. She simply wanted to avoid at all cost the semblance of forcing herself on Ulrich by pressing the issue too hard and making their relationship tenuous, although the tension between them created by the lack of tangible progress in building a family was palpable and perhaps foretold what was to come. Yet for the moment her frustration remained only a tempest in a teapot, but it had the potential to grow into something much bigger.

Helga was lucky enough to land a ten-month yearly teaching position for three years at a middle school in Stuttgart, only after she had the prospect of getting a tenured position depending on her job performance. Ulrich had also managed to find a temporary research position at the medical center in the nearby town of Tubingen. It was beyond lucky for them to have found employment so close to each other. This is how they carried on with their lives throughout the year after graduation, separated but fully committed to each other and meeting each weekend. Their salaries didn't allow them any unnecessary luxuries, but they still afforded a decent life without having to balance the ledger daily. Most importantly, they still loved each other. They often envisioned themselves as two eagles flying together, soaring ever higher, each helping the other to overcome fatigue. Ulrich said once, "I hope you will never lose your wings, but if you ever do, I will give you mine." He meant it; Helga was simply everything for him. Another time he added, "When the time comes, we'll have a wonderful marriage. I know that we're inexorably entwined." Poetic words like this from Ulrich pleased Helga and helped her stay patient, for the time being.

In the meantime, Ulrich applied for and was granted an exceptionally long four-year fellowship to complete a PhD program at the Mayo Clinic in Rochester, Minnesota. At first, Ulrich was somewhat lukewarm about his application and filled out the forms only as a routine. He, like many others in Europe, and even in the two coasts of America, pictured Minnesota as a windswept and godforsaken state where winter is long and cold and the summer is short and humid. But others who had visited Minnesota convinced him that it is in fact a nice place, and since this was a once-in-a-lifetime possibility for him to start something very important with his life, he decided to take this chance at the Mayo Clinic. "You can't possibly rebuff this opportunity"—this is what everybody told him who had some knowledge of science in the US and who knew the reputation of the Mayo Clinic. His boss was even blunter; he let Ulrich know that behind the scenes, he had put in a lot of efforts to get this fellowship for him, and it would border on impudence if he rejected the offer. Since his application to other places had not succeeded, at the end he really had no other choice. Hearing the news, Helga at first was quite indignant because she imagined only trouble ahead, but with time, as Ulrich explained all the good things he had heard from others that could come out of this adventure, and after a little nudging and goading, she began to give in. The fact that Helga didn't have a stable position in Stuttgart helped her, albeit reluctantly, to slowly begin contemplating a move to Rochester. But that move still also depended on Ulrich.

In Tubingen, Ulrich had gotten involved in a certain kind of diabetes research that he would have liked to continue, but his position there was in limbo starting the next year; in that sense, his and Helga's situation was similar. If he were to lose his job in Tubingen the next year, then their savings would sag dramatically and quickly, and he would have to find a new job as a medical doctor in any case. Working at Mayo for four years could save an angst-ridden despondency next year and would allow continuation of his research. All things considered, even though Minnesota still appeared simultaneously forbidding and enticing, eventually Ulrich accepted that the opportunity at the Mayo Clinic had come at the most opportune time possible. After several more days of deliberation, Helga finally also came around and agreed with Ulrich's conclusion that for the time being they were stymied in Germany and that this foray might push them into a new direction with more opportunities. She just hoped that she would never regret her decision.

Helga was a thoughtful young woman who didn't like to make decisions in a hurry; she was even-keeled and not at all the starry-eyed type. But when she had finally recognized that Ulrich's job at the Mayo would make it possible for them to enjoy a comfortable life for at least four years, she stopped wavering. They couldn't foresee the perils and vagaries associated with that opportunity. At first, the news that Helga would leave for America bowled her parents over, but with time they also began to see the good things that might come out of this transatlantic experiment. After all, if she got into trouble, she always had the opportunity to return to Stuttgart and start her life all over. They also trusted Ulrich and believed wholeheartedly that he would take good care of their daughter.

Around that time, some studies had begun to surface that suggested, though without definitive proof, that diabetes could lead to pancreatic cancer. This would explain why in recent decades the number of pancreatic cancer cases had appeared to increase parallel to the number of diabetes cases, but the medical reason for it was far from clear. The diabetes research group in Tubingen and another group at the Mayo Clinic were at the forefront of diabetes and pancreatic cancer research, and at research conferences, they routinely met to discuss their projects in depth, although they had not yet exchanged researchers. Both groups worked along the hypothesis that diabetes and inflammation are causally related. At that time, this was a relatively new, still tenuous theory with quite rickety experimental scaffolding. For example, even such basic questions as whether diabetes causes inflammation or vice versa were not clear. It also seemed to be difficult to combat the prevailing view that there was no connection between diabetes and pancreatic cancer. In a nutshell, their hypothesis was that inflammation, caused by specific proteins called inflammatory mediators, also triggers changes in the pancreatic cells that lead to genetic changes eventually resulting in the generation of malignant cancer cells.

Ulrich was working on a special substance that was known to cause inflammation and that incidentally was also increased by a certain agent derived from bacteria. This added a new element to the puzzle, because these bacteria were also thought to contribute to obesity, which was generally considered to be the most important risk factor for diabetes. This provided the framework for Ulrich's proposed work that both groups agreed upon. Sensing the beginnings of a competition triggered by the allure of this new and important field, they wanted to jump-start their work before other groups managed to upstage them; so not much time was left for Ulrich to procrastinate. Ulrich was supposed to start working sometime during the fall of 2004, and he did. After scraping the money for an airplane ticket together, he had been fortunate to arrive at Rochester in early September of 2004. At the airport Helga kissed him goodbye with trepidation. She had again begun to develop a bad feeling about what the future might hold for them in a completely new environment, but she didn't share her misgivings with Ulrich. But Ulrich saw in her eyes that she was worried and reassured her with a wan smile that everything would work out just fine. That moment left such indelible mark in her brain that even if time to time it submerged in the unconscious, it resurfaced just at the right time.

Ulrich's new boss, John Pierce, picked him up at the airport and took him to a reserved apartment. During the ensuing dinner, they already began to talk about research, and Ulrich quickly realized that although John hadn't yet reached the pinnacle of his career, he was working on it very hard. In contrast to his boss in Tubingen, who was a frail and sullen but very determined person with a hint of avuncular gentleness, John was a well-built man whose face radiated optimism and energy. He was also very talkative but always seemed to retain a certain focus and a high level of intellectual acuity. Ulrich got the impression that while John had a clear vision of how to go about his research, he wouldn't be offended or find it irreverent if someone challenged his ideas based on solidly factual grounds. In fact, he made it clear that he was all for thorough discussions of research problems, with input from all his lab personnel. He didn't want Ulrich to simply become his assistant. He explained that he wanted Ulrich to do his job as independently as possible and be able to stoutly but flexibly defend his research plan, not only in front of him but also the research group. He was the kind of boss who is deeply enmeshed in the initial planning but not in the subsequent execution. As a few days later, one of Ulrich's colleagues put it this way: "John is the kind of chief who will be attentive to your needs, but you shouldn't expect him to spoon-feed you, and you never have to cringe in front of him if you feel that your idea has a good chance to be correct."

John had two major five-year NIH grants that allowed him to maintain a relatively large research group that included seven postdoctoral researchers and four assistants. However, he couldn't flaunt his grants for too long because he had to renew one of them within three years and the other one in four years. Fortunately for him, Ulrich was paid by the grant of the longer duration. At first blush, the remaining time on both grants seemed plenty long enough and should have provided a reasonable level of stability to the lab personnel. But in fact, everybody needed to continue to work hard because preparations for a grant renewal start well before its expiration date. The most important papers derived from the work will preferably be published by then, and this procedure alone can easily take one year or more. In Ulrich's case, this meant that by the end of the third year, he was expected to finish all his projected work and then publish several strong papers—the more the better. John made all this clear to Ulrich, adding to what he already knew from hearsay. The work ethic was also very high in the German lab, so for Ulrich it was not daunting news that he was supposed to put in quite a bit of work. This is exactly what he expected from himself as well.

After a few days of flinging himself into the necessary administrative work and settling in, the next immediate big hurdle for Ulrich was how to find a job for Helga in Rochester. The events of 9/11 were followed by a market swoon that made this task extremely difficult. Helga's attributes were that she had received strong training in the basics of biology and chemistry, her English was good, and she had an impeccable academic record. But on the downside, she had no research experience. A teaching position for her was also out of the question because that would have required obtaining a certificate in Minnesota, which is a lengthy process. Besides, the State of Minnesota had begun cutting rather than adding teaching positions.

However, after he had been searching for jobs for quite a while, his boss informed him that the Mayo Clinic was looking for nurses. As a basic requirement, Helga first had to complete nursing school and then she could start working under the supervision of a principal nurse until she had both completed additional requirements and gained enough experience to work more freely. The upside of this arrangement was that during her school years, the clinic would employ her part-time as an assistant in a laboratory where she would handle blood and urine samples. "Now we are in a fine predicament," Ulrich thought; he could bet that Helga wouldn't like the idea of becoming a nurse. But not having any other opportunities and with the clock ticking fast, he heaved a deep sigh and called Helga. To his surprise, Helga said yes. "I do not want to be a vagrant in Rochester," she said. "Besides, I'm not skittish to change course. Becoming a nurse is just fine with me." Finding a substitute teacher for her school was only a matter of a few hours; there was a large pool of applicants for that job. Thus, the school didn't insist that she fulfill her contractual obligations.

Filling out the paperwork, including translating all the school papers and getting permission to stay in the US for training for her advancement, took only a few weeks. Thus, despite the stricter travel conditions following 9/11, Helga followed Ulrich to Rochester in December of 2004 in a sanguine mood. After her arrival, everything went on almost like she and Ulrich had planned. Despite the cold weather, the prospect of living in Minnesota was now far less forbidding to Helga than it had seemed just a few months ago. Although getting a work permit for four years was a little more time-consuming than she had expected, Mayo's needs and reputation helped once again; and in a couple of months, Helga could start working as a part-time assistant. The remaining time was sufficient to study for the credits she needed at the nursing school. Her incredulity about their choice to come to America had completely disappeared; she suddenly felt liberated. There seemed to be no foreseeable obstacles for them in the coming years. The situation that they were in didn't seem like a contingency anymore. Now they only had to apply their attention to do good work at their respective jobs, and if they did things right, the almost four years they still had would be enough for them to lay a solid foundation for their family life. "Here we are in a providential situation. You shouldn't worry about us," she told her parents over the phone.

One reason Helga felt so liberated in Rochester was that they were able to adjust to the lifestyle in this nice town quicker than they had expected. They struck up acquaintances with several people working at the Mayo, and these benevolent and helpful people gave good practical advices about health-care plans, yard sales, where to shop, and how to get a Social Security number. They indeed experienced the Minnesota-nice demeanor of legend everywhere they went, and this soon made them feel right at home. Initially they lived in the same apartment Ulrich had rented earlier, but in just a couple of months, they bought a lovely house in a nice neighborhood in the southwest part of town, within a walking distance of less than fifteen minutes from the Mayo Clinic's central buildings. At the time the lending crisis was still far away, and they were able to procure mortgage credit relatively easily.

Both Helga and Ulrich had the propensity to spend as much time outside as possible, just like they did in Germany. A nice walk in that quiet ambience exhilarated them before they separated to start the day. Besides, living within walking distance of their workplace was a particularly important advantage in Minnesota. During the long cold winters, snow and ice storms are frequent and can make driving a dangerous experience, particularly for people coming from places where winter is milder, like southern Germany. As an added convenience, the downtown area also has a vast tunnel system that connects various Mayo buildings to each other and to parking lots around the campus. This is in addition to the network of skyways that connect the second or third floors of almost all major Mayo buildings with hotels, a downtown mall, the public library, the Civic Center, and a nice Barnes & Noble bookstore, which was still open at that time and served all sorts of tasty coffees and delicious snacks.

Despite all this comfort, the occasionally merciless winter leaves its marks on human behavior; by around the end of February and mid-March, after being through the fifth or sixth wave of frigid artic air, which brings subzero temperature to the area, even people with normally impeccable behavior begin to behave like petulant teenagers. But by the middle of April, when spring is usually in the air, although still not there, people again return to their nice moods. This also signals the start of yard sales, a quintessential activity of many households in America, where people try to get rid of all the paltry and useless items that they collected over time, falling victim to the false belief that they would ever need them. The next year a large portion of items purchased usually find their way back to yard sales to reenter a seemingly perpetual cycle.

But there was a rub for the Germans that couldn't have been better contrived. From almost the very beginning, Ulrich was under tremendous pressure at the Mayo Clinic. Later, when he thought about this more often, he had to admit to himself that the pressure was to some extent self-inflicted. But in large part the pressure also came from the time limit to construct a winning research plan and collect the amount of research data needed to reach a reasonable and plausible conclusion leading to high-quality publications that would have made the trip over the Atlantic worthy. Simply put, even though he had barely started his job and still had almost four years, he was already apprehensive of his and Helga's future after that. In this regard, he was diametrically opposite to many people, the optimistic type, who would have started worrying only during their last year of appointment.

Ulrich's data was to be built into his boss's grant renewal, and it had a definite deadline that was only about three years away. As his boss often reminded him, his extension depended on the success of the grant application in the first round. However, very few grants are awarded after the first round; grant proposals are sent back almost inevitably to the applicants for revisions, which means many months of delay. Considering all this, Ulrich had only a little more than zero chance that he would be extended even if he did everything he could do to get the grant renewed. Unless he wanted to return to Germany after the end of his term, he had to start looking for other possibilities in addition to the hard work required of him to produce high-quality papers.

In contrast to Ulrich, Helga received all the guidance that she needed for her work and studies, and she found it difficult to comprehend his situation, that he was almost solely responsible for the success of a research plan that required an enormous amount of extra attention and work. Ulrich made the tremendous mistake that he didn't take the time to explain all this to Helga and that he didn't bring Helga mentally into his research. "Make your partner deeply involved in your work so she becomes interested in your success," his boss once told him, but it was a lesson Ulrich unfortunately learned only years later when it was too late.

Ulrich was one of those aspiring young scientists who, despite all the potential difficulties ahead of them, chose a very demanding task. Unlike many good students who became scientists simply because they had to do something with their lives, Ulrich chose to be a scientist because he was passionate about helping people who struggled with disease. This, in his mind, required taking risks in his research. One evening his boss stopped by the laboratory and, seeing that Ulrich was alone, went over to him and asked him how he was doing and if he did not mind taking on a very risky project. Although the question was intimidating, Ulrich was not fazed. After a short silence, he took what he considered the high road and responded that in his view, "no real breakthrough has ever occurred in science without someone taking a risk and carrying on with research that clashed with the prevailing dogma."

"You put it nicely, and I agree," John replied. "A scientist with a great idea may have to go against the grain, despite being completely alone in their beliefs. If all young scientists chose the same path that you just did, science would have progressed much further by now. But, you see, you may not know that at the same time taking on a risky project in a grant application is a gallant, almost suicidal, effort. Often a piddling grant application is more rewarded if it follows the main trend of the day than a pioneering application. The main question I tackle whenever I write a grant proposal is this: What percentage of risk, meaning novelty, should be in it? In the proposed experimental plan, there has to be a fine balance between novelty and what I call 'old stuff' where you can't go wrong."

"Is there any hope that this may change?" asked Ulrich.

"Yes, there is. Occasionally I see opinion articles in scientific journals urging peer reviewers and fund providers to support more high-risk but also potentially high-payoff or transformative research, which can revolutionize fields. Based on this acknowledgment by many people in science, it is likely that in the near future, higher-risk projects will get more support. I am optimistic about the future, although some risk-averse reviewers might still refuse me funding for my next few proposals."

Ulrich, a novice to US science, wasn't aware of that "dark side" of the grant award process and was thankful to John for explaining it to him. He would have to reckon with this reality during his independent science career if he ever got there.

After getting through the first hurdles like developing a working space with the appropriate tools and research materials, the next issue for Ulrich was how to fit into a lab community that had already worked out certain rules. The reasons for potential conflicts among workers in a lab can be endless, particularly when the new person who arrives wants to introduce new kinds of research programs and methods into the lab. Fortunately, Ulrich had an endearing manner that helped him to rapidly gain the sympathy of most people in the group except a rather boorish man who had already been ostracized by the rest, and an extremely unfriendly lady who instantly lashed out whenever someone entered her comfort zone. Overall, Ulrich was lucky because he was clearly under the protection of his boss, and except for a few occasions, he didn't have to go through the usual remorseless scrutiny that some less-fortunate postdoctoral fellows are subjected to.

Ulrich himself was relatively new to research work, and he had to learn quite a few research methods. This requires a lot of time, at least half a year, but there are a few different ways to do it. Some learn the new techniques as required while working on the project. As Ulrich learned later, psychologically that is a better approach. However, Ulrich decided that it was better to make a list of the techniques he had to learn and then go through it as soon as possible. Once he had armed himself with all the necessary techniques, he would be much more efficient with the actual projects. In principle this sounded fine, but in practice this proved to be an inefficient use of time.

Learning new techniques was crucial to his job, and he had voraciously learned a great deal of them up to the point of surfeit; however, eight months had already passed without him performing a single real experiment, which was what he had really come to the Mayo for. The learning period was much longer than he had initially expected, although that was generally the norm. This made him more and more anxious; at times he started literally panicking. The only way out, he thought, was to work harder still and spend even more hours in the lab. This was precisely the worst thing he could do, but he was too young and inexperienced to know that. He wanted to do great strides in too short time, and in science this isn't possible. Older scientists know that a little less work with some intervening periods of calm but deep thought, which can be done anywhere, is the key to success.

One evening his boss saw him toiling away and had the feeling that Ulrich was on the wrong track. He stopped by and said to Ulrich, "Listen, I appreciate your hard work, but I have some advice. Don't forget to have a life too. You see, spending all your time in the laboratory with manual work distorts the eyes. You may see a lot of trees but not the forest. To really see the big picture, you occasionally must step out of the forest to have a distant view. A good scientist sees both—the trees and the forest. The synthesis of analytical findings requires abstract thinking, and for that you must stop working in the laboratory for a while. Collecting research data without a frame to put it in is like having a hundred bullets but no gun."

Later Ulrich thought back and wished that he had accepted this advice, but again he didn't listen carefully enough to his solicitous boss.

As if his manual work wasn't sufficient to fully occupy his life, Ulrich soon noticed that in the lab he was missing something essential, and this was the Journal Club. Scientists, usually working in a narrow field, are bombarded every week by hundreds of papers that they are supposed to read to keep up with the newest developments in their field; but almost nobody has the time or energy to do that. Not all papers published in any given week have equal significance either; some of them stand out because they are especially important not only for a specific field but also for a broader group of scientists. A clever way of disseminating the most important papers is to organize a Journal Club within a research group or the entire institute. Journal Club meetings are usually held at lunchtime once a week; one scientist presents the chosen paper or papers at sufficient detail to spark discussion that, in turn, leads to deeper understanding and perhaps even applications of the ideas. The Tubingen group had such a club, and Ulrich liked it very much. At one of the lab meetings, he raised this issue and the boss strongly supported the idea. In fact, the group used to have such meetings before, but the organizer at the time recently resigned and they could not find a new volunteer to restart and direct the program. After a brief discussion with his boss, Ulrich volunteered to be the organizer of the Journal Club.

Although the club gave him even more work to do, and it didn't always go smoothly, this extra activity also earned him more respect within the group. To set a good example and establish a template for how to do it, he gave his first presentation on the possible role of insulin in cancer. New studies had started to appear suggesting that diabetes is not only a risk factor for pancreatic cancer but also for colon, liver, bladder, and breast cancers. Some of the papers that Ulrich presented to his colleagues discussed the potential mechanisms that could also lead them in new directions in their work, although they all agreed that a sheaf of questions still need to be clarified. At the end of the meeting, they also discussed who would be the next presenter one week later. One postdoctoral scientist volunteered to present the potential of using implanted stem cells to produce insulin in type 1 diabetic subjects who had lost their specific insulin-producing cells. The man promised to focus on how it might be possible to perform some genetic engineering to avoid the risk of the transplanted stem cells becoming cancerous, which was a major roadblock in developing such medical technology. This appeared to be a daunting task for most scientists involved, but a few attempts were already published that seemed encouraging and that were worth discussing. All in all, Ulrich managed to put the Journal Club on track.

After a little more than a year at the Mayo, Helga and Ulrich applied for a green card on the basis that their work was significant and useful to the United States. Because they weren't a married couple, they had to apply separately. The lawyer's fee was quite substantial, about $8,000, which completely sapped their flimsy reserves for quite a while; but they thought that this investment in their future was worthwhile even though at that time they didn't plan to stay in the US forever. Probably due to the high respect immigrant officials generally have for the Mayo Clinic, both received their green cards faster than usual, by the end of February in 2007. That year they were fortunate enough to celebrate Easter in Stuttgart with their parents and friends without the hassle foreigners often go through at the border when they leave or reenter the US without a green card. By early June, Helga had finished her training at the nursing school. This and the green card allowed her to both move to a part-time assistant nurse position and continue her studies in the nursing school for two more years to get her bachelor. She did all this to make sure that she would have a stable job by the time she and Ulrich decided to marry, although Ulrich still was quite reticent about that prospect.

Helga hadn't really tried her best to understand Ulrich's situation and his need to spend more time with research, which cut into their time together. Because of her lack of understanding, she found it more and more difficult to accept Ulrich's absence during the evenings that had become the new normal. On occasions, Helga would have liked to bellow long and hard, but at the end, she always managed to control herself. In Helga's mind they were supposed to be living a much better life and start living like a family in Rochester, but this wasn't happening. And the longer it didn't happen, the more she craved for a change. So the tension between them kept building, slowly but steadily, although they never had any sort of loud arguments. For a while Helga followed a "least said, soonest mended" policy about marriage, although it was only a question of time regarding how long she could control herself.

Occasionally they managed to take a break from their stressful lives and made small trips to interesting places around Rochester. They particularly liked to visit Wabasha on the Minnesota side of the Mississippi. A scenic highway with bluffs on both sides led into the town. They were advised by friends to have lunch in a nicely decorated and fancy traditional restaurant in that area, which to their sorrow went out of business a couple of years later. Visiting this restaurant, the Anderson House, which also served as a hotel, became a sort of tradition when they went in that direction. The food was simply superb there. They would then walk down the street hand in hand to the mighty Mississippi River, which never failed to bestow on them a soothing, bracingly clear and peaceful mind-set. Watching the lazy billowing water on a windy day made them feel like courageous survivors after another hard week, and that helped them, though only temporarily, to reset their priorities and renew their faith in their relationship and future. This is at least what Ulrich thought, and because of that, he felt safe; he would never lose Helga. Just a few more weeks or months to get caught up with his research and then he would have much more free time to spend with her.

After spending some time at the river in idle contemplation, they would drive through the bridge to the Wisconsin side of the river and go to the hilltop near Alma. The sight from there was simply breathtaking, and it was always hard to leave that spot, which inspired unfettered imagination in them. These trips and their visit to Stuttgart for Easter temporarily halted Helga's simmering internal struggles and her pent-up desire for change, namely, for a committed married life. But as they slipped back into the daily routine week after week, she began to see with ever-greater clarity that she didn't want to put marriage in abeyance and didn't want to continue her life with Ulrich if that was what it took to change her life.

But Helga loved Ulrich, and breaking up wasn't easy. She tried to postpone the inevitable, still hoping that Ulrich might change. Time and time again she returned to the topic of placing their relationship on more solid ground; Ulrich was always able to put it off even further, citing some legitimate sound reasons like his all-consuming work and their shaky future, but never admitting to the one true reason: he was still simply not ready to commit to marriage even though he was indeed crazy about Helga. Hiding the real reason was a mistake; had he explained this instead of trying to show some stubborn side to his personality that he didn't even have, Helga almost certainly would have understood and given him more time to figure things out. Instead, Helga became convinced that she might never get the commitment from Ulrich that she desired so much. That led her to the only logical conclusion that Ulrich was not a reliable partner for her in the long run. She began to lose her attraction to him, and thus the strong bond that had once blossomed between her and Ulrich rapidly began to wither. This first led to small quarrels and then to open fights that were usually initiated by her. Helga ineluctably signaled to Ulrich that on her part, it was over.

For his part, Ulrich was appalled at the thought of separating from Helga; he regarded such an action as complete nonsense and out of the question. Yet with every passing day, his feeling that it was inevitable grew stronger and stronger. Just like a spaceship that is about to achieve the critical velocity required to leave the gravity of the earth, Ulrich felt that Helga was very close to achieving the motivation needed to abandon him.

The way Helga behaved was inscrutable for Ulrich, but he decided that if Helga wanted to leave him, he wouldn't fight it. As time passed and he was placed under even greater stress, he began to believe that there was not much point in continuing this charade. Ulrich now had a better grasp, almost crystal clear, of what was happening between him and her, but he also saw that they had passed the point of no return, and he could not do anything about it. His view was that the angst Helga struggled with was, in fact, entirely of her own invention but that it couldn't be reversed. In retrospect, Ulrich realized that he had also made mistakes and that he could have been more sensitive to Helga's internal strife. But most of the time, he just felt jaded to the point that the only desire he had was to go to bed whenever he could manage it. This was for him the easy solution as opposed to fighting Helga's view of their situation. Eventually Helga sensed a crucial turning point at which Ulrich resigned to letting her go without trying to change her mind. Ironically, after it became clear to both that separation was inevitable as soon as their financial situation would allow it, Helga became much calmer and easier to get along with. Ulrich could again focus on his work, and Helga began to contemplate how to change her life for the better. Although Ulrich's job was to be terminated on October 1, 2008, his boss found a way to extend him by six months. This was very good news, and it reenergized him. The peace at home, whatever price he had to pay for it, came at the perfect time.

In June of 2008, Helga found out on the Mayo's website that Doug Lowry was searching for a technician, and she jumped at the opportunity, even though she still had one more year to go at the nursing school. Her superior assented to her request without any hesitation. Doug was looking for someone with a background in biology and chemistry. Helga seemed to be an ideal choice for this position because she had some nursing background and already had insights into how the system worked at the Mayo Clinic. Her initial duties were to learn all the necessary analytical techniques in toxicology and then help to determine drug levels in people who were apparently overdosed, poisoned, or seriously intoxicated.

In Doug's lab Helga worked with Joseph, the principal assistant who had a bachelor's degree in physics and seemed to know everything about analytical instruments. He had a jut-jawed face and in general was a little clumsy in his personal appearance; he often showed up in a rumpled suit, but he did not seem to be bothered by it. However, in personal relationships he was altruistic, and he also turned out to be an excellent teacher who never lost patience. His explanations were concise yet thorough and easily understandable. He could make a seemingly daunting task look easy. He always made sure that Helga had a grasp on the essence of the techniques before dealing with the small practical details. He was polite, and when Helga made obvious mistakes, he was never derisive or grouchy unlike many people in similar positions. He often emphasized, however, that the arduous work they were doing was very important and couldn't be done in a perfunctory manner. When it came to quality of work, he was as inflexible as his boss, Doug. For increased emphasis on the importance of work ethic, it was a daily recurrence that he cited some ominous examples, each coming from other laboratories, when subpar analyses led to disaster. He didn't want to inspire fear; he just wanted to make sure that Helga developed an unswerving commitment to do a thorough job.

One day Joseph would teach Helga how to use a mass spectrometer to analyze organic compounds. Another day he would morph into a different kind of expert and demonstrate how to prepare lipid samples for analysis with gas chromatography. Yet another time, he would prepare small chromatographic columns to separate proteins from salts and lipids. Next, DNA sequencing would be the subject, and so on. Helga loved learning these new techniques even if it often required her to stay in the lab for much longer than eight hours. In some way she was now in the same situation as Ulrich, with the major difference that she was not on her own and she did not have to develop or be responsible for a research plan. In a matter of three months Helga had become quite familiar with all essential methods used in that lab for toxicology, although to use them routinely and without error took her a few additional months. Her first independent work, although still under Joseph's supervision, was to provide DNA evidence for a forensic case. DNA analysis was Helga's favorite method; she was fascinated by the world of DNA. She happily relinquished her free time in exchange for mastering DNA techniques; this was especially easy because Ulrich was usually not at home before eleven. So arriving home at later and later hours did not cause her remorse, and she did not have to be afraid of being berated.

CHAPTER 2

Doug Recounts His Discussion with Peter about Religion and Science

Poker can be played very seriously and with full concentration, but for the friends, it was just an excuse to come together and discuss weekly events and tell stories. So far only one topic had been off-limits in their conversations: religion. Miraculously, although all of them were Catholics attending the same church, they seemed to shy away from bringing up the question of religion. Perhaps they just thought that it was a touchy subject, or maybe they figured that there was simply nothing to discuss because they all accepted the same teachings of the Catholic Church without doubt.

Then suddenly, while they were still talking about Peter, Doug surprised everybody and began to tell a story very heavily touching on religion.

"A few days before Easter in 2007, I called Peter and asked if I could visit him after dinner, because in the lab they had a problem installing software that would allow rapid reading of viable cell numbers using the data taken from a connected instrument. Although Peter was a business expert, through working at IBM he had picked up some essentials of software development. At the very least, he could give me some advice on whom to turn to. After examining the disk, he immediately knew where my assistant had made the mistake. He had not provided a crucial piece of information to the computer about how to scale the diagram. When we were through and we were just sitting there chatting, I noticed two books on his computer desk. One was entitled *Origins: Fourteen Billion Years of Cosmic Evolution* by Neil deGrasse Tyson and Donald Goldsmith, and the other was *Big Bang: The Origin of the Universe* by Simon Singh. I literally cried out and told him that I had no idea that he was interested in cosmology, and that I also was interested in that subject.

"As we began to talk about these books, Peter told me that both were relatively new, both published in 2004, and he had bought them a few months ago in the local Barnes & Noble because he wanted to have some idea of how solid the scientific arguments against God's role in Creation were. So I asked him what conclusion he had arrived at. He answered that he wasn't sure that he understood everything in the books but eventually figured out that essentially the scientists who don't see God's hand in Creation and those who believe that God created the universe must both come to the same conclusion: our universe apparently arose from nothing. He then added that in his opinion, whether something came from nothing, then providing the basis for the big bang, or the big bang itself was the event of creation from nothing, is almost beside the point. Neither science nor religion is pliable enough to admit they have a common ground except the role of God in Creation.

"Then Peter further elaborated that theoretical physicists, including Stephen Hawking, argue that although no physical objects in the universe either in the form of energy or mass can be created from nothing, the whole universe could have been. Some of these physicists go so far as to claim that due to some quantum instability, many universes, perhaps numbering in the billions or even infinite, continuously arose and are still arising out of nothing, but most of them cease to exist because certain physical constants aren't appropriate and after reaching a critical size they collapse. However, after innumerable previous trials, our universe survived because by chance all the important physical constants were at just the right balance, or as physicists say it, fine-tuned. According to some cosmologists, it would likely take the creation of about ten-to-the-one-hundred-and-twenty-third-power universes before one like ours could have been born. Peter then concluded that in his view, this idea was flawed, because if today the number of universes is infinite and tomorrow another is created or destroyed, then the number of universes would be either one more or one less than infinite. To Peter this wasn't possible, and I felt inclined to agree with his logic, although I must add that mathematicians have a very different view of infinity, which does in fact allow adding onto infinity. In any case, I agreed with Peter that if there are indeed multiple universes, then their number cannot be infinite so long as universes are born and then cease to exist in a continuous cycle. On the other hand, if we accept that the number of universes is infinite, then we also must accept that the whole cosmos is static. Universes aren't born or destroyed."

"I guess you and Peter were correct on that one, although, as you pointed out, everything hinges on the definition of 'infinite,' which I'm not so sure about either," Mike said. "Perhaps infinite is what we cannot count?"

Doug drew a breath and then continued, "Anyway, after that Peter continued his summary of his readings. As far as he could judge, he saw not a shred of concrete evidence for these secular theories. So he concluded that since the universe had to be born from nothing either way, some prodding from God was required. Eventually, both the God-less and God-based theories of Creation come down to simple blind faith, he told me with a smile on his face, and so he figured that for him it would be simpler just to stay with the conventional teachings of the church. Thus, God for him remained a spiritual super-intelligent being able to communicate with us, and He could perhaps even change physical laws and perform miracles if He really wanted. But then Peter added that he did have a little trouble understanding why God would want to change physical constants once He had established them exactly with the purpose of providing stability and thus to make them predictable. So Peter seemed to doubt that God would ever want to change anything, thereby admitting that what He had created was imperfect. In fact, changing the laws of physics would be a malevolent act, because in nature everything is connected to everything else, and a change would reverberate throughout the universe, engendering very bad consequences.

"Then Peter asked me what I thought about miracles. I answered that until most recently I had thought that God occasionally performs miracles just to prove to us that He, a supernatural being, exists. But later I changed my mind, and now I believe that God does not need to prove anything. And thinking about it a little deeper, I realized that He loves us. If He were to make life unpredictable by changing our laws of physics every now and again, it could really make life miserable for us. A change just in one law, even just once and even just a little bit, would have wide-ranging butterfly effects on the world. This would likely affect our very balanced life on Earth adversely. Consider what climate scientists and biologists predict could happen if the average temperature on our planet were to increase even just two or three centigrade. So in principle Peter and I agree that God, at least after Creation, has never performed miracles that would have required changing the laws of physics. I added that God still might well influence us by means that doesn't require any change in any law of physics."

"This is amazing!" Mike exclaimed, laughing. "I had no idea you guys were so philosophical!"

"And I'm far from finished," Doug continued. While he was talking, he realized that he was playing poorly that would cost him some money, but by now he was too engrossed in his own story to stop. "After talking a little more about miracles, I sprang my next question on Peter. How do you explain the assertion that Jesus was able to perform miracles? Peter admitted that he had been struggling with this apparent contradiction for some time now and that he didn't really know where he stood on this issue. He said that it appeared illogical to him that while God never performed miracles except creating the universe, His Son did. But then he said that since the belief that Jesus performed miracles is so deeply ingrained in him, he would go along with it for the time being, so that he could go to church without guilt. And then he asked me about my opinion. I said to him that for me, once I had accepted that God didn't perform miracles, then, well, the same should apply to Jesus as well. He asked me if I had ever thought about the possibility that God could perform miracles if he wanted to but chose not to on principle, while Jesus made few exceptions. I told him that Jesus could have performed so-called miracles through a subject's mind that didn't involve changing any law of nature. Since then I have given a lot more thoughts to this possibility.

"Peter figured that I must have put a lot of thoughts into these questions, and I confirmed that I was indeed in the process of developing several of my own theories about what God might look like, how He might have created the universe, and if He relinquished his control over the evolution of the universe and over the events right here on Earth. But I begin to sense that you guys might be getting bored of all this stuff."

"Not at all," both Mike and Steve assured Doug almost simultaneously. "Go on, we would really like to know how the rest of your conversation with Peter went," Steve demanded. "We now know Peter's opinions, but only partly yours?"

Doug searched for quite a while for a suitable opening. He wanted to come across in front of his friends as someone who would like to strike an objective balance between opposing views. At the same time, he wanted them to understand that he was at the very beginning of formulating his own ideas, which were almost certainly full of many conceptual and factual errors. But eventually he decided to follow the same line that he presented to Peter. "You, just like Peter, have put me in a difficult position," he said while finally laying down his first winning card of the day, "because my theory has really flowed and ebbed and still continues to do so, and I really do not know at the end of the day what, if anything, will remain standing. However, I am beginning to gravitate toward one specific set of beliefs that seem to me the most plausible among the many other rather convoluted possibilities that have been taking shape in my mind. If you still insist, I can talk about that direction, but I warn you that at this stage, I'm very far from presenting it to you in detail because I have not developed it fully."

"It's perfectly all right. No need to make more excuses. Just go ahead please. We aren't equipped to understand complicated mathematical formulas in any case, so we can't belie you," said Steve in an encouraging tone.

"So here is what I think about Creation, and this is what I also told Peter, more or less. The center of my beliefs is that there is a supernatural being that for simplicity I call God. But this God is different from the one described by various religions. My God is a dual being, at once spiritual and material. I emphasize that I am still having trouble to define the 'spiritual God' as a fully matter- or energyless entity, and this is something that will require a great deal of mental effort before I can come up with a satisfactory concept to explain it. In any case, I believe that we will not be able to understand God or his ability to create the universe and influence our consciousness until we recognize that he also has a material side, although not really material in the traditional sense. I see this statement is controversial, so I explain it. As I see it, one possibility is that the material being provided the prodding to create the universe, perhaps along with a parallel universe, from nothing through several steps that preceded the big bang. These initial steps were part of the grand design that eventually provided the template or blueprint for how exactly the universe should look like. I can give you an example, but first I have to ask you. How much do you know about dark matter and dark energy?"

"Well, I have to admit that I know nothing about them," said Mike.

"I heard about them," chimed in Steve. "The reason they are called dark is that they are invisible to astronomers, and therefore they have no way of knowing what they really are." He then added, "I also know that dark matter represents six times more mass than visible matter, and its best-known properties include gravity, the absence of electric charge, and its inability to interact with ordinary matter to any significant extent. I also saw a program on the public channel about dark energy where the scientist said that it appears to fill all of space, and it's the force behind the exponential expansion of the universe. If I remember correctly, dark energy represents about 73 percent of the entire universe's mass-energy. But how does all this relate to the big bang?"

Doug heaved a deep sigh and then began his "learned disquisition" to sketch out one possible scenario that could describe the events that had preceded the big bang. It came in handy that previously he had already trained his thoughts on the subject during his conversation with Peter. "First there was God, a spiritual and physical being at the same time, as I have just explained, and I emphasize that neither the 'spiritual' nor the 'physical' side of God can be understood in the traditional sense. Spiritual and physical sides of God are a kind of duality the nature of which we may never understand, although with time we might grasp at least some aspects of it. We can never understand, only guess, the nature of the spiritual side of Him, because that we would equate with '*nothing*' or '*nothingness*' as far as our observational capabilities are concerned. But *nothing* is again another term which I still need to deal with quite a bit to understand it, because I have the feeling that it is in fact something, and I wish I knew for sure what it is. Both the spiritual and material sides of God existed for eternity and are omnipresent as well as omniscient with absolute free will. That much I'm also sure of. My guess is also that at the time when God decided about the creation of our universe, His material side served as the executioner of the plan.

"I know that dualities are hard to imagine as two sides of the same coin, in this case, God being simultaneously represented by two beings. But strange, hard-to-fathom dualities exist in nature as well, for which the root cause may be, in fact, God's dualism. For example, wave-particle duality is a basic feature of quantum mechanics, meaning that quantum objects, like an electron, manifest both wavelike and particle-like properties. This not only stretches the credulity of everyday people with little training in physics, but even the most brilliant physicists are stunned and even feel stupefied struggling to understand this phenomenon, which has been proven beyond doubt by extensive experiments. Or an even more common example is a river during the winter that may have a thin ice coat on the surface, yet the chemically identical water flowing underneath is liquid. When the weather is warm, then the liquid water evaporates and eventually forms a cloud, representing yet another phase transition of the same H_2O molecules. Conceptually, it is relatively easy to understand the phase transition between these physical states because all of them are chemically the same and only their structures are different.

"In case of God, His phase transition from a spiritual nonmaterial being to a material being, as I envision it, is a lot more difficult to understand than, say, the transition from liquid water to solid ice. But for me it is still more realistic than imagining God as a solely spiritual being who is not able to act through physical means. However, the point that I really want to make is that in my belief, time started with God, and therefore it makes no sense to ask what was before God. I am not considering the option that God was created from nothing, for the main reason that from our perspective, His spiritual being is *nothing*, and therefore, He always existed. I know that this sounds confusing and probably implausible, but as I said before, many people have many different ideas about how our world was created and what role, if any, God played in it. Mine is only one of them, and of course I cannot be sure if it is true or not. That's why I call my ideas science fiction. Still want to hear more?"

"Absolutely," responded Mike. "Frankly, as a friend I can tell you that your story is starting to sound a little weird, indeed like some science fiction story as you suggested, but it is fascinating," added Steve. "I am curious about the rest of your ideas. How far will you go with debunking our image of God?"

"A big question, conceptually nearly impossible to comprehend, is what *nothing*—perhaps I should call it 'the Spiritual'—actually is. If it is dimensionless and we can't attach any attributes to it, then it is *absolute nothing*. This term is very difficult to deal with, but recently more and more scientists are considering the idea that the universe came from *absolute nothing*, which just sounds preposterous to me. I believe that this is a misconception, and I hope I can defang it. The *nothing* that I'm imagining is a matterless vacuum containing a special '*something*,' the physical nature of which we almost certainly will never understand. But I would argue that this kind of *nothing* is in fact *something* with a dimension and presently undetectable attributes that equal God.

"The other idea difficult to digest is that *something* has always existed and gave rise to one or perhaps many universes, although more and more scientists favor this concept, except that most of them do not include God in this picture. I should emphasize that I heard this second interpretation of *nothing* from a scientist, and all I added to his concept is that *nothing* equals God.

"I said that this vacuum contains *something*, the nature of which we may never understand, although it is knowable. My starting point is that if the spiritual side of God exists, He must contain all the information needed to produce and maintain anything, including our universe. Now, what if we consider that 'information,' or the spiritual God, is the *something* that is present in the vacuum? In that sense, I consider 'information' to be a kind of sovereign entity, just like matter and energy. The material God, then, is related to the spiritual God. Accordingly, the backbone of my hypothesis is that the universe sprung out of *something* that equals the totality of information, or in other words, God."

"If I understand you correctly," Mike chimed in, "when you say the word 'spiritual God,' you do not actually mean that God is spiritual in the traditional sense, that is, devoid of any form of energy or matter. Or am I missing something?"

"Yes, you have a point, and a few minutes ago I tried to explain this," Doug responded. "So again, if my reasoning is correct, then the spiritual God is a sort of entity that is in fact not really spiritual in the traditional sense because it has at least one attribute—information. However, since information is probably dimensionless, we can still consider it spiritual, even though it is the Creator of all energy fields and matter particles that exist today. With all this in mind, I will keep using the terminology of 'spiritual God' and the inseparable 'material God,' both corresponding in unity to the totality of information. Because it is a centrally important issue for me, I want to reinforce the idea that there is no disjunction between spiritual God and "material God". You should see that here I am left with a huge conceptual quandary: how to view information as an independent entity, and how exactly it relates to both the spiritual and material sides of God, and perhaps even to quantum events when subatomic particles pop out of apparently nothing and almost instantly decay back into the nothing."

"Do you mean that information can give rise to quantum objects, and vice versa? I find this a fascinating possibility," commented Mike.

"Indeed, it is. The way I think about it now is that this transition of information into pairs of subatomic particles might be the bread and butter of the existence of everything, but I have no clue how these extremely unstable and random quantum events, some call them quantum jiggers, can be tied to the stability of atoms, molecules, and macro objects," admitted Doug.

"I have read it somewhere that the whole universe could pop out from nothing, exactly as the result of a giant quantum jigger. I wonder if you have an opinion about this theory," asked Mike.

Doug had confronted this issue many times before, and he gave his best answer of the moment. "My problem with this theory is that quantum jiggers or any quantum events require the existence of vacuum space teaming with energy. The momentary formation of subatomic particle pairs, whether they are formed from information or not, must borrow energy and time from this vacuum space. In our macro-world, there are universal laws, such as the laws of conservation of momentum and energy, which cannot be broken. However, in the quantum world, these laws apparently can be broken, albeit only for such a short time period that newly formed subatomic particles cannot be detected directly. Now, those who presume that the universe popped out from nothing via a giant quantum event, and such big bang created the space, cannot explain how such quantum jigger could occur without a preexisting space and how the law of conservation of energy can be broken even for a very short moment. The reason, I guess, is that acknowledging the existence of any preexisting space with some form of energy in it could call for the role of a creator. Atheist scientists would need to find a loophole which would help to avoid such conclusion, but they have not found one, and in my opinion they never will.

"After this little detour, related to Mike's last question, allow me to continue assuming the following hypothetical case concerning the pre–big bang period, which obviously remains as unprovable as any of my other theories. At this point, let us put aside how it was done and just assume that God created an expanding space from the ever-existing information-filled vacuum that for simplicity may be equated with dark energy. By the way, I call this space the parallel universe, which I believe is interwoven with our universe. This expanding space is essentially responsible for the observed acceleration of the universe's expansion. Dark energy is very different from the matter that our known, visible part of the universe is composed of. The greater the density of matter, the greater the attraction, or gravity, is between matter conglomerates. If dark energy didn't exist, gravitational pull would slow and eventually reverse the expansion of the universe, resulting in the concentration of all matter in a very small area. Due to the influence of dark energy, scientists have now ruled out the possibility of this hypothetical process, known as the big crunch, to ever occur.

"In contrast to matter, dark energy appears to have the strange property of causing the space of our universe to expand and even accelerate its expansion. As the space of our universe expands in all directions, it appears, quite amazingly, that more and more dark energy in the form of an ever-increasing vacuum force is continuously coming into existence, thus forcing the universe to expand faster still. The consensus of scientists is that since the birth of the universe, there has always been a tug-of-war between expansion and gravitational pull, and some billion years ago dark energy won, the implication being that our universe will continue to expand forever. Scientists don't understand this phenomenon, but I have a relatively simple explanation for how dark energy could have won the competition. I think that our universe is expanding into a larger space, which was created before the big bang, and which I have just referred to as the parallel universe. Since this parallel space itself is expanding, it gets proportionally larger every minute relative to the space of our universe. Since the two spaces are interwoven, the ever-increasing vacuum force of the larger space causes an ever-increasing pull, resulting in acceleration of the expansion of our universe. Thus, dark energy may be nothing more than the ever-increasing vacuum force of that other larger space. Creation of this other space, the parallel universe, could have been the first step of the creation process, or if you like, the first big bang."

At this point Doug pulled a sheet of paper and a pencil from his pocket and drew three pairs of double circles, which he showed were increasing in volume, to compare the sizes of the preexisting space and the space of our own universe five billion years ago as well as before and now. The latest data obtained with the Hubble telescope detecting large numbers of supernovae and other features of the cosmos confirmed that in the universe's existence, there was an earlier slowdown period followed by the speedup period. The transitional state between slowdown and speedup occurred at about five billion years ago; this was the reason Doug included this time point as well.

"I see what you mean by the changing correlation between the two spaces and how that could lead to accelerated expansion of our universe," said Mike. "But what do you mean by 'the first big bang?' I thought there was only one."

"Yes, this is the consensus of the scientists, but there are some alternative theories. My main problem with the prevailing big bang theory, shared by many others, is that it doesn't give satisfactory answers to many problems. For example, why do the galaxies take the form of one of several different shapes, like elliptical and spiral? And what is the nature of dark energy? And why does the universe have so much matter and yet so little antimatter? And many related questions just keep multiplying. Also, to my best knowledge, there is no firm evidence whatsoever for the general assumption that dark matter and dark energy were formed during the big bang. As for the formation of galaxies, scientists think that within a very small fraction of a second after the big bang, a very fast expansion, called the inflation period, took place, which made the universe extremely smooth, except for very small densities caused by quantum mechanical fluctuations. These small densities of matter then gravitationally attracted cooled clumps of dark matter and condensed gas that provided the seeds for galaxy formation."

"This seems to me to be a reasonable hypothesis," Mike chimed in.

"Yes," responded Doug, "but a key question is how inflation was started and why it stopped within such a short time period. To many cosmologists, inflation can make sense only if the parameters of the inflation field had been extremely defined with just the right shape to suppress gravitation waves. Otherwise, once inflation started, one would expect that random quantum events would render it an uncontrollable runaway inflation event with the potential of spawning an infinite number of universes, each quantum event serving as a seed for a new universe. I know that this is complicated stuff, but for the sake of our discussion, just please accept that there are serious problems with the present inflationary theory and the related big bang theory. In fact, several prominent scientists, including the mathematician Roger Penrose, flatly state that the idea of the inflation theory is misconceived."

"Okay, we appreciate you saving us from the details, but I assume that in your view the first big bang would solve these problems. Am I correct?" Steve asked.

"Conceptually it might solve some of them, and here is why: Let's assume that information is a new entity in an all-encompassing vacuum that can become anything and that in that realm quantum fluctuations also occur. For the time being, let us again ignore God's role in that. Let us also assume that quantum events led to a sudden enormous concentration of information leading to its transformation into another entity with a greater resemblance to energy and matter. That event then resulted in a huge explosion and the subsequent formation of a special space with specific constituents like dark matter, the Higgs field and perhaps others. This event might have created further quantum instability in the information field that, after a second concentration and transformation process, instantly triggered a second explosion. This is what is generally known as the big bang, but what I think of as only the second big bang.

"Because the information field is very heterogeneous and the concentrated fields can be very different, the second big bang led to the formation of different energy and matter forms compared to the first one. The new space and whatever it contained went through a rapid inflation period, probably because the inflation energy derived from the big bang was so dense that gravity acted as a repellent rather than an attractant. But the expansion rapidly slowed down, most likely due to the increasing influence of gravity. From that point on, for billions of years, our universe and the parallel universe grew together in a relatively balanced way, the parallel space always growing proportionally slightly bigger than the space created by the second big bang. However, about seven billion years ago, this difference in favor of the parallel universe became so significant that it generated enough vacuum force to overcome the force of gravitation and begin to accelerate the expansion of our universe."

"Explain to me one thing," asked Mike. "Was the universe born inside the preexisting parallel universe, or outside of it? In the latter case, how is it that the parallel space didn't leave our universe behind, thus preventing any interaction between the two spaces?"

"This is a very good question, one that Peter also raised. There are two options, as you said. In the first case, the universe was born inside an unstable region of the parallel space, the latter always growing bigger than our universe. However, as a second option, if the universe had been born outside, then rapid inflation would have provided an opportunity to catch up with the preexisting space and interact with it. This would have required not only that inflation rapidly increased the size of our universe, but also that during that period, our universe was propelled into the parallel universe, perhaps aided by the gravity of the latter. If we assume that the information in the vacuum is the reason and the basis for both big bangs, then I believe the second option is more likely."

"Why do you think that there were only two big bangs?" asked Steven. "If quantum jiggers in the information field of the vacuum led to the formation of these two coexisting universes, why did only two universes come into existence? Why not billions more?"

"Brilliant! Your question is simply brilliant," exclaimed Doug. "I can explain this only by assuming that God directed the initial quantum events to make sure that the concentrates of the information field were of the right quantity and quality to yield these two universes in the correct sequence. God could do that because He is the totality of information. In that sense, the universes derived from Him."

That was too much for Steven to digest, and he asked, "But if God is nothing more than the totality of information, how could He direct the concentration of the right amounts and qualities of information to trigger a big bang? This would be like fire regulating itself."

"Well, I think the key is that increased quantity of anything at some point can yield a different quality," said Doug. "For example, unicellular and multicellular organisms represent different qualities. In a multicellular organism, cells don't behave the same way as single cells do, without having to interact with others. In that respect, your analogy with the fire is incorrect. No matter how big a fire is, it is still limited in scope and will always remain a fire. However, if that total information is a very large, perhaps infinite, quantity, it may assume a very highly organized structure that we can call God. So I'm taking the view that the spiritual God is the sum of all information, and the material God is the corresponding highly organized information structure, capable of rearranging and moving information around at will. As I said earlier, He can do this because all bits of information are part of Him. He has total control of what happens to information. The relationship between spiritual God and material God is somewhat comparable to our brain, in which the millions of nerve cells are required but alone not sufficient for higher mental functions. These require the nerve cells to be highly organized into specific structures. The totality of nerve cells without an organized structure would be incapacitated to regulate any function of the body."

"Okay, I am starting to understand your idea about the dual nature of God," said Mike. "But going back to the relationship between the two big bangs, how do your theories explain the different shapes of the galaxies?"

"Well, a likely scenario for me is that in the parallel universe, random quantum events yielded dark matter conglomerates in different shapes, which then became gradually attracted to the small densities of visible matter formed in the second explosion. This would explain why the pattern of the location of galaxies follows the distribution of extra heat, and thus extra mass, in the early universe. The initial deformities, I believe, were then further enhanced by accretion of more matter including galaxy mergers."

"I always thought that from the very beginning, the universe had been homogenous," interjected Mike again.

"Immediately after the big bang, there were probably many lumps in the tiny space due to ongoing quantum fluctuations, but inflation smoothed that out, leaving behind only minute differences in the distribution of heat and mass. So a coarse picture of the early universe, provided by the measurement of background microwave radiation, indeed shows homogeneity of space. However, at higher resolution, very small denser heat spots show up. To most cosmologists, these slightly denser spots were suitable to serve as the starting points for galaxy formation by attracting dark matter conglomerates and clouds of visible matter."

"So that explains the mystery of galaxy shapes," Steve said. "But how do you explain that the universe apparently has no antimatter left?"

"To my best knowledge, nobody really knows that for sure. Sakharov, Hawking, and others have tried to explain this phenomenon by assuming that during the very early period after the big bang, the particle physics symmetry was violated, and antimatter particles were annihilated at a greater rate than matter particles. However, results obtained with particle accelerators so far have failed to confirm such a scenario. My hunch is that soon after the second big bang, the already available dark matter somehow helped to remove antimatter particles from the equation, either by helping its selective elimination or, more likely, by attracting them through gravitation and thus making them unavailable to contact matter particles, which would result in the mutual annihilation of both kinds of particles. This could occur if dark matter's gravity slightly preferred antimatter over matter. I think no one else thought of such scenario. Another possibility is that antimatter may have a slightly greater propensity to return to the information state. Although most physicists think that matter and antimatter attract each other, there is no proof for that. In fact, some even think that matter and antimatter repel each other. If so, it's also possible that the universe has two large pools of galaxies separated from each other by vast distances. One is composed of dark matter and visible matter, and the other of dark matter and antimatter. But, in my view, such a scenario is possible only if dark matter with sufficient gravitational pull was already present in the parallel universe at the time when the second big bang occurred." Doug then leaned back in his seat, suggesting that he was done. "So anyway, for the time being, that is my idea about Creation. Of course, I could be wrong, like perhaps anybody else."

Doug wanted to get back to the game at hand, but Steve's interest was piqued. "I know you wanted to be done with all this scientific stuff, but I have to ask. You casually mentioned the Higgs field. What is that?"

"Yes, I mentioned that the preexisting space could also be endowed with the Higgs field, which is the source for the famous Higgs boson. Physicists have been struggling for decades with the question of what gives mass to the various types of fundamental particles. Their inability to solve this mystery has been impeding progress in developing the standard model, which is a quantum field theory that tries to unify the three forces by which matter particles interact: the electromagnetic force that holds the atom together, the strong nuclear force that binds quarks into protons and neutrons, and the weak nuclear force, which is responsible for radioactivity. Peter Higgs and some others came up with the idea that a certain boson could provide mass to fundamental particles. Not surprisingly, since their idea was published in 1964, there has been a frantic and still ongoing effort to prove the existence of the Higgs boson. If indeed this is how fundamental particles like quarks, gluons, and electrons gain mass, it seems perfectly logical to me that the Higgs field had already existed at the time of the big bang and that it helped create order in the formation and association of matter particles soon after the blast. In fact, the Higgs field could even provide mass to the neutral dark matter particles."

"And how do the spiritual and material sides of God come into all this?" Steve asked.

"I assume that the material side of God established the initial precise conditions, a sort of blueprints for both big bangs that functioned in concert to shape the development of our universe. As I already said tonight, since in a sense God is the totality of information, and our universe was born from information, everything in the universe is part of Him. Thus, we're part of Him as well."

"Very interesting," said Steve. "So let me make sure I understand what you're saying. You believe that the material God provided the blueprint for the birth and evolution of the universe, and then it left the universe to govern itself?"

"Yes, in a sense my present theory seems to be consistent with that," said Doug approvingly. "But I'm also saying that since the physical world is part of God, the universe has never really been left alone."

"So I'm God?" asked Steve half jokingly.

"I can assure you that you aren't, so you can relax," answered Doug with a laugh. "But you are part of Him, if it makes you feel better."

"Honestly, I'm still confused about these two Gods and the two big bangs," admitted Mike.

"Well, Peter was completely lost as well. I suppose I'll give you all the same presentation I gave him." He once again took his pencil in hand and drew a scheme outlining the theories he had just explained.

Total information (Spiritual God) ↔ highly organized information (Material God)
↓
Separation/concentration of two pools of information
↓
Intermediate energy forms
↓   ↓
1st Big Bang 2nd Big Bang
↓   ↓
Parallel universe   our universe
↓   ↓
Dark matter/energy → visible matter
Other components   ↓
Galaxies
(visible matter + dark matter)
or (antimatter + dark matter)

CHAPTER 3

MORE DISCUSSION OF RELIGION, SCIENCE, AND EVOLUTION AT THE POKER MEETING

It was already 10:00 p.m., but none of the friends wanted to stop playing. Though they didn't want to admit it, playing and talking eased their worries about Peter, which were starting to creep in on them. Peter was many things, but forgetful or unreliable weren't among them. Finally, Steve suggested that perhaps they should just call Peter. "Anyone got a cell phone? I forgot to bring mine."

"I'll call him right away," replied Mike, who felt that as the host, calling Peter was his duty. He walked over to his landline and punched in the number. With each successive ring, their hearts sank a little deeper, until eventually they heard Peter's voice mail. "Pete, where are you?" Mike said after the beep. "Call me when you get this." He hung up, and an eerie silence fell over the room. Mike couldn't help but voice his concern that something unusual was going on with Peter, but the others assured him everything was fine.

"After all, how many times have we been worried about someone, and how many times has it turned out that they are just fine?" Steve reassured him. "There is probably a perfectly trivial explanation for Peter's absence. Maybe Helga visited him again. Or maybe he couldn't resist going out with another woman and he just forgot about poker." They didn't know that Steve was not far from the truth when he speculated about another woman. In fact, just a week ago,

Steve had heard from a patient who worked at IBM that according to some rumors, Peter might have developed a relationship with his secretary. But rumors can come and go fast, and Steve didn't think this was the kind of information that he should share with anyone.

After a short pause, Doug broke the silence. "My bet is that he's on an airplane," he said. This short sentence had a soothing effect on all of them; of course, this was the most likely explanation. He must have been summoned by IBM on short notice to fly somewhere. After a few minutes of brooding over Peter's unpredictable flying schedules, Doug again interrupted the silence. "Well, how about we stop worrying and contriving all kinds of scenarios for his absence and keep playing for a while? Maybe we will try to call him later too, and if we cannot reach him, we will try again first thing tomorrow morning." That appeared to be a reasonable solution, and they continued playing, though always hoping that at some point Peter might call them with some excuse. For more than two years they had played together every single Monday evening, and someone missing without letting them know in advance had just never happened before. It is true that Peter was occasionally absent because of his business trips, but whenever he could, he arranged his travels for other weekdays, and if he could not, he always warned them well in advance. Peter knew that they usually played until midnight, and so there was still time for him to call them.

As the night wore on, Steve began teasing Doug about whether he had any more ideas for how the Catholic Church that all of them attended should change its teachings. Doug had just won the previous pot, so he was in a good mood and was ready to play along with Steve. "You know, when you mentioned Helga's name, it reminded me that I also had an interesting philosophical conversation with her and her previous boyfriend, Ulrich. The topic of that conversation relates to your question."

"We definitely want to hear about that too," said Steve. "We still have a lot of time to kill before midnight."

"Well, as you may know, I occasionally invite other scientists to my home for dinner. I'm not saying that what we serve is exquisite cuisine, but it's usually sufficient to create a pleasant environment for discussing a wide range of topics, from recent developments in science and politics, to simple gossips circulating around the Mayo Clinic. However, I should emphasize that before meeting the Germans, I had never ventured into the topic of religion with my guests, just as I've never done so with you either. I particularly used to shy away from initiating any conversation about the relationship between science and religion. Up to that meeting with Helga and Ulrich, my conversation with Peter had been the only exception, and it had occurred only because I had seen Peter's interest in the topic. Discussing with Peter my budding cosmological theories was a watershed moment for me. From that point on, I have devoted more and more of my spare time to the issues that I discussed with you tonight. Sometimes I stayed longer than I had to in my office, where I could be alone, to consider various options to describe, or at least imagine, what the pre–big bang period might have looked like and how God fits into the equation in a way that is consistent with scientific laws. In short, the things I have already told you about. I must confess that it almost became my obsession, a kind of hobby that not even Sarah knew about until very recently. I was always afraid that she would tell me I was wasting my time, developing theories that are necessarily built on intuition and guesswork. Sometimes I feel that all I am doing is piling false ideas on the top of even more, even falser ideas. But no matter how irksome and frustrating this is, my approach was and still is that even if only 1 percent of my theories ultimately prove correct, my efforts would already be worthwhile."

Unfortunately, what Doug failed to see was that in the process of working on his theories during the long evenings, Sarah became increasingly annoyed because she didn't get the attention from him that she desired. But she still loved him enough to ever let her simmering dissatisfaction show; besides, she thought that it was only his work that occupied him overtime. Very likely it would have been better if she had aired her frustration instead of bottling it up; this would have forced Doug to come clean.

"Well, let me summarize my conversation with Helga and Ulrich," continued Doug. "After Helga's arrival to my laboratory, I soon learned that her German boyfriend also worked at the Mayo in the recently built Guggenheim building, which houses a fairly large number of excellent research groups. Through my nutrition and obesity studies, I also became interested in the cutting-edge diabetes research that was going on there and elsewhere. Thus, it was only natural that I invited Helga and Ulrich to lunch on the following Saturday. I expected to discuss a lot of things, so our agreement was that they could stay for dinner as well if they wished. Sarah volunteered to prepare a Minnesotan specialty for dinner, a plate with Valley fish and steamed vegetables. I knew that she had really been looking forward to this party because for months she spent very little time with me, and now for almost a full day, we could enjoy each other's company. I bought a few bottles of white wine produced in the Rhine River area that complemented not only the fishplate but also the occasion to welcome our German guests.

"Helga and Ulrich arrived around noon in good spirits. Nothing even remotely suggested to us that they were on the brink of parting ways. This obviously didn't mean much. They were too intelligent to show their true feelings in our presence. After all, we were important hosts for them. It was a nice day, and by the time the guests arrived, I had already laid the table outside on the deck away from the street. I still remember that the weathervane was almost motionless and the temperature was pleasant. On weekends, when the weather permitted, Sarah and I liked to dine out on the deck regardless of whether we had guests or not. We still have this habit. We enjoy the short-lived peace that we can find there. The children like it too. They are usually excused in the middle of the meal and let them play around the deck. Anyway, I led Helga and Ulrich to the table and offered them aperitifs, but finally we all settled with the white wine. Ulrich noticed that the wine was from Germany and asked us how we liked it. We all agreed that Germans are also very skilled at producing fine wines, not only the French and Italians. In fact, I just had a great idea. I think I will bring some white German wine to our next poker party.

"In any case, I already knew through Sarah how they had ended up at the Mayo, and we did not waste any time going over this again. Instead, while Sarah was serving lunch on the deck, I remember jumping right into the topic of Ulrich's research and asking a lot of questions during the meal. I was circumspect not to ask personal questions. I was surprised to learn from Ulrich that inflammation is important for both obesity and diabetes and that fat tissue and the macrophages in it are not at all innocent bystanders. I always used to look at the fat tissue as if it had no significant physiological role aside from storing fat and therefore wasn't worth studying. But recently this old view has dramatically changed, as some stubborn scientists have kept discovering more and more key regulators of inflammation and metabolism that originate from the white fat tissue. Ulrich explained to me that he was close to finding a suitable intervention to safely turn off a master regulator of inflammation in the white fat tissue, which would curb weight gain and diabetes, and hopefully the risk of pancreatic cancer as well. But his time at the Mayo would soon be up, and he would have to find another means of earning an income in a hurry. Despite this outlook, he didn't seem to be flustered. He said that he was contemplating founding a biotechnology company and thus employing himself. I assured him that this was indeed an amazing opportunity, but I added that it must be done right because the failure rate is quite high.

"'I agree with you. If I don't do this right, it can easily lead to a financial debacle,' he said.

"As we carried on with the conversation, we also touched upon Helga's and Ulrich's expectations in the US, and that was the only occasion when I saw on Sarah's face that she might have instinctively sensed a rift between the two Germans. Ulrich was aware that their future was going to be beset with difficulties. However, he didn't try to hide this fact of his life and didn't show any sign of being downcast. He was clearly ready for the challenge. He illustrated their situation with debonair wit. Helga sounded a little strained and didn't seem to share Ulrich's optimism. When Sarah asked if they missed anything in Germany, after a little pause Helga answered her with a definite yes, but Ulrich said, 'Not really.' But then he added that one thing he sorely missed was to watch matches in the soccer league there, particularly when the Stuttgart team played. In his words, 'once you have been caught in the thrall of soccer, it stays with you forever,' and he was one of the many strongly committed fans. I agreed and said that I also used to watch English teams play on the weekends. It was difficult not to notice at this point that there was something off about Helga's tone when she answered questions about Germany. 'I miss a lot more than soccer,' she said once. 'I just can't expunge some good memories.'

"Seeing the different reactions of our guests, at that point I sensed that I might have unintentionally touched a nerve and that it might be better to change the subject of our conversation, from the personal to the more general. I said I assumed that they did not have a lot of time to read books. Ulrich responded that in general I was correct, although he had just finished reading *The God Delusion* by Richard Dawkins. Although reading it took him almost a year, he had finally finished it. Ulrich was clearly excited about the book and elaborated on his keen interest in evolution from an atheist's point of view. He had a high opinion of Richard Dawkins, a scientist who indeed has a good record and excellent reputation in this area and who is presently one of the most effective and intelligent atheist opponents of religious beliefs. I agreed about the qualifications of Dawkins, though adding that he still can be as ruefully direct in attacking religious people as many other atheists, for example Bill Maher, a comedian, but at least he can back up some of his arguments against religious dogmas with serious science. This book entices those, I said, who believe that evolution must rule out creation of the universe by God.

"I also told the guests that I knew Dawkins's work very well and that I had read some of his other books too, including the *Selfish Gene* and *The Blind Watchmaker*. I couldn't help but further elaborate that with all due respect to Dawkins as a scientist, I find his approach to religion, namely, the intolerant way he discusses it, unacceptable. For example, it seems to me that in his mind, only easily gullible and essentially crazy people are religious. I think that Dawkins fights religion because in his view it's something malicious that only serves to give people the illusion of hope. I don't see it that way. For me, the existence of God, even if I see Him differently than traditional religions do, is not an illusion. I am appalled when someone tells me with utmost confidence that my belief in God is an illusion, implying that I'm stupid. Contempt of anyone's belief, be it religious or atheist, by a righteous person always bothered me."

Steven nodded in agreement and then commented, "It sounds like you have other problems with Dawkins as well."

"Oh, I have plenty of problems with him and atheists in general, as I pointed out to my guests," Doug continued. "For example, Dawkins thinks that people are religious only because they were indoctrinated during their childhood. Thus, somebody with an open and free mind would automatically have to become an atheist. In Dawkins's opinion, people become liberated when they have the courage to leave their religious beliefs behind. I can't help it, but Dawkins's strident blasphemy page after page and the way he summarily characterizes religion as a fallacy simply irritate me. It's one thing to believe in something, in Dawkins's case that there is no God, but it's another thing to label other people as unintelligent just because they don't share your opinion. I consider this an extremely arrogant viewpoint. After all, what gives him the right to think that he knows better than anybody else? Unlike him, I believe that I have no reason to think that my beliefs are more correct than anyone else's including his."

As Doug continued his account of his conversation with the Germans, he raised his voice without even noticing it; while talking about Dawkins and rebuking his approach to religions, his passion had clearly been stirred. Fairness was everything to him. He would have been equally zealous against any intolerance toward atheists. "How can anyone make any solid statement with any certainty about how and from what the universe was created? Everybody should be humble when talking about these hugely important issues, and I hope you can see that I am trying to be humble too. First, we still know very little about the universe. We know a few things, about maybe 4 percent, of the visible universe but practically nothing about the rest except perhaps that dark matter has gravity and mass but no electric charge and shows no signs of any of the other forces. But really this is all, and who knows what other mysterious particles and energy forms might be out there that we are not even aware of? Scientists are always discovering new and baffling things about the universe, which they can't explain.

"So, in my view, you may have theories about our universe and how and from what it began, but to say anything with certainty does not indicate to me very deep thinking. The same applies to how life was created. All scientists and theorists will have to take a step back, and instead of rushing into judgments and lambasting religion, they should think things through more carefully and, above all, respectfully. We all should be prepared to be more open-minded because new discoveries can change our views rapidly. Remember that not too long ago, materialists claimed that our universe and even matter had existed forever and incessantly criticized and even ridiculed the Catholic Church's stand on Creation. Now that creation of the universe, whether via the big bang without God's involvement or another event, has become established, the internal harmony of materialism has been compromised. So presently the main questions that most scientists are busying themselves with revolve around the nature of the *something* from which creation started and how this *something* was detonated by a presently mysterious event.

"At this point, Ulrich noted that it was quite clear to him that I was religious, or at least I believed in the existence of a god, but Helga and he weren't, and as far as he knew, most Germans weren't either. To his surprise, he had observed that most people in Rochester and elsewhere in America regularly went to a church, and from that he drew the conclusion that religion must have something to offer to the believer. But he found it hard to fathom why this big discrepancy between Germans and Americans existed. Then he asked me what gave me the push toward believing in the existence of God. Was it indeed childhood indoctrination as Dawkins would say, or did I come to this conclusion on my own? Nobody had asked me that question before, and I remember pausing for some time, pretending that I was temporarily engrossed in drinking tea from a nicely decorated mug Sarah brought with her when we had married.

"Now that I have brought up Ulrich's question, I would say that Sarah and I have had equally long and eventful evolutionary journeys until we finally developed our similar religious beliefs. When we met, we didn't have the same idea about God, but He was still always a major linchpin between us. Sarah had accepted God's existence as a child, but very early in her high school years, she concluded that the Bible should be taken merely as a beautiful piece of literature that provides time-tested essential guidance for how to live one's life. She believed that the Bible might bear God's signature, but it was intended for people who lived centuries ago in far less advanced societies with limited knowledge of natural laws.

"During the period when Sarah was still dating Peter, she went further in doubting the church's doctrines and had a temporary bout of full denial of God's existence. Then, during her college years, we began to have a series of frequent discussions on the subject. For quite a while she lurked between belief and nonbelief, but after a time, she finally became convinced that there must be a god. Although Sarah and Peter also used to talk about religion, she didn't like what she heard from him. At the time of their dating, Peter was also a church regular, and although he didn't seem to have very strong beliefs either, he didn't show any openness to new ideas about religion. Of course, Sarah couldn't know that recently Peter had become interested in explaining God and religion from a different perspective and that this strengthened his faith. The difference in their attitudes toward religion certainly played a role in Sarah's decision to break up with Peter. At least this is what she told me.

"The German guests listened intently without interrupting me and without asking who Peter even was, which I had forgotten to explain. Thus, I continued speaking. If I really want to flesh out the reasons for my faith, I should start by saying that I also was a kind of theological rebel in my youth, perhaps even more so than Sarah. I was not quite sixteen when I wrote an essay in high school about the beauty of evolution. In it I cited many examples of how new discoveries in life sciences proved evolution to be true and how they disproved the religious views of the creation of earth and of life. And you may not believe this, but I even read some books from Lenin like *Materialism and Empirical Criticism*, in which some of the basic tenets of the materialistic worldview, including disdain for any religion, were laid down. I even attempted to read Marx's *Das Capital* in the original German, but after reading about twenty pages, I gave it up. Both the text and the language were too difficult for me."

"You did a lot of things in high school we didn't know about!" interjected Steve. "I thought it was mostly football and biology you focused on. But in any case, go on with your story. We still have some time left."

"Well, I then talked about how and why I had become interested in envisioning how the first cell might have formed. During my high school years, our biology teacher presented this as a simple process with no need for God's involvement. Many millions of years ago, the first cells were formed from molecules that were put together from basic chemicals present in the hot primordial soup on earth's surface and the oceans, and then these cells evolved to meet the various challenges the environment imposed on them. As you know, a cell is defined as the smallest functional unit of living creatures, and it is composed of an internal viscose fluid that is kept together by a more solid structure called either a membrane, as in animal cells, or a cell wall in plant cells.

"After a relatively basic introduction to cells in high school, at Harvard I was soon immersed in disciplines such as biochemistry and cell biology, and these courses began to have a serious impact on me. I began to appreciate the enormous complexities of cells. I soon learned that when someone tries to understand how the first cells formed in the context of evolution, the devil is indeed in the details, big time. I truly venerate the scientists who deal with the processes that might have led to the birth of first cells, but I believe that most of them take an unrealistic view on this. I have never really talked to any of you about this part of my life, but it was critically important to me in shaping my relationship with God. I refused to accept the oversimplification of how the first cells formed. It really annoyed me when I heard simplistic explanations one after another. But aren't you guys bored yet?"

"No, we aren't bored at all! At least I'm not," exclaimed Steve. "Please go on with the story. It's just getting more and more interesting. We still have a good hour to play if we want to stick to the midnight schedule. Besides, if you keep talking, we have a greater chance to win."

"I admire your resilience. In that case, this is how I continued my explanation to the guests: Well, the science courses I took first introduced me to the molecular building blocks of cells with more and more layers of complexities added later. It was amazing to me to learn that a single cell has many small organelles composed of tens of thousands of proteins, lipids, carbohydrates, DNA, and RNA, all working together in mysterious concert with millions of highly organized and balanced interactions among them. And then cells also undergo incredibly fine-tuned complex changes for which they are getting cues from both their environment and their genetic makeup.

"The myriad of biochemical events in a cell that underline all cellular functions really captured my fantasy. My recalcitrant, know-it-all attitude began to change, and I was less and less willing to view atheism as the only valid view of life. My harsh criticism of religion abated. I slowly began to realize that life, even a single cell, is far from being as simple as I was led to believe earlier by ostentatious teachers and even scientists. The same complexity of life, which changed many scientists from believers to atheists, forced me to gradually step away from my rigid materialistic view of the world and move instead toward belief in a higher authority who somehow had a hand in arranging this incredible complexity so that there may be order to life and everything else. It was simply too hard for me to believe that within the time constraints of earth's existence, the immensely complex cell could have formed and then evolved without some divine intervention providing some key organizing principles. Such completely autonomous self-evolving scenario became even less likely when I considered the added complexity of organs, composed of many different cell types, and their multilayered interplay in the human body.

"I then slowly but surely accepted God's existence and authority, and after I made this decision, I have never veered from that path. Somehow, I found solace in my newfound faith. At the same time, I also agreed with Sarah that God wasn't interested in controlling people's lives in detail. God must be infinitely intelligent, and an intelligent God wouldn't possibly want to subjugate anyone. As you recall, this is exactly what Peter also told me when we talked about this subject. We also agreed with Sarah that the Catholic Church's view of the role of God in the creation of humankind is very confusing. Slowly we arrived at the conclusion that while God had a hand in the creation of man as well, His role was indirect, and He chose the process of evolution to accomplish that feat."

"You really whittled down God's role in the creation of humankind," noted Mike. "If I may advise you, don't talk about this idea in our church because it won't burnish your reputation as a Christian. They will accuse you of wanting to shear the church of its identity."

"I know, I must exercise prudence in this," agreed Doug, "particularly because I don't even know yet if anything I am saying has any merit. But you are my friends, and I see no reason to hush up my thoughts in front of you."

In the meantime, they had started a new round of poker, which Doug followed only superficially; he was once again a biologist, sharing with his friends his take on cells and evolution as explained to his German guests. "The more I immersed myself in life sciences, the more I began to wonder if my teachers were biased or ideologically too committed to not see the folly in their explanations of nature. But evolution and natural selection were such strongly interlinked theories that I understood why most scientists didn't believe in God's existence and His role in the creation of the universe and humankind. However, my criticism at that time was aimed not only at atheist scientists. I also rebelled against the long-term untenable positions of the Catholic Church that almost voluntarily separated itself from science, and I believe that it continues do so except for minor, though encouraging, recent positive steps it has taken. I believe that science should complement, rather than contradict, God's plan. Theologians should be on equal footing with materialistic scientists in attempting to define the universe. Why can't evolution and natural selection be part of God's plan? Why should science and God be mutually exclusive instead of mutually inclusive? These college years were formative for both me and Sarah, and our frequent discussions helped us both to return to God, even if it was in an unconventional way. Thus, on

Sundays I still often go to church, always together with Sarah, because of my strong faith in God that I received through a deep scientific soul-searching. And the fact that my view of God deviates from the God described in the New Testament doesn't change that. But this is today. Tomorrow I might conclude that my forays into these new ideas were in vain and that the church was correct all along."

"And your reputation as a true Christian will be restored," said Mike laughingly.

"Yes, but I didn't finish. I also can't totally exclude the possibility that after the end of my journey, I will have no other choice than to conclude I can't reasonably argue for the existence of a god. Right now, however, my internal pendulum strongly tilts toward God."

Then it was Mike's turn to chime in again. "It seems to me that you and Sarah have problems with some of the church's fundamental doctrines, and the two of you pretty much support each other in that. Am I correct?"

Without hesitation, Doug was ready with his answer. "I would say that an upshot of our continuous search for God is that by now we are at least in coarse agreement about where we stand on religion. For example, we both think that He is not the kind of God who ostensibly requires submission of the people He created. To give an example, we believe that our God wouldn't require us to do things like go to church every Sunday, or ever for that matter, even though such a practice is a positive experience and has value. But the urge to go to a church service must come from the inside. In our view, this and other church's routines were entirely invented by people without divine intervention. We can converse with God in many ways, and church attendance is only one way of doing that. We don't necessarily have to rely on the priest's role as a messenger, although good sermons can teach us, or at least make us think about valuable things. Thus, when we begin to miss the anodyne words of the priest, we go to church. Otherwise, we stay home and think about Him there."

"I wonder how Sarah can keep up with all these theories of yours," remarked Steven incredulously.

"Frankly, I haven't really been totally honest with her, and tonight I already explained why. I haven't told her that recently on many evenings I stay in my office after hours thinking about these ultimate questions. I simply want to be alone and think undisturbed about these things, but I'm often home very late as a result. When I think of it, I reproach myself, and I'm in the process of changing this habit, but I'm afraid I have already robbed Sarah and my children of valuable time we could have spent together, which I'm not at all sure that Sarah would understand. The other thing is that since Sarah's opinion is very important to me, I don't want to risk damaging her opinion of me with ideas that might seem baseless or even idiotic. I think I'll only tell her once the internal logic of my theory becomes infallible in my eyes, even though I know that factually it can never be proven. Once a few months ago I tried to talk to her about the possibility of the dual nature of God, and her answer was, 'Doug, I believe you would do better if you just stuck with medicine.' This remark made me think twice before prematurely sharing my ideas with her, although indirectly I just did that when I explained some of them to the German guests."

"You started to talk about the German guests a couple of times already, but you didn't really finish the story yet, and it is already half past eleven. Will you finish the story some other time?" asked Steve.

"No, I will finish it right now. No more side stories. So to respond to Ulrich's question about my faith, I needed to go back to Dawkins's morally repugnant position that religious people must be stupid. His arrogant view is a stretch because many good scientists, indeed thousands of them, are religious. It is sufficient to mention Francis Collins, who headed the Human Genome Project, and there is talk that he will be the head of NIH. Nobody can say that he is a less accomplished scientist than Dawkins. With that I only want to say that whether someone believes in the existence of God or not, it has nothing to do with the person's mental capabilities. Two persons with the same mental power able to analyze the same information provided by science and the church's teachings may come to opposite conclusions about God.

"I mentioned Collins. In his book *The Language of God*, he argues that a caring God created humankind through an evolutionary process. The Almighty created the universe with its natural laws and chose evolution to be the process for the creation of humankind. In Collins's mind, science doesn't exclude the existence of God. In fact, it argues for it. I agree with Collins and disagree with those who believe too literally in the events described in the Bible. Like Collins, I also disagree with proponents of the intelligent design theory who claim that God occasionally intervenes in the process of evolution to perfect it. Collins thinks, and again I agree, that this is a serious underestimation of the exceptional intelligence and supreme nature of God. He put this universe in motion with all the laws required for the evolution of the universe and of life to take place, and it would be an admission by Him that He created something imperfect if there would be a need for occasional adjustments. God isn't interested in subjecting Himself to regular self-aptitude tests. After creating the conditions for the Creation, whether it was the big bang or something else, He might have intervened only one more time to create cells, but only in a way that didn't require Him to change any of the natural laws. But I must admit that the more I think about it, the more I doubt that He was involved even in the creation of the first cell.

"I allowed the German guests some time to digest what I said and drink some more wine, and then I added that what I had just said was far from what the Bible teaches. But in my opinion, it would be better if Christians started doing their part by fully embracing science, although only good science, while in the process confronting both atheists and themselves. Finding harmony between faith and science should not be too difficult. All it would take would be to accept the evolutionary process as part of God's plan to create all living organisms and humankind, although I'm aware that such message even today would appall many Christians and engender resistance. But I argued further. Consider the huge gains that such a new approach would bring. For one, all great new scientific achievements would become yet another celebration of the still ongoing creation process, both in the universe and here on earth with the evolution being a major part of it. The present confrontational relationship between faith and science would become much simpler, in fact bracingly clear, free of any rancor. Instead of working against one other, scientists and believers could come to the same side of the fence, or at least people of faith wouldn't have to be on the defensive anymore. The most significant outcome would be that atheists could no longer lay claim to the evolutionary process, which would dispel the myth that evolution demands atheism. Atheists would have to give up their current perch as the zenith of reason particularly because their ideology is also full of controversial issues. Such a move by the religious would level the playing field of science, providing equal access to it by all sides. But if such a change doesn't happen, Christianity and other religions will keep floundering as nothing more than obsolete ideology in the face of atheism. I know that many people who believe in God might think that the danger I'm describing is a figment of my imagination, but I think it is very real even in America, not even mentioning Europe and other parts of the world."

"Do you think that God would allow the downfall of Christianity, or any other religion for that matter?" asked Steve.

Doug firmly responded, "No, God wouldn't allow this to happen. He wouldn't forsake us. In fact, Ulrich asked me practically the same question, and I remember answering that in my view, although God wouldn't tinker with the arduous process of evolution He started, I believe that He does have the means to direct us toward saving the church by helping us put the church's teachings on a new and more credible foundation by fully incorporating science. I told Ulrich that all this might not directly answer his question of how exactly Sarah and I became believers in God's existence, but perhaps you can see the thought process we went through. At this point the children understandably grew bored and went inside the house with Sarah. I refilled the empty glasses, and with the German guests we stayed outside. Quite honestly, after Sarah left, I became more liberated, almost reckless, in talking about my theories. This was such a nonpressure environment to debate Ulrich, an atheist, and hone my ideas."

"I can imagine. It must have been an interesting experience to argue with another smart scientist who had a completely opposite viewpoint on God," said Mike.

"Yes, it was." Doug smiled as he recounted that interesting encounter. "In any case, I continued to criticize Dawkins's approach to religion, which I consider disrespectful and unforgivably arrogant. It particularly upsets me that while he scolds religions for indoctrinating young minds, he conveniently forgets the atheistic indoctrination of everyone in socialist countries where atheism was the ruling ideology. As an example, I told him about one of my older colleagues in a different department, Joseph Hiller, who finished his schooling in socialist Hungary before coming to America to do research. He told me many stories about such brainwashing, starting in his elementary school and continuing until he graduated from a medical school in 1980. Dialectic materialism, based on the teachings of Marx, Engels, and Lenin, was the only sort of 'philosophy and economy' that was taught and considered accessible for the young minds in socialistic countries. All other philosophical and economic systems were taught only briefly and through the critical prism of dialectic materialism. Officially, religion in any form was ridiculed and despised, and it was regarded as a poisonous fallacy that was dangerous to the socialist workers' state.

"At some point, back in the seventies, Joseph became interested in Western philosophy, but the only relevant book he could find in the library of an institute fully devoted to teach 'philosophy' was a book on Kant in German that had somehow escaped the attention of otherwise very guileful authorities. Communist atheists in those countries did a nearly perfect job of eliminating books that didn't fit their ideology. They were probably better at it than the Nazis, who similarly eliminated certain types of books in the thirties.

"Anyway, on that day I was really in my zone, and in retrospect I should acknowledge that I let almost no one else talk, perhaps in part thanks to the good wine. Here's how I continued my stance:

"Apart from the Nazis and Communists, and apart from the duplicitous arguments and convenient oversights of the atheists regarding who indoctrinated whom, Dawkins and other atheists are too simplistic when they explain the emergence of humankind with natural selection as the sole driving force of evolution. This is where I beg to differ from most biologists and even Collins. While

I also argue for the role of evolution and natural selection in the development of life, I consider contribution by other processes as well. For example, Stephen Wolfram, a theoretical physicist with very original ideas, recently put forward an additional possibility intertwined with natural selection. He has made lasting contributions to mathematics, computer sciences, biology, and cosmology with his theories. In his book *New Kind of Science*, he reports on hundreds of computer simulations, each demonstrating that by applying a few simple rules to an initially completely random system, it can become increasingly complex and surprisingly well organized. Such amazingly effective self-organizing principles alone could, according to Wolfram, account for the evolving complexity of living organisms without necessarily needing the involvement of natural selection as originally proposed by Darwin and fully backed by thousands of scientists including Dawkins. While I believe that Wolfram goes too far with his conclusion, his idea is really something worth considering. I can imagine that once we have a clearer idea about the initial organizing principles, we can then better explain how self-organization based on initial simple rules, in combination with natural selection, had directed the creation and evolution of life forms.

"At this point Ulrich interrupted me and asked how such simple initial rules could lead to the development of highly complicated, self-replicating organisms. For him this was the major question: how the very first cell with the ability to divide and transfer coded information into the resulting new cells came to be. He said that even though he was a materialist and understood the plight of cell biologists, he had to acknowledge that they were still far from providing a coherent hypothesis with sufficient depth to solve it, although not for lack of trying. He felt that somehow evolutionary scientists were held back by old ideas that didn't seem to be relevant for determining how the first cells were developed. Then Helga joined the conversation, stating that she had recently become more of an agnostic. She doubted that scientists would ever be able to solve this conundrum of the first cell, just like they would probably find it impossible to determine if there was anything before the creation of the universe and what exactly consciousness was. To Helga, these were the three most crucially important ultimate questions that even the most advanced science would never be able to solve.

"I thought that Helga was right on target by identifying these three key questions, and I agreed that in this present stage of science, it is indeed difficult to have a substantive debate about them. But I disagreed with her view that they could never be known, because I thought that these were all in fact knowable and the humankind should never desist from trying to understand them. When Helga looked at me with doubting eyes, I further qualified my statement. What I really meant was that due to future technological advances, we might, in principle, be able to find answers to all the ultimate questions, but this doesn't necessarily mean that we will. Then I added that I had some ideas about what might have been before the big bang, although this is again something that we may or may not ever know for sure.

"I really didn't want to talk about the big bang period and what might have been before it, because at that time I still didn't have any good conception of these events. Therefore, I shifted the conversation back to how the first cells might have formed. Ulrich was very much interested in this topic because, in his words, he was a great believer in the theory of evolution, and he was keen to hear any idea about how life might have started. I assured him that for quite a while we wouldn't know what the innumerable intermediary steps were after the formation of the first organic compounds from carbon, derived from supernova explosions as well as oxygen and hydrogen. But I was ready to share my ideas with him and Helga, although I warned them that there were still many holes in my theory, partly because we still know so little about what earth was like two to three billion years ago, and partly because I was still an amateur in the science of evolution."

At this point, Doug once again paused. "It's really starting to get late. Do you all still want to hear how our conversation about the first cell went?" he asked.

"I would love to hear it," answered Steven, "but I can't speak for Mike."

"It is alright with me," responded Mike, sweeping his latest winnings toward him with both arms.

While Doug prepared a fresh set of cards for the next round, he continued his account of the mystery of the first cell or cells, which for him was the most fascinating and important event after the big bang. "I began to think about how the first cell might have formed only after I read Wolfram's book, and although some key information was missing, I still came up with an idea. Indeed, some of the missing information wasn't even relevant, like how earth had acquired water. For the story of the first cell, it's enough to assume that having water on earth was part of God's plan. Water always had many simple chemicals dissolved in it, but some mechanisms were needed to get them concentrated, allowing continuation of the chemical reactions to yield more complex compounds. Some sort of closed structure could fulfill this condition by ensuring that water and the chemicals inside couldn't escape but miniscule compounds could still get inside. In other words, there was a need for water-repelling or 'hydrophobic' compounds to be synthesized by chemical reactions that could form a stable semipermeable membrane layer around a water droplet to stabilize it. I call this primitive structure a *skinned liquid droplet*.

"These *skinned liquid droplets* were probably simply waterdrops enclosed by hydrocarbon molecules, which prevented the water from dispersing. This enclosure later evolved into lipid molecules endowed with both water-repelling and water-attracting parts. In fact, today these lipid molecules form the bulk of the cell membranes of real cells. The hydrophobic part of the lipid repeals water, and therefore it doesn't mix with it except under extreme conditions. For example, strong mixing of water and a lipid or lipid-like substance by a strong swirling force or ultrasound in a laboratory results in micelles or liposomes, wherein the hydrophobic part of the lipid membrane surrounds a liquid interior, transforming it into a membrane-surrounded liquid droplet.

"In ancient times, waterfalls or hot vents could have provided the strong mixing strength required for the formation of these skinned liquid droplets from suitable hydrocarbon molecules and water. However, initially the skin around the watery interior had to be semipermeable. In other words, it had to be impenetrable from the inside while somehow penetrable from the outside. In addition, the skin or membrane had to behave as a molecular valve, initially letting simple water-soluble inorganic and organic compounds to flow only from the outside inward. However, in these closed structures, the various chemical reactions generated waste that somehow had to be removed to the outside. Thus, the skin around the liquid droplet further evolved, now allowing selective movement of various substances in both directions without making the structure too leaky. These droplets began to resemble today's cells more and more closely. The Germans were with me up to that point. Are you too?"

"I guess we're," said Steve. "We'll tell you when to stop, or maybe our snores will let you know. But have you noticed that Peter still hasn't called?"

"Yes, I did, but I still hope he might. Midnight is still fifteen minutes away. Let me keep you entertained with my cells until then," said Doug.

"So where was I? Ah yes, I was explaining Wolfram's ideas and my ideas about cells to the Germans. Now, applying Wolfram's findings, the number and nature of the initial mixture of compounds in the enclosed liquid droplet equals the initial random condition. However, in a way the chemical nature of these compounds can also be viewed as providing some predetermined programs, because certain chemicals will react only or preferably with certain other chemicals in special ways. Thus, the initial composition of the skinned liquid droplet determines what other successive chemical reactions can take place and what type of new compounds can emerge with time. Then new chemical elements and simple compounds, already present in the outside medium, can enter the skinned liquid droplet, which then may contribute, as building blocks, to the chemical synthesis of more complicated molecules. Since many chemical reactions require energy input, and maintenance of an organized structure against entropy-driven disorganization also requires energy, further synthesis of more complicated molecules and evolution of the skinned liquid droplets must have occurred only in and around hot vents like some geysers in Yellowstone or in Africa. Certain elements or simple inorganic compounds that were specialized in speeding up the chemical reactions were probably also incorporated into these liquid droplets. Today we call them catalysts, and chemists often use them to optimize chemical synthetic reactions.

"Of course, one of the big questions is, How could the primitive skin around the liquid droplet let the molecules move only inward? A while ago I thought that the skinned liquid droplets had the chance to be stabilized and further develop only if they were submerged deep below the surface of the water so that water pressure allowed the flow of material only inward but not outward. But this is probably a very naive view that only an amateur like me could ever consider. It is particularly difficult to visualize how certain elements, like phosphate as well as carbon, zinc, manganese, magnesium, potassium, sodium, and the compounds derived from them, could become sufficiently concentrated in an ocean so that all of them could be simultaneously present in the closed skinned liquid droplet. The first *liquid droplet*, in my view, could only be formed in shallow lakes derived from geysers. A shallow lake that went through cyclical drying out and refilling could serve as a concentrating mechanism for these elements and simple organic compounds. The liquid droplets that were formed by the strong force of hot geysers could be dispersed in the adjoining lakes, rich in these elements. All this would have required that the surviving liquid droplets' interiors became semisolid, resembling the interior of an egg, and the compounds inside stuck to that structure, preventing their outward movement. Hence, there was no need for a high-pressure environment, such as what can be found deep down in the ocean, to hold the structure together.

"But if the skinned liquid droplet was left alone, it was still very vulnerable to external forces. Therefore, I believe that almost simultaneously many similar liquid droplets were formed in a small area of the shallow lake, which then became attached to each other forming a more complex structure that provided more stability and which ensured free exchange of minerals and primitive compounds among them. If one droplet managed to chemically synthesize copies of a new compound, they were shared with the other droplets, which probably also synthesized their own unique compounds shared among the community of droplets.

"It seems conceivable to me that later some new molecules transported to and remained in the skin and became transporters of small molecules from the exterior to the interior. Like every other function, transporters later became more complicated, assuming selective functions, and some of them were also able to remove, via the process known as secretion, unused or waste products that would have otherwise put a brake on further evolution of the cell, or cell conglomerate. Of course, there are many other questions that will have to be addressed if this theory, based on new research, will be taken seriously and not treated as crapshoot business. But in this frame of thinking, which many other scientists are also following in several modified forms, I believe that all steps involved in the creation of the first cell or cells are scientifically and intellectually addressable. One of them is how the skin material was formed and how the essential substances used for the construction of internal organelles were retained. Yet another big question is whether these skinned liquid droplets, which we might even call *primitive cells*, were able to divide or not. A point on where I disagree with many scientists searching for the origin of life is that while they argue for the role of the oceans in the formation of the first cells, I argue for the role of hot geysers and their associated shallow waters. I also have not found reference for the lipid droplets or pre-cell structures to form conglomerates to ensure stability and faster evolution.

"Then Ulrich asked me how these primitive, but already somewhat structured, liquid droplets could have divided into separate droplets that were different than the original ones, thereby starting off the process of evolution. My answer was like this: When a primitive droplet broke up to form two new droplets, which in turn divided into two daughters of their own, eventually, somewhere down the chain of further divisions, there was bound to be a descendant droplet that was a little different than the parent. Statistically speaking, after many generations, eventually it had to happen. And this difference, no matter how small, would generate a myriad of new opportunities for the sequentially forming new daughter droplets to evolve in different directions, with the process eventually resulting in real cells as we know them today. Evolution also certainly required the distribution of either these stable skinned liquid droplets, somewhat developed cells, or primitive multicellular organisms into the ocean and land. The lakes would become swollen after heavy rains, making seasonal connections with rivers and the oceans, but they would also occasionally dry out, helping the distribution of these primitive structures onto land. In these different environments, many new opportunities for the evolution of cells could arise. The influx of new kinds of elements and primitive compounds would change the direction and magnitude of chemical reactions, thereby propelling evolution in many diverse directions. However, what sometimes makes me despondent is that why such primitive droplets have never been found in water samples, regularly collected from small and large water bodies and analyzed by powerful microscopes. Perhaps only the ancient earth's environment was suitable to form droplets, or fish simply keep removing them? Or perhaps my theory is wrong? I may never know that.

"At this point, Ulrich had another question. 'Asteroids hitting the earth often contain simple amino acids and even nucleotides that make up DNA. Do you think that asteroids played a role in the development of the first cells?' he asked. I answered that my problem with this theory is the same as what I had just explained in connection with the ocean theory. I do not see how molecules present in a meteorite can mix with the multitude of other molecules needed to begin the process of cell formation.

"Then Helga asked a question that almost gave me a headache. It went something like this: 'If there was such a long series of events that presumably took place prior to the formation of a new cell, do you think that all life is derived from one single particular cell, or from many different cells that arose at different times each following a particular path of evolution? This may account for the incredible varieties of flora and fauna that exist today. And to take it even further, have you ever considered that some primitive new cells might still be arising today and spawning ever new chains of evolution?' I think she missed what I told them about why droplets are not found today, perhaps because she was too deep in her thoughts.

"I answered that to my best knowledge, many specialists think that all living organisms are derived from one single cell, although there is a lot of debate about it. However, DNA analysis indicates relatedness among all living organisms, which would imply that even if initially there were many attempts by nature to transform the primitive *skinned droplets* into real cells, and even if some of them were able to achieve some higher level of complexity, the daughter cells of the most adept cell probably crowded out the offspring of the less viable ones via the process of natural selection, and this is the cell from which the entire modern living world is derived. The basis for this cell's and its descendants' victory over their competitors was probably their ability to pass on genetic information to offspring without error and very rapidly like modern bacteria do, and to be flexible enough to develop many different cellular forms without dying. But to produce such a surviving and rapidly proliferating cell, the skinned droplets had to go through many evolutionary steps, always the best ones being able to adapt and survive. At this point, I reinforced the idea that the skinned droplets survived and further developed only within a colony and only after the genetic material formed could individual detached cells survive and proliferate on their own and start many new branches of evolution. While in the colony, the participant cells had the same genetic makeup, and they became divergent only after they became detached from the colony. In that sense, all living might derive from the many cells derived from the same surviving colony and not from a single cell.

"After detachment from the surviving colony, some of the surviving cells would undergo further changes in their inheritable genetic information due to changes in the environment and by fusing with each other just like the skinned droplets might have done it. Furthermore, cells occasionally engulfed other cells, like bacteria, and this also added new genes and structural elements to the cells. All this would ultimately yield dramatically altered paths of evolution, resulting in organisms much different from those that live today. Many of the evolutionary paths that did come into existence weren't viable in the long term and eventually died out, while others, including the one that eventually led to the development of humankind, survived. In that regard it's amusing to think that we could have assumed a quite different physical form had a different evolutionary path become dominant due to changes in the environment. Or consider this: According to scientists, humankind could not have developed at all if the large meteorite that impacted the earth about 65 million years ago wouldn't have killed off the dinosaurs. Because of the many, often haphazard environmental conditions that influenced evolution on earth, I strongly believe that if similar evolution from single cells into intelligent beings took place on any other planet, we would probably not even recognize these beings just by looking at their physical appearance. In fact, the building blocks of the cells may even be composed from different elements. Silicone might replace carbon, or arsenic might stand in for phosphate. Perhaps even oxygen and nitrogen could be replaced with other elements in a very different physical environment. I believe that if we ever find life elsewhere in the universe, it will flummox even our most creative and intelligent mind.

"As I finished this last sentence, Sarah appeared in the door and reminded us that it was almost five and perhaps we should prepare for dinner. Helga and Ulrich were at first reluctant to stay, mostly out of politeness, but I easily convinced them that we should continue our conversation. We all agreed that since we had eaten less than half of the food prepared for lunch, we would simply warm up the leftovers for dinner and eat inside. This arrangement saved more time for our discussions. As soon as we settled down in the living room, Helga immediately asked me again if I thought that God had had a hand in the creation of the first cell. She refused to let me off the hook. She clearly hoped that I'll expunge God's role from that process. I told her that I could think of two options. The first has something to do with the big bang. Whatever the source of the big bang was, it had to be extremely well organized and programmed to ensure that all subsequent events occurred in a sufficiently orderly manner to produce the exact types of energy and particle forms needed for the creation of a stable universe that would eventually lead to the development of life as we know it. According to this first scenario, the source of the big bang contained all the information, as assembled by God, necessary to achieve all this through evolution including the eventual rise of humankind. Thus, after the big bang, there was no need for God to ever again physically intervene in the process of evolution.

"It didn't take long for Helga to guess that the second scenario I had in mind was that God had indeed played a role in the creation of life. My response was that I also could imagine that God had had a role in the assembly of the first skinned liquid droplet or primitive cell, but only if that somehow happened without violating the natural laws. Just as He might have provided the blueprint for the organization of the universe, He also might have done the same for the development of cells. For example, He might have provided seed material in a template, containing sufficient initial information, for the assembly of a skinned liquid drop, which then enabled cell formation. Before they could ask any more questions, I pleaded with them not to ask me what such a template would look like and what organizing principle could have been used to bring about clusters of elements and compounds with a path to forming a skinned liquid drop and then a cell. But I told them that the one thing I felt strongly about is this: if a template existed, it was composed of matter, even if it was a kind of matter unknown to us, something that God has always had a grip on.

"Helga wasn't entirely satisfied with my answer and asked me which scenario I thought was more feasible. I told her that I would go with the first one because at this point, it can be more easily explained, not very well, but better than the second one. For example, there is more and more evidence to support the idea that dark matter played an important role in galaxy formation and is still needed to keep stars from flying away. It has also been established that the vast stretches of what appears to be empty space between galaxies are filled with dark matter. From these new discoveries, it's just one more step, although admittedly a big one, to say that concentrations of dark matter in preexisting space served as molds for galaxy formation. But frankly, I can't imagine how dark matter or anything else could have served as a template for a cell, even if God was involved. All things considered, my gut feeling is that providing a physical template for a cell would have been too much intervention by Him.

"I still vividly recall Ulrich complimenting me by saying that my idea about God's very indirect role in the creation of life was quite interesting and he was intrigued by it. He said that until that day, all he had ever heard was that science and faith are as antagonistic as water and fire and that they could never be reconciled. But, in his words, I seemed to be confident that such reconciliation was possible. He didn't stop there. He wanted to know if I had any idea what practical steps could be taken and by whom to bring science and faith together. It seemed to him that in the Catholic Church, for example, such a course of action would require some major prodding by high-ranking people including the pope and key cardinals. In the case of other religions, such a process might be even more complicated.

"'Listen,' I said to him. 'In all honesty, I have no idea how this could be done. But I am fully convinced that faith and science not only can be reconciled but must be reconciled to prevent the further decline of Christianity and other religions as well. I can't see the practical steps ahead. But if it will ever be done, it will take frank, far-reaching discussions within faith communities and with nonbelievers too, and the process will include a lot of give and take. Atheists, fully confident in their intellectual superiority, are understandably not much interested in dialogue. For atheists, and I'm afraid you two are included in that category, domination of science over faith is a done deal. To break atheists' domination, the community of the faithful first must fully embrace good science. For me, this should be the first step but will also be the most difficult one. It requires the faithful to adopt a new approach to the concept of God while still preserving the essentials of the Bible. However, the situation isn't as hopeless as it first seems to be, because Christian teachings and Christian values are constantly evolving and have been for ages. Perhaps the Catholic Church should advertise more the often subtle but important changes it has made in its positions on science.

"'I should add that several denominations of the Christian Church subscribe to some forms of evolution under the terms of *theistic evolution* or *evolutionary creation*, essentially advocating the idea that evolution of the universe, including biological evolution, is God's tool to eventually develop human life. According to these theories, astronomical evolution, geological evolution, chemical evolution, and biological evolution each evolve naturally within still ongoing creation. Theistic evolution theory is close to my idea except that I introduce the term *material God*, while the God of advocates of theistic evolution, like the God of the Catholic Church, is a completely dimensionless, omnipresent being.'

"I saw on my guests' faces that they didn't really believe in the changing attitude of the Catholic Church toward science, so I tried to come up with some convincing examples. 'Look, for example, already decades ago Pope Pius XII endorsed the big bang, and since then the church has at least tolerated the scientific view of the creation of our universe, albeit with the caveat that God was behind the whole thing. Obviously, the Catholic Church and practically all Christian denominations will continue to see God's hand in Creation, as it should. Or, as another example of the church's changing views, in 1992 the Vatican, after an inquiry that lasted for more than a decade, admitted that it was wrong to declare Galileo's sun-centered cosmological view as heresy. And John Paul II visited the birthplace of Copernicus in Poland and praised his scientific achievements, which again centuries ago the Catholic Church had fiercely repudiated. All this indicates to me that the church is in the process of getting more familiar with science and imparting more significance to it. I sense that some new ideas are approaching the Vatican, and I hope that these will eventually lead to the acceptance of well-established scientific facts, particularly those related to evolution.'

"All this again took Helga and Ulrich by surprise, and some time passed in silence before Helga asked me if I was optimistic about religious people and atheists someday uniting under the same big tent of science. I had to admit that everyone involved in these debates knows that arguing the case for the reconciliation of Christian beliefs with science is an extremely daunting task. For atheists, religion equals superstition, or worse, madness. But on the other side of the equation, many Christians reject the idea of mixing religion with science and are unable to accept, for example, the key role of natural evolution in human development. And this is a very bad start for an honest debate. Thus, the very first task Christians face is to clearly separate the Christian religion from elements that can be regarded as superstition. This would force nonbelievers to give up an important piece in their argument and would also make the Christian belief system more credible.

"'But this is what I don't get,' interjected Ulrich. 'You are talking about the Bible, the key document around which the entire belief system of Christianity revolves. How can someone reform the belief system of Christianity without messing with the Bible? And what would become of Christianity without a strong reference point like the Bible, the source of numerous hallowed traditions?'

"'Yes,' I replied. 'This is why I told you that, as of right now, I don't see a way out of this conundrum. It is possible that the Christian churches will be content with a few core believers who don't want to change anything, and those who would like to see changes but will remain with their church no matter what. But if you really want to know my very private opinion, a very far-fetched idea indeed, I can tell you about it. Just as the New Testament was a great improvement compared to the Old Testament, God might eventually guide someone, or more likely a group of the learned faithful, to further advance the church's teachings and compose a "Newer Testament," which would be the Third Testament. This would accomplish two important things: it would use the language of the third millennium and incorporate science as much as possible. Despite the poor odds, stemming from religion's resistance to change and atheist scientists' arrogance, I can still imagine that something like this could happen at a propitious moment of future human history. For millions of those fresh out of school, where they acquired so much knowledge contradicting certain Christian doctrines, there is a great need to line up the church's teachings with modern science without contradicting the absolute essentials presented in the previous testaments. In this way, the Third Testament would give a more advanced interpretation of some of the teachings of the New Testament in exchange for placing Christianity firmly in line with modern science. And who knows, other religions might follow suit or maybe even join around the new central idea! After all, the "basic tenets" of the Bible became rooted in Christianity well after Jesus's death, and while they were intuitive to people who lived centuries ago, they may not apply as well to modern society. Now, the Third Testament would have to come up with new basic tenets, but they would ride on the coattails of both science and the Bible, and it wouldn't contradict either of them. What a riveting possibility indeed.'

"At that point I was starting to regret even bringing up the idea of the Third Testament, but I couldn't retreat. I knew that more questions would follow, because this was clearly so unexpected and radical for my guests that it took quite a while for them to decide whether it was entirely ridiculous or if it was worthy of further discussion. Ulrich finally seemed to decide the latter, and he asked, 'What exactly do you mean by lining up the church's teachings with modern science? How can you put thousands of years old teachings into the context of modern science while maintaining the authenticity of both?'

"'What I meant,' I explained, 'was that the Third Testament should retain the key elements of the Bible, including that God is a loving being who created the universe perhaps via what we now call the big bang, Jesus was His extension on earth, and that He is omnipotent and thus has the ability to communicate with anyone He wants to. Science will never be able to touch these key points. On the other hand, the Third Testament would have to incorporate everything that good verified science has to offer. For example, it may give a new interpretation to the idea how God created humankind—that is, He created us by indirect means, namely, through evolution. Also, the Third Testament might give a different interpretation for if and how God performs miracles. Events that seem to be miraculous can almost always be explained by natural causes without bending the laws of nature God created. Not that He could not perform miracles if He really wanted to, but it might seem unlikely to many people in the twenty-first century that He would want to.'

"'Maybe He might want to perform miracles to stoke enthusiasm in people,' suggested Helga.

"'I don't see a chance for that to occur,' I cut her short.

"'Do you think that so-called miraculous events that can be explained by natural causes still occur?' she asked.

"'Sure, they do,' I answered. 'There are many examples in the medical practice when someone is very sick because of psychosomatic reasons and doesn't respond to drugs, but as soon as a psychologist begins to work with the person, he or she begins to feel much better. In some cases, people who were clinically dead have been brought back to life by treating the psyche, or I should say the mind, rather than the body.'

"'Speaking of the clinically dead, some of them claim to have been in heaven before coming back to earth, which, in my mind, would qualify to be miraculous events. Do you think that there is any merit to such claims?' asked Helga. Without even thinking, I answered her, 'Nobody can be sure about whether those claims should be believed or not, but instinctively I believe them.' I didn't add that my belief stemmed from my view of the materialistic relationship between soul and body, which I had no time and desire to elaborate on.

"Then Ulrich shifted the direction of the conversation again and said that he doubted that there ever would be a pope willing to authorize the construction of a Newer Testament, thereby freeing himself and his flock from the restraints of the church's age-old teachings and essentially creating a new church. In his words, 'this would almost certainly result in a revolution inside the existing church and the pope's forceful removal from office. On the other hand, no one would attempt to do something like this in contempt of the pope's wish. Although there are precedents of large-scale reformation, like the great schism instigated by Martin Luther, but what you are proposing would go much farther than anything before.'

"I agreed with him that this seemed like an almost impossible idea, and indeed nothing may ever become of it. But I ended on a note of hope. The Old Testament was an old covenant of God with mankind, and the New Testament was a more recent one. Obviously, Christians firmly believe that both testaments were written under divine inspiration. Thus, a Newer Testament must also be thought of as a new covenant between God and humankind. Since the pope, who is God's ambassador to earth, would be preferably the one who initiates the writing of the Third Testament, the group of people writing it would be doing so through divine inspiration. It wouldn't be a revision of the Bible because the Bible needs no revision. Every basic principle presented in it is valid. The Third Testament would just be an addition to the Bible, a third chapter restating the basic tenets of the Bible and then adding new ones based on modern science. For example, Creation could be described in the most modern scientific terms, by specifying that the natural laws reflect God's commanding role. The creation of humankind could also be described as the result of evolution, starting with the big bang under God's direction, and eventually resulting in the appearance of the first human and so on.

"Then I approached the relationship between the Third Testament and the previous two testaments from a different angle. 'The goal of the Third Testament would be to improve upon the Christian teachings without reducing their meaning. For example, think of the Old Testament as the way the human body is taught to children in elementary school. Then the New Testament is like a document describing the same human body in more details for high school students without contradicting what they had learned in elementary school. And finally, a much more detailed description of the human body at a medical school would correspond to the higher level of teachings found in the Third Testament. The same thing can be looked upon at various levels depending on the general knowledge of people at the time. As our understanding of nature grows deeper and deeper, we should not hesitate to go deeper and deeper into our knowledge of God. Two thousand years ago, people had a very minimal knowledge about the world. You couldn't explain to them the concepts of cells or the big bang. Now we have a better understanding of these and many other concepts, and we have a better chance to relate them to God. However, I emphasize that those who are just fine and happy with the New Testament shouldn't be forced or even encouraged to change their views. A new interpretation of God and His church would be only for those who have already left or are about to leave the Christian churches, those who once felt close to God but became disillusioned with the church's teachings. The number of these people is already alarmingly large and steadily increasing. They, including the two of you, should be given a chance to remain within the Christian community.'

"As I finished speaking, Ulrich noted that he had a different opinion about the people who lived two thousand years ago. Those people were far less educated than we are, but that didn't mean they were stupid. He thought they could have understood any scientific concept if someone explained it to them the right way. Since none of us had any direct experience in the issue, we couldn't settle it.

"Helga told me that she was fascinated by these radical ideas, but she was also confused. She wanted me to elaborate a little more on the other new central elements of the Third Testament, besides the existence of God. I answered that in my opinion, a central change should be that God chose evolution as His main instrument in creating humankind. Another thing I would like to see is that He established the physical laws of the universe at the very beginning of Creation, but He chose to never intervene in the subsequent development of the cosmos. A third central element could be that He has retained the means to communicate with us within the confines of natural laws.

"Helga objected, saying that the ability of Jesus to perform miracles was an important part of the Bible. I replied that in my interpretation, the miracles described in the Bible are secondary compared to Jesus's teachings for how we should behave and relate to God. Besides, it is my belief that the Old Testament and the New Testament had to talk to contemporaries in a language that they understood, which meant using a lot of parables and referring to miracles for greater emphasis. The Third Testament would talk to modern people in a language that they would now understand, thanks to widespread dissemination of the great progress science has made in revealing the secrets of nature. But the trick would be to make the Third Testament an organic extension of the Bible that builds on the spirit of both the Old Testament and New Testament instead of replacing them while also tying them to science. In other words, the Old and New Testaments would remain essentially spiritual books designed to guide our lives and relationship with God, while the

Third Testament would help the faithful to fully embrace science and build an even stronger tie with God, now involving not only our heart and conscience but our brain as well. The Third Testament would recognize that the teachings of the Christian churches also undergo evolution and that this occurs in full compliance with God's plan. In fact, the Third Testament, as I envision it, will be open-ended so that it can continuously incorporate new elements as science progresses and our understanding of God deepens.

"Then Ulrich asked, 'Should the Third Testament point out the flaws in the previous testaments and address them, or it just should reluctantly acquiesce to all of their teachings?'

"I was very determined in my answer. 'The flaws you are referring to don't really exist. What appears to be flaws to some people are only reflections of the fact that those testaments were written for and by people who lived many hundreds of years ago.' Although Ulrich disagrees, the fact is that people back then had a very primitive knowledge about the world around them. Just think about it. A thousand years from now people will probably consider us as primitive people, and they will probably have just as hard a time to understand our way of life and our motives, which now we treasure so much and are so proud of.

"It was getting late, but I wasn't yet done explaining my ideas to the guests, and they didn't seem to show any signs of boredom, so I went on. Going back to previous examples of reformation in the church, during the first few centuries after the death of Jesus, the initial loosely knit Christian religion went through a great many revisions and improvements when select people appeared to receive new instructions from God. The first major documents to help the emergence of a dominant direction of Christianity were Paul's letters, written to Christians and would-be Christians around AD 50–65, followed by the Gospels and other writings in the first and second centuries. A standard Latin translation of the Bible was first made only around AD 383–405. Thus, there would be nothing un-Christian about maintaining this ancient process of growth, in which a Newer Testament would be only the latest alteration. If this won't be done, it will irremediably reduce the allure of Christianity for many people.

"Seeing that they were still not convinced, I told them I had one more thing to add. The *Newer Testament* would make the existence of all the various religions we have today simply unnecessary. The Newer Testament would make it abundantly clear that Catholics, Protestants, Baptists, Muslims, Hindus, and others can't all be equally correct in their view of God. There is only one truth, and this truth must conform to established science. It might take centuries, but I'm hopeful that eventually such unity might be realized. The Newer Testament shouldn't only apply to the Catholic Church—it should be universal. Wouldn't it be nice if the inter-religious bickering could be replaced by peaceful religious unity?

"The ever-doubting Ulrich reminded me that the Catholic Church's history shows that it doesn't treat dissent leniently. This is what he said: 'And let us pretend for a moment that Jesus visited earth once again and left behind the Third Testament. Deprived of doing miracles, how would He be able to convince people of the truth of the new teachings and prove authenticity of Himself and the new Bible? And there are over a billion Muslims who are equally convinced that their religion is superior. How could you possibly convince them that what they have believed in all this time is false?' All this came out of Ulrich at once. My guess is that he was shocked to hear all these religious ideas from a scientist at the Mayo Clinic, which, in his mind, should be a bastion of atheism. I believe that in his eyes I began to comport myself as a theologian rather than a scientist.

"I assured him that I considered all his questions valid and added that I would like to analyze them one by one from my vantage point. First, let us deal with the fact that there has been an exodus of young people from the church. You can't plug a leaky dam with a wad of gum. Likewise, you can't stop such a massive exodus from the church with small measures. If it wants to survive, the church will soon have to start thinking big. Presently, the church is woefully out of touch with young people because they are getting inadequate answers to their questions.

"As for Jesus having to perform miracles, once God created the universe governed by natural laws, I doubt He would allow anyone, including Jesus, to perform miracles requiring changes in the laws of nature. I know that this is a very strong statement to make, but this is what I think now. And I think that it is illogical to assume that he performed miracles for another reason too. Just think about it. He wanted to live as a human and sacrifice himself for us as a human being. Humans don't perform miracles, but if he did, his life as a human being would have lost authenticity. However, Jesus could almost certainly influence people's behavior and even health by His strong personality and most probably exercising mind control. Take for example the case when He resuscitated a young man who was believed to be dead. Now, in the absence of sensitive instruments, two thousand years ago it was impossible to distinguish between a dead person or a person in deep coma. Thus, when Jesus was led to the young man, He could recognize, through sensing very shallow vital sign, that he was in deep coma and not dead. He then might have used His mind power to wake the young man up.

"I believe that Jesus wasn't quite literally God in the flesh, but He was endowed with the power to speak so powerfully and convincingly that His followers simply believed Him. This strong belief then could even lead to the healing of some sick people through the so-called placebo effects. Today we know it very well that a pill that doesn't contain the medication, just some ballast material, can be 30–40 percent as effective as the real medication, provided the patient who gets the placebo pill is convinced that he received the medication. But meditation and some other methods that enhance the body's ability to fight off disease can also have healing effects.

"Convincing many people to believe in you and follow you is not necessarily godly. Napoleon, Gandhi, and even Hitler had many faithful followers, and we can agree that none of them, particularly Hitler, could have claimed himself to be God's messenger. The difference, I believe, is that Jesus was God's messenger—that is, God specifically empowered Jesus, a human, to be His mouthpiece. In fact, perhaps God took full control of Jesus's brain and body. One might say that while the Bible is very specific about the fact that God came to earth as the Son of God, it's also possible that God chose a man, Jesus, and made Him the Son of God. In the former case, God had to perform miracles against physical laws, while in the latter case, He could take over the mind of an existing human by some physical means without performing an act that we view as miracle. These two options are very different.

"My last conclusion led Helga to ask about my thoughts relating to Jesus's birth and His childhood. And she also suggested that in her view, there were too many things that would have to be said differently in the Newer Testament. This might draw some people to the church, but perhaps many more would leave it. In short, it seemed to her that the church could not win; it would be in the predicament and lose followers either way. However, I was ready with my answer. Do you want to hear it, or should we all pack up and go home now?"

"If it is not too long, please finish your story, but at around one, I really want to go home," said Steve.

"I second that," rejoined Mike.

"All right then, I promulgate your wishes. This is how I responded. 'First, nobody knows for sure which biblical events described in the New Testament did or didn't happen. I tend to think that Jesus's sermons did happen and that they are as valid as described in the Bible by God's inspiration. But I also think that events that involved breaking natural laws were put into the Bible without godly inspiration as an attempt to give more credibility to Jesus's godly status. This uncertainty is one more reason why I believe the time is ripe for the Newer Testament. It could clarify these issues without insulting anyone who accepts the Bible word by word. As I already explained, I don't consider these issues to be flaws in the Bible. There is no evidence to debunk the stories of the miracles Jesus was said to have performed, provided no laws of nature were broken.

"As for Jesus's death, my private opinion is that until the moment He died, He lived His life as a human being and was part of this physical world, having no supernatural qualities. I also believe that it would have been against God's laws to raise Jesus from death and take Him into heaven in His physical form. For me it would seem more likely that when Jesus reached adulthood, God communicated to Him what to do and how to do it. He also provided Jesus with the means to fulfill His mission, in the form of knowledge, willpower, and a kind of hypnotic personality with which He could select and convince the first twelve people, the apostles, to follow Him and essentially lay the groundwork for God's Christian church. I know that many people would find such a description of Jesus contemptible, but this is what my logic dictates.

"I also believe that if the Catholic Church wants to embrace science, it can't escape dealing with the Trinity issue and rephrasing it. I think that the first pillar of the Holy Trinity is God, the second pillar is Jesus, initially the most important prophet and communicator of God's will who eventually became the Son of God, and the third pillar is the Holy Spirit, essentially God's means to communicate with humankind. It is the Holy Spirit that is present in our everyday lives, exactly as the church teaches. I would say that today my view on the nature of the Holy Trinity is really very close to the church's view, although tomorrow I might think differently.

"Helga's next question was whether I really meant that God literally provided the knowledge, willpower, and hypnotic personality to Jesus. If I did mean that, she saw a huge contradiction in my statement right there. Once God had established the natural laws that He then obeyed, any communication between God and Jesus or between God and anybody else had to have been a miraculous event with no material basis. She wondered if I had ever noticed this contradiction that was as clear as daylight to her.

"I told her that this was an excellent point, but it is beyond my capacity to explain how God might be able to communicate with us via material means, except that I believe that the messaging has to have a material basis. Our brain might be able to decode God's material-based messages, just like it can sense many other environmental cues like light, smells, sound, electromagnetism, pheromones, and many others. At that time, I thought that our brains might have some chemically marked flag posts that could specifically sense the incoming signals and transform them into a memory or an idea via additional chemical changes in our brain. I added that they shouldn't think that this idea is far-fetched. For example, I recently saw an abstract at a scientific meeting that gave strong evidence that such flag posts, made up of chemically tagged neurons, play dominant roles in the creation of long-term memory. Could our brains contain such marked neurons that specifically respond to God's messages? Once we develop a reasonable idea of what to look for and a few influential scientists begin to give their backing to this idea, this kind of communication might be approached by scientific methods, however incomprehensible it might seem at that moment. But, as someone once said, 'it's not worth doing something if it's not hard.' It behooves us if we want to choose the easy road or the hard-to-travel road.

"Of course, I explained my position further. There is a difference between things that are knowable and things that we will never know. For example, no matter how close scientists get to the moment of big bang, even as close as one nanosecond or less, I strongly believe that they will never be able to figure out what came before and how it happened that our universe apparently came from *nothing*. This essential fact is true regardless of whether God was involved or not. In principle, the story of what came before the big bang is knowable, but in practice, I believe it will always resist direct scientific confirmation. Theoretical physicists and theologians may always hunt different prey, but I do hope that at some point they will come together.

"At this point we once again paused and drunk a little wine, which gave the German guests some time to think over what I had said. It was getting late, and Sarah took the kids to the bathroom to prepare them for bed, but Ulrich and Helga had clearly decided to stay for a little while longer. They were curious about what else I had to say. Helga was the first to interrupt the silence, and searching for the right words, she stated that she might accept that God could have the means to communicate with us, but if no laws of nature can be broken, how can one imagine the afterlife? Just as our universe was created from *nothing*, can we also become *nothing* and still enjoy God's company in heaven? She added that such a scenario didn't make any sense to her if God had a material side, as I had suggested to them.

"That was a very hard question then, and it is still now, and probably will remain forever, as I said to Helga. This is perhaps the question that I meditated over the most during the evening hours that I spent in my office. I'm not sure, I told them, that I've gotten much closer to solving it to my satisfaction, but I believe this much: If I assume that there is a material God, I think I must also assume that we have a material copy of us that we can call hologram or soul that survives past our lives and is able to recognize and interact with the material God. But I can't even guess what material our soul would be composed of, although I'm sure that physicists could come up with some wild ideas. For example, according to some theoretical physicists, every elementary particle has a corresponding complementary particle that we can't detect. Who knows, maybe these complementary particles are suitable to form a surviving soul of us into which all life experiences we have accumulated are imprinted. For all this to make sense, individual souls should survive the owner until eternity. But of course, this whole idea might be totally ridiculous, and I will be the first to admit it. These are only playful thoughts. I don't want to advocate something that may completely contort and subvert science, thus undermining intellect.

"When I hinted that perhaps I had talked too much, they insisted to continue our dialogue. It was Ulrich's turn to shift the conversation in a somewhat different direction. He asked, 'If Jesus wasn't God, at least not initially, then how is it that He died for our original sins to redeem us? As we would say it today, Jesus didn't have the "authority" to do this for us.'

"I answered that even though Jesus was human until He died, it was quite acceptable to me that, through the Holy Spirit, God elevated Him to the status of Son of God, thus making Jesus part of Him and one of the Holy Trinity. But I also acknowledged that his question had merit because perhaps Jesus didn't die to liberate people from their sins, whatever they might be. It could even be imagined that Jesus was murdered because He didn't retreat from His enemies who didn't accept His new way of thinking about the relationships between God and humankind, as well as among humans. Possibly He also wanted to be exemplary in demonstrating to His disciples what might be waiting for them if they remained steadfast in spreading the new faith. By becoming a martyr, He gave enormous credibility to the power of God and strength to the disciples as well as the thousands of followers who accepted this new religion within an amazingly short time span.

"It's also possible that Jesus might not even have been murdered the way it was described in the New Testament. But this may not be that important. Long story short, I would have no problem accepting Jesus as a prophet whom God chose to declare and confirm His existence and who taught people what God expects them to do, and above all, love and help one another as well as treat everyone equally and fairly. If you think deeply about what Jesus taught, you can't find one single sentence that in good conscience you can object to. He had such deep wisdom that it could only have come from God. If you pressed me, I might opt for the idea of Jesus indeed being the Son of God, one of the inseparable pillars of Holy Trinity. But the way I think of Him now, He became one pillar of the Holy Trinity as the result of a process. This was a kind of duality, being mostly human and part God, enabling him to be God's credible messenger while on earth. I still added that spiritual God and material God are inseparable in the Holy Trinity.

"My explanation led Ulrich to ask my opinion about sins. I told him I had never had a high opinion of how the church treats sin. I think most Christian churches place too much emphasis on continuous contrition, even for minor misdeeds. I believe that God has no desire to see us living our lives in constant guilt or to scold us for exiguous offenses, but He is probably not in the 'respite-care business' either. He's interested in our sins only for the simple reason that only good people can truly live in lasting peace and harmony with one another. His love seeks no recompense. I cannot stress this enough. I don't believe that God is pleased to see the wars and the many wrongs that people are willing to commit sometimes in His name, and that in the process the meaning of life and the opportunities that He has provided for us are reduced. The way I see it, Jesus was in a way the tool of God used to create the conditions necessary for the spread of Christianity, which has the power to make the world better. That's why He had to deliver God's message as a human. For sure, the church has gone through many turbulent times, but no one can deny that it has also delivered on many goals that Jesus had set for His followers. I firmly believe that without the Christian churches, our world would be a much worse place, because the many good things they have done eclipse the wrongs they have committed over the centuries.

"'Okay,' said Ulrich. 'Let us assume that the Catholic Church or even the other Christian churches accept the notion that there is a need for a Newer Testament to align Christian teachings with modern science. How would other religions relate to this urge for change? If your logic is correct, they are also in need of modernization.'

"'They might consider it and I think they should,' I replied, 'but if history is any indication, it will be a very long and arduous process. But I stand by my belief that what I have said about the Catholic Church is basically also true for other religions. Today they all suffer from the same fundamental problem of falling behind modern science. I can't help but keep coming back to this crucially important point, although I can only speak for Christians. Also, soon there will be no room for the many, in fact too many, Christian churches to remain separated. Thus, I want to reiterate that there is a chance, however small it might be, that the Third Testament might serve as a focus point to unite all Christian churches and maybe even other religions under the same tent, with the purpose of becoming one single religion, worshipping the same God and adhering to the same principles. And if God wants this to happen, it will happen. He may not want to intervene with the process of natural order that He initiated, but I strongly believe that He has the power to influence us through our minds and souls. The question is if He really wants to do that, and if He does, how can He still secure our freedom of choice, which He gave us?

"'Therefore, God might simply trust humans to settle this issue on their own via exercising their free will. I rather like this latter option even though free will can lead to bad things too. We shouldn't minimize the importance of the fact that humans are undergoing constant evolution that, in my view, will help propel the renewal of

Christianity forward. We also shouldn't forget that of the thousands of different gods that mankind has worshipped over the course of the human history, only a few remain today. This means to me that the evolution of religion will eventually arrive to a point when the union under only one God, and with only one interpretation of Him, can be achieved. Asking people to switch their religious allegiance seems to be an impossible task. However, I believe that a compromise could eventually be made. How, you ask? I trust the collective capacious mind of humankind and the extensive give-and-take in the Third Testament, placing emphasis on commonalities rather than differences among major religions, to lead us there. But the first step is to achieve peaceful coexistence of various religions instead of recalcitrant opposition to each other's ideas.'

"'Do you see any chance that the evolution of the human mind will eventually lead to the unification of humankind under atheism?' Ulrich asked.

"'That would be a sad outcome for me, but it is possible. Although, I would like to add that the trend I sense now is that people are increasingly turning to various kinds of mysticism, which is just one step away from recognizing a god."

At this point, Steve interrupted Doug. "Can I ask you a question about Sarah?" he said hesitantly. "Wasn't she surprised to hear about all these ideas of yours? I mean, your explanations sound rational after you explain them, but at first blush, they do seem a little, well, outlandish."

It seemed to Steve that Doug avoided his eyes as he prepared to respond. "Well, as I have already said tonight, I hadn't talked to her about some of these specific things before my conversation with the Germans. But even during it, she heard only small parts of the conversations because she had to deal with the dinner and the boys. But when she was there, I noticed that she seemed to be slightly irritated by what I was saying, even though earlier on in our marriage we had talked about God a lot. Whether she was irritated because I had not discussed these ideas with her as I previously used to, or she just thought my ideas were stupid, I really don't know. Perhaps the latter, because I remember her remarking that significant alignment of the Christian churches with science at the expense of rejecting any doctrines laid down in the Bible would be met with enormous resistance despite the pent-up demand even by churchgoing young people for change. She also pointed out that even if there were a consensus that the church needed to move closer to science and shed any semblance of superstition, one big question still would remain: Who would carry out such a reformation, and how? In her opinion, church leaders are too timid and not yet convinced to lead this process. 'Obviously, such a reformation process would need to be organized from the bottom up,' she said, although she had no idea how. This was clearly contrary to my idea of change coming from the top, but we have never had the chance to discuss this topic between the two us since then. At that party I only reflected, somewhat yielding to Sarah's concerns, that in Jesus's time and particularly after that, no one could predict how Christianity would evolve. Similarly, no one today has the capacity to predict how and when the kind of evolution I am hoping for will take place. We simply can't see that far ahead, so we will have to accept a certain level of uncertainty.

"Anyway, going back to the Germans, we sat for a while in silence, sipping our wine slowly. The subject we were discussing was too complex with too many dimensions for quick mental absorption, considering that it all started as a light conversation about books we had read. Obviously, I could not be sure, but I believe I had a good idea of what might have been going through my guests' minds. Ulrich and Helga, like most atheists, thought that equating religion with superstition is a no-brainer. Society and their own bad experiences in church had planted in their minds the belief that modern science could refute even the most sophisticated arguments for the existence of a spiritual God. For them the eulogy for religion, any kind of religion, had been written long ago. It was a finished business on which there was no need to waste any more words. In their minds, the time was very fast approaching when religions would no longer be able to beguile people with their dogma, and they would wake up from their delusion, never again returning to it. The world would finally be free of this opprobrium called religion. Science-based reason would finally triumph over superstition.

"But now I had challenged them with the possibility that God's interactions with us may actually have a material basis. That probably put the whole issue of the relationship between faith and science, however nonsensical it might have sounded to them, into an entirely different perspective, that is, the interaction of a somewhat known physical world that still hides many secrets with an invisible and totally unknown, but probably knowable physical world that includes a god with a dual nature. Even an afterlife in the form of a surviving soul carrying the memories of a lifetime might be explained on a material basis, although I had to admit to my guests that this was the weakest point of the many other weak arguments I had made. But the idea that an invisible but material world exists is no more fantastical an idea than, for example, string theory, which claims that strings are the fundamental units of matter and energy without any proof that we may ever be able to physically detect them. But to be fair with string theory, it has serious mathematical background although based on unprovable assumptions, while my science fiction 'theory' has no mathematical background at all. The mathematics of my theory needs to be worked out to be taken seriously.

"I was sure that my guests were getting the idea that what I was trying to do was just another effort, attempted by many others before me with varying success, to save a religion under increasing pressure from scientific evidence that questions the word-for-word meaning of its teachings. It would have been completely understandable if they thought that. After all, it should have been conceivable to them that I and the many others who think that there is something beyond our material world must feel that we are making a last stand, and if we lose, nothing will protect us against self-destruction as far as our religious, or in my case semireligious, beliefs are concerned.

"I also wondered if they truly understood my main point, so I explained it further. You know, I said, I have a natural tendency to protect the physically and intellectually weak, and right now I can detect serious weaknesses in the messages of Christianity and, in fact, of all religions. My remedy is that instead of being stuck in a state of denial, believers must recognize that in the long haul, intellectual laziness, sticking to the old dogmas without any attempt to refresh them, will not work for their advantage. Believers must be active participants in change. Atheists have gotten overconfident with their success in using science against faith, which could make the task of reformation easier.

"I also wondered if they understood a second main idea I tried to convey, so I further elaborated. By and large, the Christian churches and scientists both operate on the principle that issues of faith and issues of science are completely different, representing two separate realms with no crossover. I believe Christians must challenge this idea, because otherwise, more and more people will equate faith with superstition, which is just one step, indeed just a small step, away from delusion. So Christian scientists must start working to bring science and faith closer together, based on the principle that scientific laws are God's laws. I would love to see a fair competition between Christian and atheist scientists. Reviling each other should have no place in such competition."

"That was a very nice ending to your story about your German guests. With that I guess we can go home now," said Steve, somewhat desperately. They finished one final round, which Steve won, and they agreed that it was pointless to wait any longer for Peter's call. Mike volunteered to try to contact Peter early the next morning, no matter how long it might take. The day ahead was a little easier for him; he only had to start teaching at ten in the morning, and he could return home early in the afternoon.

CHAPTER 4

DOUG COULDN'T STOP RELIVING HIS CONVERSATION WITH THE GERMAN GUESTS

Doug went home with the firm decision to sleep for at least six hours, but it seemed like it was just not meant to be. The more he tried to force himself to sleep, the less sleepy he became. His thoughts kept circulating around his conversation with his Germen guests. There were more things he had discussed with Ulrich and Helga that he had no time to mention to his friends that night. Ulrich was particularly interested to know more about how Doug viewed Jesus and how the Catholic Church might respond to any ideas that would alter its teachings.

As he lay there awake, Doug replayed in his mind the rest of his conversation with Ulrich. Ulrich had protested against his ideas, saying, "History has shown that the Catholic Church and almost all other religious groups have a dismal record of tolerance when it comes to the kind of really free-thinking theories you are talking about. The way I see it as an outside observer, religious leaders tend to endlessly mull over the problems they face and move forward in only very small, very cautious steps instead of trying to solve them all at once. For example, Christian teachings are based on the basic tenets that Jesus is God and He had the power to perform miracles against the principles of natural laws, if He so chose. Now if anybody tried to take these principles out of the religion and relegate Jesus to no more than a man, that person would basically be trying to destroy the entire Christian belief. The persons who wanted to do that would be treated at the minimum with pungency, and more seriously they would be labeled as evil and extremely dangerous, not only by the church's hierarchy, but also by most of the true believers. Not even the pope would be an exception, should he by some miracle ever entertain the possibility of leading an effort to construct the Third Testament. Don't forget that many churchgoers are very comfortable with the separation of faith and science, and they will resist change. At least this is what I experienced when I used to go to church with my parents. They will label any new idea that anyone might propose as blasphemy. No, church leaders are extremely wary, and my prediction is that they will strongly oppose any significant addition to the Bible. They will say that God has led the church to survive and grow for two thousand years, and they don't have the slightest doubt that this will remain so for thousands of years to come."

Doug remembered his response almost exactly: "This is why I said earlier that if someone is happy with the New Testament and the belief that Jesus could perform miracles, then all the best for them. What I am talking about is a higher-level approach to God for people who are still within the gravity field of Christianity but don't accept certain aspects of the Bible that sound too far-fetched for them. These are the people who think that the Christian churches provide dawdling responses or no responses at all to their many questions concerning, for example, Creation, development of human race, evolution, and so on.

"Listen, the point is that nonbelievers are steadily gnawing away at the foundation of religion. Something must be done to retain the sanctity of faith. Obviously, whether it will be a pope or a prophet, anyone chosen by God to advance the idea of reform in the *Newer Testament* must have the power to imbue people with this new kind of spirit. When adding on to an old building, you must begin with a strong new foundation to make sure that it will be able to hold the new structure but will not weaken the old one. This new foundation for the church would be the principle of evolution, which God chose as the driving force for the development of the physical and biological world, including a steady progress in our understanding of the relationship between faith and science as well as between God and humanity. The oldest and strongest foundation for Christianity is the existence of God, who wants us to be good people for our own sake, and that shouldn't be destroyed, but cherished and strengthened as much as possible. The key to all this is that the common walls between the new and old buildings should be strong enough to hold the entire superstructure.

"Continuing the metaphor, it is equally important that there are wide, always-open doors connecting the two sections to allow effortless traffic between them. You see, if Jesus had told His followers how important evolution was, they would have never understood what He was talking about. He had to use a different language, one that people at the time understood. Today, the concept of evolution is widely known, and most people are also aware that natural laws that don't change govern the universe. Therefore, someone who will be chosen to advance God's new covenant with people should be able to convincingly expound the significance of these God-chosen principles in shaping our physical world and driving continuous improvement of Christianity by introducing new science-based ideas and by erasing superfluous dogmas."

"That sounds interesting, but going back to my earlier question about the flexibility of Christian churches and of religion in general," interrupted Ulrich, "I still believe that they are afraid of making the same mistake that Gorbachev made during the late eighties in the Soviet Union. He saw that unless he allowed some limited liberalization of the economy and political life, his country with its rigid system would not be able to compete militarily and economically with the West. However, once he let the rouge spirit of liberalism out of its bottle, people didn't stop at small reforms, and eventually this led to the breakup of the entire Soviet Union. Gorbachev didn't have a contingency plan how to deal with that new situation. The Christian churches are probably afraid of adding something new to the Bible because they might lose their identities, which may eventually lead to the creation of even more churches, separated from the mainstream ones."

"Yes, this has basically already happened during the turbulent years of the Reformation, and it may happen again," responded Doug after a short silence. "But I don't wish that those turbulent times be repeated. I would rather like to see a smooth and controlled process, led by the pope and his trusted scientists and theologians, that would eventually lead to the unification of Christian churches and perhaps later other religions as well. But undoubtedly, the renewed church would resume a new, and I believe much stronger, identity by undergoing the alignment with science that I envision. And I should add that this process should be open-ended. It should result in a system that allows seamless incorporation of new ideas that fit in with both established science and key church doctrines, thereby allowing the church to get closer and closer to a full picture of the truth.

"Obviously, you can't expect that all Christians would immediately accept the new science- and evolution-oriented ideas—far from it. But I would bet that eventually the Catholic Church and even other religions would unite behind a pope-led reformation process if they would see it as the fulfillment of God's wishes. You see, a true atheist scientist is ready to change his or her mind if new evidence convincingly challenges the old theory. For example, most of them had no problem with shifting from the idea that our universe always existed to the idea of the big bang. And most recently, many of these atheists have no problem saying that probably there was no big bang after all, or what we call the big bang was the formation of our universe from a black hole or a previously existing universe that had contracted. Widely different materialistic theories about the formation of our universe abound. So I acknowledge that it is easier said than done that the *Newer Testament* will have to conform to science, but in the case of Creation, which science should it adopt?

"Just like materialist scientists, a Christian should also be able to change his or her mind about previous ideas if continuously updated new scientific evidence requires it. As soon as we Christians will proclaim that our search for God is a continuous process, with only a few basic truths written in stone, I believe that we will firmly win the hearts and minds of most of our fellow believers so that they will never have to even contemplate atheism. Instead, they will seek unity with other believers. The number of young people turning away from religion is alarming to me, although I know that you are probably happy to see it.

"If I want to summarize my thoughts once again, I would say that what I expect from the new addition to the Bible is that it would contrast a godless material universe to the one that also includes a material God and an interwoven space dimension, invisible but material. Right now, we have the tools to study only the more simplistic and smaller visible material world, but this doesn't mean that we should be fastidious and not to deal with the presently invisible part of the universe. Our efforts to understand the remaining 96 percent of the universe where God may or may not reside are in principle no different from trying to understand the visible world. I know that most people, particularly those who are scathing about any form of belief in God, would find my ideas farcical and would burst into laughter, but at this moment, this is where I am in my thought process."

"But scientists are trying to understand the nature of invisible part of the universe like dark matter and dark energy," Ulrich countered. "The reason they so far have come up empty is not because of lack of trying."

"This is true," I said, "but they are studying the universe from a narrow perspective without considering that God is part of it, who used an elaborate plan for Creation, which can be understood only through a new way of thinking. If scientists keep ignoring the role of God, they will never understand how the big bang happened and what is the nature of dark matter and dark energy among others."

Having replayed the entire conversation in his head, Doug now looked at the clock on the wall, which showed it was five past two. He went out to the kitchen where he found a bottle of vodka that he rarely touched. He drank three mouthfuls directly from the bottle and went back to bed. Before he fell asleep, his mind once again drifted back to the conversation with his guests. He remembered that while he was finishing the last sentences for the German guests, Ulrich began to fidget like someone who needed to leave but was reluctant to admit to it. Indeed, it was already nearly midnight, and it was high time to think about going to bed. But that wasn't the reason Ulrich was a little nervous; he admitted that he still had to finish an experiment and for that he had to be back to the lab in around thirty minutes. He only had about fifteen minutes of work to do, but the timing was important. Doug remembered that before leaving, Ulrich asked if he and Helga could invite them to their home next weekend. He suggested that perhaps that time they could talk more about diabetes and science, instead of faith. "But perhaps we can talk about faith healing as well," he added. "Personally, I don't believe in it, but I would really like to know your opinion on it, whether you think it's real or just another forlorn hope people in need are drawn to."

Ulrich appeared to be sincere with his invitation, and Sarah was happy at the prospect of seeing the Germans again. Later, she told Doug that Helga seemed to be surprised, in fact momentarily nonplussed, when Ulrich offered the invitation, suggesting to Sarah that there was some sort of strife between the Germans. Perhaps for that reason, the planned meeting had never taken place, and the invitation was eventually completely forgotten.

CHAPTER 5

Helga and Ulrich Make Big Decisions

After Helga and Ulrich had left, their minds were still whirling from the day's events, and they didn't quite know what to make of what they had heard from Doug. Ulrich was the first to break the ice, saying, "I don't know about you, Helga, but in my opinion, your boss is either a genius who spends his time thinking about things that we couldn't even attempt to comprehend, or he was simply playing a big elaborate prank on us. Well, there is a third option as well: he might just be crazy. But considering his reputation as a very smart guy, I think it is probably the second. It was probably a mistake to invite him to our house."

"Well, I think he explained quite well that what he was telling us was just one of many possibilities," Helga replied. "Who knows what else is in his mind? I am sympathetic to what he is trying to do. Essentially, he is trying to figure out how the Christian churches can be saved from irrevocable downsizing and finally destruction. I think this is the idea that impels him. And he might be correct in that the only way of doing so is to convince the faithful but less orthodox believers that the church must get closer to science and accept the existence of evolution. He is right—science has clobbered the Christian faith system to the extent that soon something must give. The church simply cannot keep its present staunchly conservative platform. I agree with you that Doug is a very smart guy, but I also believe that he is a sincere one, and I cannot imagine he was trying to trick us.

"Just look at us, Ulrich," Helga continued. "We both abandoned the church for the sole reason that we did not believe in miracles. Doug says basically the same thing. He doesn't believe in miracles either. I agree with you that everything he said today basically sounded like science fiction, but when Jules Verne wrote his books about flying to the moon and things like that, wasn't that considered science fiction too? Doug just might be ahead of his time, and people like him are needed to promote these new ideas. Everything he said might turn out to be nonsense, but only those who try can fail. I think there is a chance that some of his ideas might actually gain some popularity." As she said this, they arrived at the front door of Ulrich's lab. It was the first time Helga had visited Ulrich's place of work; later she often thought that perhaps she should had shown more interest in Ulrich's research. As high-strung as Ulrich was, this could have helped him regain his inner composure. And perhaps this would have helped save their relationship.

Ulrich quickly finished his work, and while they were driving home, he addressed Helga's thoughts. "I still believe that regardless of whether the church needs this *Newer Testament* or not, the present ideology is so deeply entrenched in all aspects of society that the church can't undergo serious change without a very high risk of being totally ripped apart. Besides, there is absolutely no guarantee that whatever might emerge from such mayhem is going to be any better than the present teachings of the church. As far as I'm concerned, the church leaders will never contemplate any changes to the time-tested theological teachings, even after hearing Doug's ideas, if they ever reach them." Although their approach to Doug's theory was different, they were both surprised that despite their imminent separation, they were still able to debate honestly and respectfully such deeply ideological issues. This was one of the few things that kept them together for the time being, even though they knew these momentary attractions would do nothing else, just delay the inevitable.

The following Monday, Helga was once again spending the evening in the lab when Doug showed up with his friend Peter. They were about to go see their other friends to play poker, but first Doug wanted to know how things were going in the lab. He expected to get some urgently needed serum and urine analysis data from his long-time assistant, Joseph, to provide some guidance for the treatment of a seriously ill cancer patient.

Although Peter had been Sarah's first boyfriend, he and Doug remained close friends. Their friendship, free of any bitterness or hard feelings, stayed strong over the years. Their recent chat about cosmology and religion had just confirmed this close tie between them. Before that conversation, Doug had thought they had some major differences in their approach to religion, Peter being presumably more conservative and entrenched in his beliefs. But now he saw that they were essentially on the same page, fending off the basic tenets of both atheism, which rejected God altogether, and creationism, which insisted on a literal reading of the Bible. In fact, Doug was very happy to see this new side of Peter. He had already planned on holding more discussions with Peter about Genesis 1 and Genesis 2 and how the church might eventually be induced to update its view on the creation of the universe and of life.

Even though Peter was skeptical of some of Doug's ideas, the fact that they were on a similar wavelength assured fertile ground for such discussions. And they both knew that many churchgoers were indeed beginning to place more value on science. At the same time, Doug meant what he said to Ulrich; he had no plans to publicly disclose his idealistic, in some aspects almost certainly irreverent, ideas. He also knew that his theory was still too inconsistent, and he would never go as far as to challenge the church's current doctrines in wider circles. He wasn't the type of person to seek confrontation; he knew his boundaries in public relations and wasn't ready to challenge and possibly insult anyone's beliefs. He liked that he had his own personal space to formulate his own theories, into which he would allow only a few people in, and he had no desire to go any further.

Doug briefly introduced Helga to Peter and then turned to Joseph for the data. As soon as they started talking to each other, Helga sensed a strange kind of titillation in the air; later Peter admitted that he also felt the same thing. For Peter, there was something inexplicably but intensely attractive about Helga; for him it was quite literally love at first sight. Helga couldn't help noticing a sort of natural kindness in Peter's appearance and demeanor, which she found very endearing. She also sensed an air of security, strength, and determination about Peter, which she had never found in Ulrich. These were exactly the traits she was searching for at this juncture in her life, after she had just experienced so much instability. In an instant she decided that she would accept Peter's advances, if they ever came.

At that point, she wasn't interested in analyzing the depth of her feelings toward Peter. After all, love may come gradually, but it may last longer than love that comes at first sight. Her mother explained this to her many times, perhaps referring to her own personal experience. This is exactly what also happened to Monica, Helga's older sister. Monica got married just a few months before Helga moved to America. In the early stages of her marriage, Monica didn't feel anything other than very high respect toward her husband, who provided her with full security and complete freedom. But over just a couple of years, Monica's respect gradually grew into deep love. During their last Easter visit to Stuttgart, Helga couldn't help but notice that Monica and her husband had become a happy, inseparable couple, with no signs of any of the usual marital squabbles. When Helga asked her sister if they managed to successfully hide their disagreements in front of others or if they really didn't have any, Monica answered very eloquently, "My husband and I are acutely attuned to any signs of differences in our opinions, and we try to resolve them with a lot of talk and a lot of love when we are alone. Also, we don't have a lot of time to trifle away with unnecessary fighting. There is a saying that silence is golden, but we don't subscribe to this." With the story of Monica's marriage in the back of her mind, Helga was ready to give Peter a chance if he happened to be interested in her; but as it later turned out, she still had a lot more to learn about peaceful cohabitation.

As they made idle small talk, Helga began to talk about her nascent interest in art and how she would like to experiment with painting but had no idea where to start. Peter mentioned that he knew someone who might be able to help her, and this gave them a good reason to exchange phone numbers. This little act and the way they looked at each other during the exchange signaled an incipient attraction between them that foreshadowed a next step beyond Peter's promise to help her.

That evening Helga went home with the absolute certainty that she riveted Peter's attention and that the next day she would get a call from Peter. She sunk in her thoughts so deep she didn't even notice when Ulrich came home, which was probably after midnight again. She seemed to have completely stopped caring about what Ulrich did or said. Her mind was now on Peter. "Could this be the change I was waiting for?" she kept asking herself. The next morning, she hadn't even reached the lab when Peter indeed called her and invited her to dinner. She said yes to the invitation almost instinctively, and she had turned a deep shade of crimson by end of the short conversation. Although during the first meeting they were a little bit shy with each other, their relationship evolved rapidly and ended in marriage just before Christmas of the same year. To be fair to Ulrich, Helga informed him about her new relationship with Peter, which he accepted without indignation and without trying to change her mind, although he was deeply hurt.

Peter had a friend at IBM who married a German woman he had met in Berlin while he was serving in the military there. Whenever he saw them, he never failed to notice that their marriage was exemplary. Perhaps because of their different upbringing and language, they had a particularly sensitive way of tolerating the cultural differences between them. Unfortunately, this transnational marriage worked out quite differently for Helga and Peter; parallel to the breakneck speed by which the relationship initially developed between them, after only two years, their marriage took a nosedive. Nobody, not even Peter, could fathom the real reason. Once, in a desperate moment, Peter revealed to Doug what was truly going on between him and Helga, but the exact reason for it was shrouded in mystery. Peter was seized with scruples of conscience and began reading books on marital relationships to find out what he did wrong. The experts cited in these books the five ingredients that were important for a good marriage: love, trust, honor, respect, and desire. After reading these books, he was still no closer to divining where he was going wrong or what more he could do to reinvigorate his marriage.

Based on some of the examples the experts cited, it appeared that there was something, which even Helga could never really put her finger on, that had quite suddenly diminished her desire and, along with that, her love for Peter. In fact, the more Helga analyzed herself, the more she realized that she had never really felt the same deep love that Peter had for her. The trust, honor, and respect that Helga still had just couldn't make up for the lost desire and the absence of true love. As it happens so often when a woman starts panicking at the thought that she will have to continue living with someone whom she no longer desires, Helga couldn't control her irritations with Peter. This happened even on the rare occasions they met with Peter's friends. They too were flabbergasted by Helga's rapidly changing attitude toward Peter, and it was clear to them that something was simmering under the surface. Once, when Helga and Peter had left, Mike loudly began speculating what problems the two might be having. Perhaps they were simply an incongruous couple physically or psychologically, or maybe Ulrich's shadow was still alive in Helga's mind, or perhaps she missed something in Rochester, like her family or the old country, that neither Ulrich nor Peter could make up for; or, most likely, the change in her attitude toward Peter reflected a combination of all these.

Whichever was the case, the emotional parity between Helga and Peter, necessary for any good marriage, didn't seem to exist anymore and perhaps never did in the first place. Peter gave more and more of himself to keep the marriage alive, while Helga appeared to give less and less, having lost interest in maintaining the relationship. Her mind had already moved on, and unlike Peter, she had a pretty good idea of where they were headed. What she couldn't put her finger on was exactly what she was missing that Peter couldn't provide. Strangely, what had happened to Helga toward the end of her relationship with Ulrich now happened to Peter. Helga came home later and later, and on the days when he wasn't traveling, Peter often sat down at the dinner table alone brooding and wondering how much longer he could cope with this situation of being both married and alone at the same time. God's love may not seek recompense but his, as a human being, did. But for months and months he remained patient trying to contrive some solution and willing to cling to his second marriage at all costs, although his confidence that it would work was fading day by day. His ability to save his marriage narrowed to the point that he began to feel helpless; the cards were now in Helga's hands. It was only a question of time when this struggle would end.

Peter's character had two essential elements: he hated confrontation with anyone, particularly his wife, and he thought in terms of pure, sharp logic. When he realized that he could do no more and that Helga had emotionally passed the point of even being willing to listen to him, Peter decided that Helga had abandoned reason and had entered the realm of illogic, and he gave up on their marriage as well. When he finally stopped trying to fathom their piteous situation and decided to make no more attempts to understand Helga's mysterious, implacable nature, his inner peace was quickly restored.

During his last months with Helga, even though he was ready for the breakup, he still thought quite a bit about what he could be missing that caused him to keep losing his wives. His failures perplexed him. He begrudged other men for their ability to keep their women. He knew several married men who had cheated on or even physically abused their wives, and on the top of that, almost all of them had additional alcohol and drug problems. Astonishingly to him, in many cases the wife wanted to stay with her man. Peter did none of those things, yet both of his wives and Sarah he had loved had decided that he wasn't good enough for them. Possibly, deep in their hearts some women prefer aggressive and abusive men? Perhaps during the evolution of humankind, an aggressive man was better equipped to protect the family against outsiders? When choosing a partner with whom to establish a family, could this subconsciously be the most important criterion for some women, even if they would never admit it?

A lot more went through Peter's mind during this transition into impending divorce when he sat at the dinner table alone all those evenings. Probably the only tangible reason for the tension in their marriage was the ideological divide that clearly separated them. Initially they glossed over it with the noble lie that it wasn't important, although it should have been blindingly obvious to them that it was. For Helga, Peter's religious conviction, even though he interpreted the Bible's text less literally than others, reflected some level of intellectual laziness. It seemed that there wasn't much room for compromise with Helga in religious matters. The real problem presented itself when one Sunday Peter asked Helga to join him at his church service. Marrying Peter in a church was already a big enough compromise for Helga, but going to an actual mass to praise a god she didn't believe in was just too much of a stretch for her.

During the first year of their marriage, Peter often talked to Helga about faith issues and his disagreement with mainstream church doctrines, hoping to bring her closer to him, but to no avail. For a committed atheist like Helga, who found the idea of religion ridiculous, Peter's acceptance of God's existence and His exclusive role in creating the universe and humankind was an issue that kept her increasingly irritated. She explained to Peter that in her view, the belief system of Christianity is so misaligned with modern science that she couldn't see how anyone could rationally believe in God's existence, but she added that she could imagine that people of belief have their own system of rational thinking. When Peter asked her if she ever had some sort of guilt for not accepting God, she answered with a definite no. She had no contrition whatsoever. To her credit, however, Helga wasn't condescending or derisive to others' beliefs. She made no effort to badger Peter to give up his faith in God, she never ridiculed him, and she never tried to make him feel like his beliefs were less valid than hers. For her, tolerance for others' religious views was an important virtue, even though inside she was irritated by it. Her attitude and Peter's calmness were the reasons that their conversations about faith and atheism never became heated, although Helga brooked no criticism of her atheism. But after discussing these issues and seeing Helga's continuing resistance, after a while Peter decided to eschew discussing them again. He was afraid, and with good reason, that if he kept pressing the religion issue, the small fire burning in Helga that had so far been kept under control may evolve into a conflagration. Apparently, the long conversation in Doug's house almost two years ago about faith and science didn't help very much in smoothing over Helga's aversion toward Christian beliefs, even in a less orthodox form. She continued to view the church as a collection of wishful thinkers with slanted views of the real world.

Peter's first wife was a natural complainer, the sort of person who seemed to have the need to regularly get worked up and remonstrate about something. She could go off about almost anything, and for Peter her scenes almost always came out of the blue, leaving him totally unprepared. Peter was an easy target. Since he rarely went anywhere after work, he was readily available to listen to Anna's outbursts, and calming her down always seemed an impossible task. Not even flattery worked. After hearing long rants from Anna about how miserable her life was, of course because of Peter alone, he was sad but never fought back, other than occasionally pleading, "Why can't we have a little peace?" When Peter asked her what the real problem was, she would only beat around the bush and couldn't give a logical answer, probably because there wasn't one. When others looked at them separately, everyone thought of them as very nice people, who simply happened to be incompatible.

Unlike Helga, Anna was a Catholic with strong beliefs in the Bible, taken word for word. The weekly Sunday church services did some good for their relationship, even if temporarily. After the birth of their children, Anna became preoccupied with them, but she still found time and reason to squabble with Peter and instill a feeling of guilt and shame in him. Finally, Peter had had enough of Anna's aggression and unpredictability and gave up on the marriage. When he filed for divorce, Anna was terribly surprised. She accused Peter of being irresponsible, leaving her in the lurch with two small children. It was unfortunately true that Peter's decision to separate also applied to his two sons. However, he was strongly committed to see them as often as he could, and he wasn't about to compromise his relationship with them. He would look for a new relationship only if he found someone who understood him fully, who would try to solve a tense situation rather than fight about and would help him start a new life with the full involvement of his children. But he didn't actively seek out a new partner.

Peter, his heart filled with ebullient love, thought that Helga would be the woman to start that new life with. But as it became painfully clear over a short span of time, she was not. She demanded that she come first and Peter's children come second. He reluctantly agreed to that, not only for the sake of peace but because she truly loved Helga. Yet, inevitably, his decision created an ever-deepening hole in his heart, particularly because what Helga really meant was that she wanted Peter to completely replace his children with her. He tried very hard, but he couldn't make Helga understand that his love for his children didn't exist at the expense of his love for her. The opposite was true. These weren't competing feelings; they enriched, not diminished, each other. Peter always hoped that eventually Helga would understand this 'obvious truth,' but she never did.

Helga had her own take on how to relate to Peter's sons. She obviously thought that she had nothing to gain by accepting them, although she never said so openly, giving Peter the false impression that with time she might come around. Peter also hoped that once they had their own children, it might help Helga to understand his struggle. Unfortunately for Peter, soon after their marriage, Helga indicated in no uncertain terms that she had no intention of having a child anytime soon. This was something that they should have discussed in depth, along with their differences in faith beliefs, prior to their marriage. Had they discussed this, Peter would probably not have given his assent to this, and Helga wouldn't have married him. It was too much for Peter to bear that he could have no children with Helga, and at the same time, he was forced to reduce his relationship with his two children. Sometimes all he wanted to do was just bawl and cry. He couldn't imagine maintaining that arrangement for much longer.

One of the few bright spots in their relationship was Helga's burgeoning interest in painting. Peter offered to help and would have been more than willing to provide Helga with everything she needed. But Helga kept postponing her decision, and after a while, Peter stopped bringing up the issue. Perhaps a little more goading from him would have helped Helga find her true calling, and that may have diffused or at least delayed the tension between them. Helga never explained to Peter what dissuaded her from creating the art that she was once so enthusiastic about. But, in an interesting turn in Helga's life that Peter wasn't aware of, she has recently rekindled her interest in painting, triggered by observations she had made in the lab.

Helga's job included producing DNA fingerprints from tissues that consisted of many orderly horizontal lines derived from broken DNA. To her they looked like little maps, and one day, as she once again admired them, her desire to paint, this time with the eyes of a cell biologist, suddenly came back. "If I put my mind into it, perhaps I could find an artistic way to express the vast richness and beauty hidden in our cells," she thought. Each night after that, she went to bed with this thought in her mind but was too diffident to act on it. Yet this incubation time, so important for scientists, artists, and politicians before making decisions of lasting consequence, was very useful for her to hatch ideas. She didn't have any formal training in art, which made starting a painting career even more difficult. Her lack of experience, and her fear that any experiment with painting would end in total disaster, was probably the reason she kept postponing her decision for so long—so long in fact that her desire to create art was driven into her subconscious, although never fully lost.

Where did the idea to create art come from in the first place? She only recalled that in her teenage years she often had vivid dreams about painting abstract pictures, although she could never decipher what the pictures meant. Visiting art galleries was always an exciting experience for her, more so than for any of her friends, and now the thought of once again creating art filled her with the same exhilaration. But this time she had confidence in herself, although all she knew was that she wanted to create paintings that would give the viewer a glimpse of the mysterious complexity of the molecular events that our lives are based on.

The transition to becoming an artist is a long, arduous, and brittle process, requiring quite a bit of determination. It is almost impossible to get through it without some sort of support, and Helga had none; she had to rely on her own determination. By then Peter had given up on her in every respect, which was entirely Helga's making, so he was no longer a resource of strength she could tap into.

Since Helga only had the desire to paint but had no idea how to get started, it took quite a while to conceptually develop a unique personal style. This process kept her for even longer evening hours in a separate corner of her lab, where she could be alone with her thoughts. After weeks of trying out several different approaches, she settled with a semiabstract style and grew confident that she could get it right and may be able to create an attractive new brand of art. The immediate question was, What first steps did she need to take? She knew that she would succeed only through a great deal of trial and error. But to avoid falling into the trap of producing cheap artificial work instead of real art, she expended a lot of energy and thought into the planning process to come up with a strategy that might bring her success. To embark on such a radically different path, she also needed quietude in her life, which gave her the last bit of motivation she needed to divorce Peter. In her interpretation, valor won over tepidity; now she knew how she wanted to proceed with her life.

Joseph, who didn't have a family yet, often stayed in the lab after the official hours too and kept wondering what Helga was doing in her corner. It was evident she wasn't working but merely thinking about something while looking at some drawings she had made of cells as well as maps of DNA and proteins. He also noted that after work Helga read a lot of art books, as if she was preparing to teach art history. In Joseph's mind, work meant hard manual work at the bench, which hard thinking wasn't necessarily a part of. He tried to contain his curiosity for a while, but one evening he couldn't help but ask Helga if she had a problem that made her stay in the lab longer than was expected of her. This gave the opportunity for Helga to talk about her idea to start painting, the difficulties she was experiencing in trying to develop a unique style, and that despite the odds, she was sanguine about her future in art.

Helga's plan resonated surprisingly well with Joseph. He admitted that he also had some artistic aspirations and that he had already acted on them. Joseph was an avid photographer, but he was looking for something entirely new that no one else had done before, although in art and science, one can never be sure if an idea is indeed new. Once, as he stood at his window staring at the cloudy sky, he witnessed a small lonely cloud as it began forming the shape of a bear. Surprisingly, little imagination was needed to recognize the shapes of the bear's large body and head, complete with eyes and ears, all about at nearly the right locations. The bear also had a mouth, although it was somewhat dislocated. Another time he saw in the cloud the face of God as described in the Old Testament. Soon Joseph saw shapes in clouds wherever he went, and he decided to keep his camera close to him all the times in case he was lucky to find one worth recording. And indeed, over the years, he collected an amazing variety of pictures all showing clouds with interesting shapes, all abstract in a sense but still evoking the image of objects or subjects real or imagined. Most of them were portraits with misplaced eyes, lips, nose, and ears, but these grotesque faces often resembled humans in extreme pain or exuberant joy. Less than a year into his photography adventure, Joseph assembled fifty of his best photographs for an exhibition proposal that he intended to submit to the Minnesota Institute of Arts. He had no way of predicting whether the curator at the institute would like the idea behind the pictures or find it preposterous. Artists are often subjected to withering scorn before finally finding an interested gallery. Once, gallery curators used to be impressed by unusual art, but not anymore; by the time Joseph submitted his proposal, they had really seen it all.

Joseph and Helga's mutual interest in art, and the fact that they both were readying themselves to set off into uncharted waters, helped them become good friends. The other day Joseph brought in copies of his picture collection, which Helga found amazing, and they had a long conversation about it. Joseph had developed a technique to focus on the important sections of the cloud; without it, the picture would be too fuzzy to recognize shapes. During these discussions, Helga discovered a surprisingly deep intellect in Joseph.

That night Helga went home around midnight and found Peter preparing for bed with a wry smile. This was a kind of off season for him; he hadn't traveled anywhere for two weeks, and during the long evenings at home, he felt very much alone. Peter didn't ask why she came home so late, and Helga didn't feel obliged to give any explanation, so they went to bed without saying a word, but at least without any bickering. Even when things had been better between them, Helga had held to the view that in marriage, both partners should have total freedom to do anything they want, without the need to explain their whereabouts or actions to the other; the two partners should just trust each other fully. She would reject questions from Peter like "Why were you so late?" or "Who were you with?" She thought there was no place for suspicion in a marriage.

In contrast to Helga, Peter always gave his wife full disclosure about his whereabouts because he never had anything to hide, and he just thought that it was considerate to let her know when she could expect him to get home. He even called Helga frequently when he was traveling so that she knew that he was all right. It took him a while until he figured out that Helga really didn't miss that information. In fact, Helga thought that all those times Peter called, he only wanted to know where she was and what she was doing; she regarded this as an invidious intrusion into her privacy. Thus, while Peter was practically an open book for Helga, she was much more guarded.

Helga kept to herself everything that would have helped Peter understand her more completely, and she was starting to become almost obsessed with her own independence, another sign of what was to come. Even though Peter had already given up on their marriage, that night, experiencing Helga's complete lack of interest to talk to him, he was particularly hurt, and that was the point when he decided that he would take no more of this. For the second time in his young life, he had reached a point of no return. Helga's act, which by itself shouldn't have been so significant, finally put him over the edge.

Peter was surprised about his own reaction. He thought that he had already settled this whole thing in his mind, but he didn't. Now he finally realized that his efforts to keep their marriage intact were indeed hopeless. He still felt very strange about it, but his realistic mind, once again, had finally accepted Helga's intransigence and her unspoken verdict that she wanted to end their marriage. In fact, he accepted this eventuality with a sort of relief; finally, there would be no more bickering, no more rants from out of the blue, no dirty looks or snide comments. He was ready to once again live a wholesome, carefree life without Helga. It was high time to prepare for their divorce.

"Helga," Joseph said, beginning one of their many evening conversations that followed. "After our conversation the other night, I started thinking about your future artistic plans, and I had an idea. Don't you think there are some parallels between the events that take place in cells and the ones in human relationships? Just look at the 'chaperone' proteins. Most newly formed proteins in a cell don't come with the perfect three-dimensional structure needed to do its job properly. Thus, to acquire it, the newborn protein needs help from chaperone proteins, which, in turn, also need help from yet another set of proteins called co-chaperone proteins. The chaperone proteins, together with their aids, contact the newborn protein and like tailors cut pieces off here and there and sew some loose pieces together, creating the perfect size and shape. Does not the same thing happen, figuratively speaking, after a baby is born? The mother, who is like a chaperone, begins to shape the baby's mind with the help of her co-chaperones such as the father, the siblings, relatives, friends, and teachers. They cut and stitch many things in the child's mind until it grows up and is ready for the tasks in life. The only difference is that while in a cell, chaperones work like very effective robots and do their jobs in a second or less. In real life, the guidance of a child into adulthood takes much longer. But to me the principle is the same."

Helga could see the inordinate pride on Joseph's face as he figured all this out. But Joseph wasn't done yet; he came up with another extemporaneous idea.

"Now, how could one express this similarity in a painting? I was thinking that you could, for example, create a divided picture. On the left side, you could paint a semiabstract painting showing interactions among the newborn protein, chaperone, and co-chaperones with an artist's eye. I know that this is an incredibly difficult job, but with your tenacity, you could do it. On the right side, you could paint some lovely event involving the baby with his family and friends, with some indication that the formative process has already started. The right-hand picture doesn't have to be a traditional painting. It could be in a semiabstract form too. Please, before you say the idea is completely crazy, just consider it for a moment. I'm telling you, if I were as talented as you are, I would paint something like that without hesitation."

Helga chuckled a little bit and found herself wishing she were a little more aloof; the stark truth was, she hated Joseph's idea. The proteins might turn out all right. But painting a baby with a loving family? How could she do that without using the syrupy style that she so detested? But she knew Joseph was just trying to help and didn't want to hurt him; she just smiled and said, "I will think about it." What she really wanted to say was that Joseph should just stick to his clouds.

Joseph was taken aback by Helga's reaction; he thought she probably didn't understand what he was trying to say, so he continued to expound his idea. "What I really meant," he said, "was that artists and scientists shouldn't put a limit on their experimentation with new ideas. If they do, then they will fall into the trap of traditionalism, with the consequence that scientists will stop making breakthrough discoveries, and artists will not develop new artistic directions and solutions. What I have proposed could be very far from the right approach, but maybe it can at least stimulate some creative thinking." Put that way, the idea sounded a little better to Helga.

The fact that Helga was clearly not overly thrilled about his proposal failed to discourage Joseph. He remained in an enthusiastic mood and continued to develop ideas for Helga, although just minutes ago he had admitted that he had no practical knowledge of painting. "Just imagine what happens when terrorists attack a country, and how this reflects the immune system," he continued. "In response to the attack, the country, initially staggered by the attack, mobilizes its defenses and will most likely eventually capture or kill the terrorists, and with that remove the danger. After resolving this conflict and assuaging people's fear, life can return to normal. Similar things happen when we are infected with a small number of viruses or bacteria. Our body readies the immune cells to fight, and after some violent hours or days, the immune cells pass muster and eventually destroy the invading terroristic microorganisms. For most people this is nothing more than an uncomfortable temporary event that might even be helpful in preparing the body to withstand the next, perhaps more serious, attack. But what happens when a big country with a large army attacks a weaker country? Despite mobilizing all its defenses, the small country will probably lose and might be taken over. This is like situations such as serious sepsis, when the infection is so overwhelming that it results in the victim's death. Once again you could make a divided painting, which I wish I could do myself. On one side, you could depict in an abstract but recognizable way the battle between attacking microbes and the defensive immune cells. On the other side, the clashes between two opposing militaries could be painted using an unorthodox approach."

But, as he had already observed twice, Joseph knew that painting these pictures would be much harder than just talking about them, and so he finally stopped short of describing any more of his ideas for which Helga was very thankful. Neither of them was aware that others were already experimenting with the kind of art that Joseph was describing. Several years later, Ulrich showed Helga an article in a magazine, entitled "Double Vision." The article profiled a young graduate student at the University of Heidelberg who made such divided drawings, on the left side showing intricate protein structures that matched the events on the right side, like a couple dancing or people watching a bullfight.

That evening Joseph was in a talkative mood, and he asked how Peter was doing. "Doesn't he mind that you go home so late? Or is he traveling again?"

"No, he's home, but it's not a big deal. In our marriage, we've kept our independence and we usually don't have to explain all of our comings and goings to each other." She omitted that this was only her view of their marriage. Helga's answer surprised Joseph, but he had no further comments; he just packed up and left for home.

From time to time Helga got the feeling that Joseph was trying to initiate a romantic relationship with her, but she had a subtle way of signaling that this would not happen; she was simply not attracted to Joseph as a man, even though in a way she liked him. For one, Joseph, being quite a gaunt guy, was nearly not as handsome as Peter or Ulrich. To make things worse, Joseph was also very religious, almost to the level of bigotry. Her experience with Peter alone was a sufficiently strong force to keep her at a distance from Joseph. So, although they had developed a strong intellectual bond, perhaps the strongest Helga had ever had with a man, this wasn't accompanied by any sort of attraction. Fortunately, Joseph was cognizant of this situation, and he knew his boundaries.

Since Helga and Peter had decided to divorce, she made great strides to change her life as an independent woman. She moved out into a three-room apartment at the edge of the town near the airport. She also decided to give up her job at Doug's laboratory so she could even become independent from Peter's friends. By then she had completed all the necessary nursing exams and had found a position as a principal nurse in a small clinic in Stewartville, which was only about ten miles from her apartment. In her new apartment she converted the room with the highest ceiling into a studio, into which she could fit her large easel that she had bought earlier at a low price. She decided that on the weekends and in her free time she would start painting seriously. Soon after moving in, she also bought canvasses, brushes, special pencils, and paper for drawing. Although she now had the means, she was neither mentally nor technically ready to create actual paintings. She was like the people who buy fancy and expensive exercise machines but then never end up using them. She began to realize that she was perhaps in over her head in her attempt to start an entirely new occupation, without any preparation and being on her own. She clearly needed help from somebody, but for a while she was bereft of opportunities.

Fortunately, once she talked to Mike's wife, Nancy, about her problem, who not only loved art but also was formally trained in it. Even though Nancy was at home with two small children, she still managed to keep in touch with the local art community and its programs; somehow local art news kept finding its way to her. Her art degree gave her a certain status in the local art community, and before the children were born, she had helped to organize regular meetings of the Rochester art group, during which they would draw sketches of live models. Once Nancy brought Helga to one of these meetings, after which she enthusiastically continued to attend every second week.

Joining the art club was a real eye-opening experience for Helga. She learned very soon how important drawing was for a painter. Many abstract painters can't draw, and they cunningly mask their deficiencies by producing abstruse abstract works. Helga wanted to become a painter without this handicap, and she therefore made drawing every evening a priority, and she adhered to this new regular program. She also read several technical books about achieving the desired matching colors in a composition. The extensive progress she made in drawing gave her more and more confidence, and after several months, she was brave enough to paint with brushes for the first time. Her first oil painting, a mediocre start at best, showing a vase with some unidentifiable flowers in it, taught her a lot about how to develop the right mix of colors and how to use the canvass to make sure that enough space is left in each direction for the planned composition. After looking at her first creation for days without touching it, she firmly made her mind to continue her experimentation with painting.

Often, a burgeoning artist's first failed attempt saps the will to ever paint again, but this did not happen to Helga. Her next picture, an old violinist, worked out better, although she still had trouble with the space and a small part of the violin was missing from the picture. But it was really the following few pictures, where she finally tried Joseph's idea of artistically tying specific microscopic events in a cell to similar events in people's lives, that set her in a direction that she then never abandoned. The colors and the contours were still tenuous but improving with every new picture. Within a few months, painting, thanks to her obvious talent and relentless fervor, became tantamount to her primary job. The time between her first completed painting to the day she found herself handling her brush masterfully, though not yet quite artfully, was a surprisingly short period. She was inexorably on the path, an audacious journey, to becoming an artist. To do both her day job and painting required her to become very stingy with time; she had almost no time for anything else. But once she had managed to ensconce into her studio and sense a faint whiff of success, she was happy; painting gave her all the satisfaction she needed, and that was all what she wanted for the time being.

In the meantime, Ulrich's research on inflammation and diabetes took a sudden turn. Triggering of inflammation involves certain so-called regulatory proteins that tell the DNA apparatus in the cell nucleus which genes to use to produce some other kinds of proteins that boost inflammation. This was a distinctively resistant area of research, difficult to crack the secrets of DNA to glean new information. Most previous ideas had been shattered by newer data, but the resulting new theories were also on similarly shaky grounds and in constant flux. Ulrich discovered a small organic compound that in diabetic mice inhibited a key regulator of inflammation, which then led to a significant improvement in the diabetic condition. He then went on to prove in larger-scale experiments that as the diabetic condition of the animals improved, fewer mice developed pancreatic cancer. He had arrived at the apogee of his planned project.

He got the most important results at the end of his tenure, and he had to decide how to go forward within the frame of his possibilities. He didn't have a lot of choices. Staying for too much longer at the Mayo didn't appear to be an option, although another extension to perhaps six months was still possible. His boss's grant renewal application had failed at first attempt, and from that source, he had no money to keep Ulrich for an intervening period until a decision was made on a revised application. But he had some reserves and considered once more prolonging Ulrich's position for a while. Due to the recession, it was also virtually impossible to get a suitable research position at any of the Mayo research departments. Ulrich was not certified to practice medicine anywhere in the US other than California, and he couldn't get a stable research position elsewhere, despite his research succeeding beyond anyone's expectation.

At this point in his life, Ulrich really began to be apprehensive of his future, even more so than before. As he saw it, he had two options: he could go back to Germany and find a physician position, which did not seem to be too difficult, or he could develop a biotechnology company in the US based on this versatile antidiabetic, anticancer compound he had discovered but owned by Mayo. Being a German doctor, he could return to Germany at any time as a physician, so he chose to first take a shot at the latter. He simply hated the idea of letting his research in abeyance. From his remaining savings, which mostly came from selling the house several years earlier, he founded a limited liability company by himself, which he named Diabetes Solutions LLC and which he registered in Minnesota in January 2009. Since the initial technology was developed at the Mayo Clinic, a very good name in the venture capital world, Ulrich was able to raise an additional three million dollars in venture capital with relative ease to establish a small biotech facility in Stewartville. In the first step he purchased the rights for the compound from Mayo.

Each year small biotech firms are founded in large numbers with great enthusiasm but little cash, each hoping for epic windfalls after the work is done. However, most of them survive only for a short period. In general, in the biotech business, the burn rate of committed funds is very high and often made even worse by bad decisions such as starting with an inflated management group drawing large salaries. Often the CEO also operates under the belief that more money spent equals more money earned, leading to the squandering of precious funds due to excessive spending and tolerating slackers. Other times, the company does everything right to optimize spending, but then the FDA may require additional and more expensive studies that they just can't afford. At this point, venture capitalists can also chicken out because they don't see with clarity the path forward or the end game. To make matters worse, around that time, ominous tremors on Wall Street further lowered their enthusiasm to support new start-ups. Thus, at any point of time during the year, there was always a relatively large pool of scientists with industrial backgrounds looking for jobs.

Ulrich advertised three PhD and three research assistant positions for experts who were specifically qualified to plan, perform, and analyze experiments with diabetic animals. He received 124 applications, and from this large pool, he easily selected six whom he thought best fit his needs. A little while earlier, the pharmaceutical giant Pfizer began to reduce the number of its employees in Ann Arbor, Michigan, and two PhD scientists came to Ulrich's lab from this big company. In the end, he assembled a competent research team with six scientists and research assistants from all over the country all hoping for prodigious success, and with that stable employment.

Having scientists with industrial background was very helpful because they brought into the company many useful research practices that Ulrich had never heard of but that were required to push their drug through the chain of development quickly, with maximized results and minimized expenses. For example, good and relatively inexpensive methods that don't require animals are available to test whether the drug might adversely affect the cardiac muscle. This was one of the important issues that needed to be studied before even starting the very expensive animal experiments.

The significance of these kinds of toxicology studies was painfully clear to some large drug companies that had to stop development of their antidiabetic drug candidates, after they had already spent hundreds of millions of dollars, because in clinical trials, such toxic effects in the heart were found. In one well-known case, an already approved diabetes drug, Avandia, with several billions of dollars in market value, was suspected to cause latent heart problems. Serious discussions were held at the FDA about whether the research data had reached statistical significance and if the drug should be taken off the market or not. Although the drug maker's own data didn't confirm this possibility and no final decision had yet been made, the drug's reputation, and along with that its market value, was suffering.

Even for a major company, removing a multibillion-dollar drug from the market can be a staggering blow. But a similar failure could force a smaller drug maker out of business altogether. Thus, the task for a small company with limited resources is to find out at the earliest possible phase of drug development if the drug has undesirable side effects. But this can't mean that they compromise either the quality or the required quantity of data. Not being able to solve this conundrum is a sure recipe for failure. Despite being a scientist with no industrial experience, Ulrich was aware that the formation of a company is fraught with many risks and that his position was not a cushy job. But he trusted the science behind the project, and that gave him hope and confidence.

After the three months that were needed to set up the facility, Ulrich and his small team were ready to start testing their drug candidate in cell cultures. They looked for signs of toxic effects as well as if the drug elicited the correct cellular effects that could be indicative of its usefulness to treat inflammation and diabetes in animals and later in humans. Eventually, Ulrich's team demonstrated that in mice, their drug could tell the body to use less insulin for the same good effect on blood glucose. After six months of research with diabetic mice, the results were so encouraging with no reason to be leery about the quality of data that Ulrich and his coworkers decided to throw a boisterous party in celebration. By all indications, they were on the path of developing an important drug. It was also time to file a patent application to protect the compound's effects on inflammation and diabetes. To get worldwide protection, they made the application international. This is usually a long process, but the presented data and the accompanying claims are protected from the moment the patent office acknowledges receipt of the patent application. Then after the filing, usually nothing of substance happens with the application for at least a full year or two. Sooner or later the US Patent Office sends an office action in which the examiner either accepts the claims and issues the patent or finds problems with the application that the applicant may or may not be able to correct.

So-called prior art, such as a previous patent or scientific publication that resembles the patent application the inventor files for, is the most dreadful, and quite common, reason for the rejection of a patent application, which can be rarely overcome even by the best-trained patent lawyers. Unfortunately, this is exactly what happened to Ulrich's team. Several years back, a research group reported in an obscure biological journal in Japanese, but supplanted with an abstract written in English, that a drug structurally very similar to Ulrich's drug also inhibited the same key regulator of the inflammation process and as a result reduced blood glucose levels in diabetic mice. The examiner's sophisticated search program somehow found this piece, and on these grounds, the entire application was rejected. The examiner's letter ineluctably indicated that he saw little chance that the application can be revised to make it acceptable.

This was truly devastating news to Ulrich's company. It had already spent over two million dollars and started its third year with a large toxicology study underway, which now had to be stopped. Without patent protection, all their efforts were good only for a tantalizing story. It was clear that even with full patent protection their small firm would have difficulties in raising money for their drug development work, but without perfect patent protection, it was impossible.

Ulrich's original idea was that the three million dollars they had received would be sufficient to bring the project to FDA approval; they would then raise another ten million to perform their first human clinical trial. If the trial worked out, they would approach a larger pharmaceutical company either as a partner to perform additional phase 2 and phase 3 human clinical trials or to offer the technology for licensing. The value of the project at that point could be worth $50 million or more, although payments would come in smaller portions over time.

Ulrich knew that larger pharmaceutical companies were usually interested in products offered by small companies only after a successful phase 1 human trial, proving that the drug candidate is safe for human use. Preferably, at this stage the trial would already also reveal preliminary evidence of the drug's efficacy, as indicated by preclinical experiments, in treating the targeted disease. Since full development resulting in entering the market may cost up to $1 billion, the licensee wants to be sure that the offered drug candidate is not only safe and effective but also has good intellectual property protection; in other words, no one else can use it without the license holder's permission. With the previous publication of practically the same data that he had, Ulrich's dream was destroyed. After squandering most of the venture capital money and his own investments as well, he had to quickly come up with another idea to save the company and himself.

The investors from the venture capitalist group had always been keen to see how their investments in Ulrich's company were faring, and they evinced no empathy after hearing the bad news; instead, they immediately cut their losses and fled the company. That's what they do, understandably, if the future of a drug candidate is equivocal. One solution for Ulrich was to declare defeat and file for bankruptcy. But he wasn't yet ready to say goodbye to the company for which he had worked so hard. Toppling the company into insolvency would have meant that he personally would also go bankrupt because he invested all his money in the company. A frantic effort, combined with an absolutely genius idea, was needed. He retreated into his office and ruminated for a few days, taking stock of his potential options. With somewhat less than $1 million still on hand, the first order of business was to lay off everybody, except one technician, within thirty days. To create some positive cash flow, he subleased some of his laboratories and an office to a dental mechanic for two thousand a month. Finally, he had the idea that by adding another substance to his compound, he might be able to get a patent for a new composition that was equally or even more effective against diabetes. To hedge his bets, he decided to also change the structure of the compound and then see which approach would work better. He hoped that with some tweaking of the molecule he could come up with a related and still effective compound that was structurally sufficiently different to become patentable.

Ulrich's stubbornness to find an effective drug to prevent and treat diabetes soon received a new impetus. During a recent routine annual checkup, he was diagnosed with insulin resistance, the first stage of diabetes. This was very disconcerting news, which made him feel despondent at work for days. This diagnosis added to all the stress in his life that he already had to endure. As a diabetes researcher, he knew too well what insulin resistance meant. The consequence of it is that after a meal, the excess glucose, the most common sugar, lingers in the blood longer than it should because the skeletal muscle, the major reservoir of glucose, will become desensitized to the action of insulin, requiring more insulin production by the pancreas for the same effect. When this happens, higher glucose in the blood begins to exercise its destructive potential, slowly but surely, in several organs. The most damaging effect is that some of the insulin-producing islet cells in the pancreas begin to die during periods of high glucose. This means that by the next day, fewer islet cells are available to produce the same amount or even more insulin that is needed. As this vicious cycle of insulin resistance, high glucose, and islet cell death is repeated, after a meal the blood glucose value never gets back to normal, and afterward it just gets progressively higher, bit by bit, with every passing week and month. At some point, that person needs to take regular antidiabetic medication unless exercise and diet can keep his or her blood glucose in the normal range.

Ulrich also knew that the existing medications provided only a temporary solution, which could last for any number of years, but eventually many people with this condition would have to be treated with insulin, if they lived long enough. This was a very real and dreadful scenario for him, even though recently safer and longer-acting insulin products that were simpler to use and caused less complications had arrived on the market. He also knew that exercising and eating less carbohydrate could help delay the onset of diabetes. Still, he was scared, and after the diagnosis he became even more focused on tweaking his drug's structure to make it suitable for patent protection and diabetes treatment with as few side effects as possible. He hoped to sail through this new discovery period, a quite onerous one, successfully, so that he could use his own improved drug for his own treatment.

CHAPTER 6

A Tragic Event Occurred

On Tuesday morning, Mike woke up at around 5:30, much earlier than he was used to. Like Doug, he had had a hard time falling asleep. When he finally did, he had a nightmare; he dreamed that that night while he was walking home, he was suddenly set upon by an attacker, and in self-defense he killed the man. He was immediately arraigned and sentenced to death without having the opportunity to present his case or plead his innocence. All he could do was mutter some words in his defense that nobody paid attention to or even heard. Just before the final preparations for his execution were made, he woke up in a cold sweat. His wife, knowing nothing about Peter's absence the previous night, was surprised when he offered her coffee that early. Normally, at that hour Mike was still asleep, with his head burrowed into his pillow so as not to hear Nancy's movements. She sensed that something might have been out of the ordinary, but she wasn't concerned.

"What's going on?" she asked, amused. "My birthday is still two months away... or maybe you cheated on me last night and you want me to forgive you? Just admit it," she joked.

"Don't be silly, this is not a good time for it," said Mike with a surly face. "Besides, no husband would voluntarily admit to cheating. Honey, we have a very different kind of problem. Ever since we first started playing poker, it has never happened that anyone missed it without informing the others in advance, but last night that's exactly what happened. Peter didn't call and didn't show up. He's usually so responsible about everything that I just can't help but think that it was more than a temporary lapse in his memory. I am afraid something bad might have happened to him. I can't think of any other explanation. On the top of that, I just had a terrible nightmare. But what do you think, is it too early to call him?"

"No, you should call him immediately! We can't waste any time." Nancy suddenly grew very worried. She knew Peter very well and considered him a very reliable friend. "If everything was all right, I think he definitely would have called one of you before the meeting."

Mike heaved a sigh and then dialed Peter's number. The same thing happened as had happened the night before; the phone rang several times and then went to the voice mail. Having already left a few voice mails, Mike hung up. He exchanged another troubled look with Nancy, saying without words that not being able to reach Peter portends something bad. "Maybe after taking care of the kids I'll contact Helga to see if she knows anything," she said. She also started to have a bad feeling about Peter.

After a very short breakfast, Mike picked up his cell phone and called Peter again, with the same result. "Peter, are you there?" Mike loudly repeated his question several times, to no avail. But each time he called, Peter's phone rung several times, meaning that Peter was probably at least in the country, because the cell phone he used with his friends didn't work abroad, and the calls would go straight to the voice mail. As soon as he had put down the phone, Doug called him, with Steve also on the line. Doug informed Mike that they had had the same experience; Peter's phone was on, but he didn't answer. They were all shaken, but Doug reasoned that it was too early to assume something had happened to Peter and they should wait a little longer before contacting the police. "Maybe one of us should go to his house after work?" he added.

"I'll do it," said Mike.

They continued chatting for a little while about the possible reasons of Peter's absence. Eventually the three friends agreed on the idea that Peter must be traveling within the country and had forgotten to tell them; with that, they all said goodbye and went to work. Early in the afternoon they called each other again, and since Peter still hadn't contacted any of them, the thought that something had happened to him after all was starting to seem more and more real. For the rest of the day, they all struggled with the same frightening thought: What if Peter was in trouble, and they did nothing in time to save him? Were they simply too averse to inconvenience? For them, doing nothing to prevent a disaster would have been unforgivable, to say the least, although there was probably nothing that they could have done the previous evening or afterward. Or was there? They could have called the police in the morning, but none of them could muster the willpower to get to that point. They kept postponing this, partially because each had urgent tasks to deal with on that day. But later in the afternoon, what really began to worry Doug was, as he told his friends, that he knew Peter had two cell phones. How unlikely was it that Peter had lost access to both of his phones? Increasingly, their failure to contact Peter just didn't seem right, and they felt compelled to do something about it that very same day. Mike assured his friends again that after work he would visit Peter's house.

In the morning it was cloudy with light rain that fell occasionally until a little after 2:00 p.m., just enough to make the roads and grass a little wet. But by the time Mike left the school to visit Peter's house, the sun was out. He pulled over to the curb in front of Peter's house, and as he was walking up the driveway, he noticed the faint traces of a motorcycle's tires. He found this very strange because Peter didn't have a motorcycle. This was the only sign of a vehicle; the rain had obviously washed away the traces of any cars that might have been in the driveway before that morning. "The motorcycle had to have been here after the rain, sometime after two o'clock," he mused. He saw no movement in the house when he looked through the huge windows that occupied most of the front wall. Arriving on the porch, he rang the doorbell a couple of times. The fact that Peter didn't answer did nothing to dissuade him, and he tried the doorknob. To his mild surprise, the door was unlocked, and he cautiously entered the house.

Half of Peter's house was one large space with little separation between a spacious living room combined with a dining room, an office that Helga used to occupy, and the kitchen. Beautiful and rich woodwork, both on the walls and the ceiling, gave the house an especially warm ambience. A large stone fireplace, although never used, added additional charm to the place. A few years before his marriage to Helga, on a very cold and windy day, Peter tried to light the fireplace but was greeted by a big cloud of smog and ash that promptly gushed back from the vent, indicating that something was clearly wrong with it. Since then he had never attempted to clean it. Dredging ash and dirt, and becoming sooty in the process, was outside both his expertise and taste, and he didn't know anyone in Rochester who did that sort of thing.

Mike couldn't find Peter anywhere in the main area of the house and heard no noises that could provide any clue as to where he was. He wasn't in his bedroom either, and Mike was just about to leave, feeling despondent when he decided, as a last hope, to check the bathroom as well. As he walked through the door, he immediately recoiled in horror. There was Peter, lying on his back beside the bathtub with a towel halfway wrapped around him. The bathtub was about two-thirds full of water, as if it had just been used. Mike was paralyzed for a minute or so, staring agape at Peter's body. He was no coward, but the scene was suffused with some mystic darkness that filled him with a strange and overwhelming sense of fear and disgust.

Eventually he was able to slowly pull himself from his trance, and as he regained his faculties, he began to think logically about the situation. The first thing he realized was that the light was off, which he found very strange. He turned on the light and then stepped closer to Peter to see if he could still do something to help, but Peter was clearly motionless. As he touched him, he noticed that his chest was wet while the towel was almost dry. Out of desperation, he rejected the idea that Peter was dead; it just didn't make any sense. He quickly tried the usual methods he had learned back in high school to detect any vital signs, but he detected none. He tried to resuscitate Peter by breathing air into his mouth and pressing on his chest for a few minutes. Usually he felt a kind of squeamishness when he saw blood or very sick people, but interestingly, he didn't feel anything like that this time. He fully concentrated on performing his resuscitating efforts correctly, and this completely smothered his nausea. But all his efforts were in vain, and eventually he had to concede that Peter was dead. The finality of this realization sent a new wave of icy shivers down his spine, and now he could hardly suppress his urge to vomit. He was devoured by fear. He closed his eyes and prayed for a second that this whole thing was only a terrible nightmare and he would wake up any moment; but when he opened them again, the dead man lying in front of him couldn't be more real.

Then in few minutes later, he regained some control and decided that this wasn't the time to try to help or analyze the situation any further, and he called 911. He identified himself and gave a short report of what he had seen and where. He noticed that he was literally stammering, but the call taker was very patient with him. Although he himself had never been in such a situation and had no contingency plan to adhere to, he knew that he shouldn't touch or move anything because that would interfere with the police investigation of the scene. He even regretted that he had left so many footprints all over the house because even that might make the work of examiners more difficult, but he could never have dreamed of discovering such an abominable situation.

The police and an ambulance car carrying a medical examiner and his assistants arrived shortly. They were happy to hear that Mike had tried not to touch or move anything. The examiner briefly introduced himself as John O'Neil and then rushed into the bathroom alone. Peter had fixed dilated pupils, and John couldn't observe breathing or any reflexes. With his instrument, John also determined that Peter's heart was not giving out any electrical signals.

"Well, he's dead, all right," John declared gravely, "but I really can't tell exactly when it occurred or how."

His body was still relatively warm, although it was clearly at least halfway down to room temperature as they quickly determined with a thermometer. His face and other body parts already showed the characteristic discoloration of a corpse; however, his muscles had not yet stiffened. Rigor mortis only sets in about five to ten hours after death. His position on the floor and his facial expression suggested to John that he had died of a sort of convulsive condition, but at first blush, they didn't notice any tangible signs that could point to homicide.

Meanwhile, one of the policemen, Bill, who had come with John, started looking for footprints. He recognized four different ones on the hard vinyl cover in the hall; two clearly derived from men's shoes, and the other two appeared to be traces of women's shoes. He asked Mike to take his shoes off so they could identify his tracks both in the hall and the bathroom. They also found a couple of pairs of Peter's shoes, and they soon found the one that fit one of the prints in the hall. Thus, in the hall they could account for the traces of two different men's shoes, but not the women's shoes. Bill's aides further scrutinized the footprints in the bathroom and around the front door. It was immediately clear that in the bathroom, one of the woman's shoes as well as Mike's shoes and Peter's slippers had left marks, but there was no sign of the other woman's shoes. While Bill and his aides worked on finding fingerprints, hair, or any other telltale items, John also measured the temperature of the water in the bathtub; it was $76°$ Fahrenheit, although the room temperature stood at $70°$. "I have the impression," said John, "that while he was sitting beside the bathtub, he became sick very quickly, which prevented him from standing up. But at this point, the reason for that is anyone's guess—heart attack, stroke, a sudden lung failure, food poisoning, maybe some kind of foul play?"

Then Bill came by and turned his attention to the towel partly wrapped around Peter, which had tiny marks of blood on it. On a closer examination of Peter's face, they found a small amount of clogged blood in his nose as well. However, Bill and John saw no physical signs of a struggle.

"Nevertheless," Mike heard Bill saying to his assistants, "the bloodstain on the towel might indicate that someone used the towel to strangle him. This would leave no physical signs other than some blood from the nose and mouth. Since there was no sign of a struggle, Peter already could have been in a very weak position, maybe even unconscious. It seems to me that if there was a struggle, it was between him and a woman who entered the bathroom. Of course, the blood in the nose could also have come from a convulsion or seizure. We have to keep all these possibilities in mind."

At this point, in a flash Mike remembered that when he had entered the house, the light was off in the bathroom, and he could see Peter only because the bathroom door was open and there was enough light coming from the living room. He shared his observation with Bill, who frowned in confusion at this development. "Then who turned on the light? Was it you?"

"Yes, I turned it on in panic," Mike replied. "Don't be surprised if you find my fingerprint on the switch."

Bill frowned again. "Well, if you turned on the light, then somebody other than Peter must have turned it off. It wouldn't make any sense for Peter to take a bath with the lights off or that he kept the door open just to let some light in. Well, we definitely have to dust for any other fingerprints on the switch," declared Bill loudly enough to make sure that his assistants heard the order. "We will take yours too."

"I believe," Bill said, now turning to Mike, "that your observation gives us more than enough evidence to assume that somebody other than you, probably a woman, spent some time in the bathroom along with Peter. Perhaps by turning off the light, the woman wanted to prolong the time until someone discovered him. This seems obvious to me. But why would she leave the front door open?"

"Maybe when she saw Peter dead, she rushed out of the bathroom and the house as fast as she could because the scene was so scary, and perhaps she was also afraid of going through an interrogation by the police. In such situation, nobody would care about the bathroom door."

"All are valid points," said John, joining in the conversation again. "She could actually be just an innocent bystander wanting to avoid any possibility of being implicated in Peter's death. That could explain why we can't find any clear evidence of murder. But at this point, we have to move forward examining all possibilities without predilection."

These early observations placed the probable time of Peter's death within about two hours of Mike's arrival on Tuesday. To be more accurate with the timing, they would have had to know the temperature of the bathwater at the time Peter took his bath, but that was impossible. If it was lukewarm, then Peter could have died thirty to sixty minutes before Mike's visit. If it was very hot, then it could have taken more than an hour to cool down to 76° Fahrenheit. Since Peter used to go to work around 8:30 in the morning, the medical examiner noted, Peter couldn't possibly have died in the morning after taking a bath. While debating these possibilities, the medical examiner and an aide quickly prepared Peter's body for transportation and rushed him to the clinic. They decided that a thorough autopsy was needed to see if they could still get some usable blood and tissue samples for analysis. If they waited much longer, the cell death in Peter's organs would be so extensive that it would be impossible to determine if a poison caused the death. However, a fast-acting poison like strychnine could still be found in the gastric fluid and blood or even in Peter's tissues for many hours after his death. Naturally, they would also take samples of Peter's nails and hair because some poisons, like arsenic, would remain there for a long time. Predicated on the findings of the clinical laboratory, they then can either decide to close the case with a report or move further with additional investigations.

In the meantime, Mike asked Bill's permission to call Peter's office at IBM. Although it was already 5:25 p.m., he still could catch Peter's secretary, Agnes. She informed Mike that Peter was out of work that Tuesday and that he was planning to fly to Los Angeles early Wednesday morning. She added that on Monday afternoon, Peter took everything with him and wasn't planning to go back to his IBM office before going to the airport. She hadn't heard from him since then. Peter left a lot of work for her, but she had been able to take care of it all without seeking any advice from him. All this explained to Mike why no one at IBM was concerned about Peter's whereabouts on Tuesday.

Agnes asked Mike if she could take a message. When Mike gently told her that Peter was dead, first she just repeated the word *dead* with tremulous incredulity in her voice, and then she apparently broke down crying; at least, Mike could hear nothing but static for a full minute. Finally, in a broken voice, she promised to inform Peter's boss immediately, who was still at the office as well. Before he hung up the phone, Mike had one more question for her: "Agnes, on Monday, did Peter by any chance mention to you that he didn't feel well?"

"Well, actually he did," replied Agnes. "Before lunch he mentioned that his stomach was feeling strange, and he even seemed a little irritable, which is very unlike him. But he still went out for lunch with business guests and then stayed in his office until about 4:00 p.m. without any more complaints. When he left, I didn't notice anything unusual, certainly nothing that might have suggested that anything was wrong with him. I'm still in shock."

Until that point, Mike hadn't had the strength to call Doug or Steve, but now he finally decided that he must. Of the two, he could only reach Doug, who was about to leave for home. Doug was shocked beyond description, but when he finally pulled himself together, he promised to come to Peter's house as soon as he could instead of going home. He also said that he would attempt to call Steve.

Although Peter used to travel a lot, to Mike's best knowledge, he had never taken a one-day vacation prior to a short trip like this; and even if he had this one time, what could have prevented him from calling his friends on Monday evening? Or if he had by chance lost his first cell phone, why didn't he just use his second one? It almost looked as if Peter had intentionally been avoiding them on Monday evening. At that moment, though, there was no plausible explanation. Another strange thing that Mike noticed was the impeccable order in the house. After Helga's departure Mike visited Peter twice, and he never saw such order. Usually piles of books, notebooks, and fliers were all over the place, which seemed only natural because Peter often continued his work at home. He had a large table in the living room with literally hundreds of documents scattered all over. When Mike asked him how he could possibly find anything, he answered that he remembered the exact location of each document. "You know," he had said almost ostentatiously, "the same disorder which is forbidding to you is wonderful to me because I see order in this chaos." Even the floor, which used to be littered with pieces of papers that had fallen from the table, was now spotless—one mystery after another.

Mike began to think about the cell phone. Bill clearly told everyone that he didn't find anything in the pockets of Peter's clothes. So then where were Peter's cell phones? If at least one of them had been working that morning, it should be working now as well. If he could find at least one of them, it could offer some clues. Upon a whim, he yet again asked Bill if he can make calls. This was granted, and he dialed Peter's number. He clearly heard ringing from the receiver, but the phone couldn't be located anywhere in the house. Mike called again and went to the garage to see if he could hear some ringing from Peter's car, but again he heard nothing, even though the phone was clearly turned on and was ringing somewhere. Mike drew the conclusion, which he immediately shared with Bill, that at least one of Peter's cell phones was most likely in someone else's possession. Mike unfortunately knew neither the number nor the carrier of Peter's other cell phone. This already bizarre situation was becoming more and more mystifying. If ostensibly no foul play was involved, as the police temporarily concluded, why couldn't they find his cell phone?

While all this was happening, one of the police officers called the city's chief homicide detective, Carl Schafer, who was neck-deep working in a nearby small town on a domestic violence case. Despite this, he arrived at Peter's house in a remarkably short time, considering the more than ten miles between Rochester and the town he was in, and that on his way he even had to find time to contact John, the medical examiner, who was back at the Mayo Clinic by then. John informed him that there was no sign of any violence, although he couldn't rule out strangulation while the victim was already weak or unconscious.

Carl was in his early sixties, and with his pudgy face and drooping eyes, he bore a rather striking resemblance to the famous detective Poirot. He was constantly and loudly talking about his dreams of retirement, although nobody around him was sure if he was serious. One of his potential successors used to question this with thinly veiled hostility. In any case, he still had few more years to go until reaching full retirement age. He had spent most of his career in the New York Police Department chasing murderers, of which that city had far too many. It was a particularly thankless job, made even worse by frequent accusations of police brutality by human rights activists who usually found a way to contort the facts. Those activists were silent and rarely cut the police any slack even when police officers were killed on line of duty. There were many incendiary situations, some of them race related, and all this made Carl rather tired of New York, so much so that he was eager for a change of location. He met his future second wife, Rosanne, a nurse, at the Mayo Clinic where he came to receive treatment for a rare case of leukemia he was diagnosed with. The combination of radiation and a new experimental chemotherapy appeared to be successful, but he had to take three months of leave to recover. He was already almost completely bald, so at least he didn't have to worry about losing his hair. Shortly after that, he married Rosanne, and from then he never again went anywhere in a rumpled suit.

Carl was offered a detective position at the homicide division of the Rochester Police Department, which he took gladly. It didn't go unnoticed there that back in New York, Carl had boasted higher than 80 percent crime-solving rate. He was highly venerated, and everybody in the department hoped that his stellar record would continue unabated in Rochester. "What a change from New York," he thought. "I can relax here for a couple of more years before retirement." But the reality was different. Rochester was growing rapidly, which brought in many new people and, with that, more crimes.

On that day, Carl still had to find time to arraign two brothers for alleged drug possession. He usually didn't have the chance to get home before 9:00 p.m., and even after that, he was often called to various crime scenes. This explains why he was looking forward to his retirement so eagerly. However, his work in New York had driven into his brain that while on the job, he must give his absolute best effort to be successful. It came with his job that he always remained intentionally imponderable to people he contacted during an investigation. He usually answered questions tersely, letting others know that he was the one who held the power as he usually did. He knew the downside of giving out more information than necessary, even though this was often disconcerting to his associates. He wheezed up on the slight slope of the driveway to the house with a gloomy expression on his face; his attempt to get home in time for dinner had once again been thwarted.

As he entered the bathroom, Carl already had the gut feeling that this present case was not going to be an easy one, although in every investigation, there is always a sort of allure to try to make the case as simple as possible. Although Mike and the medical examiner could not find overt signs of foul play, it was difficult to accept that a seemingly healthy thirty-eight-year-old man would suddenly just collapse and die without any previous health issues. Mike informed him that Peter had been in the early stages of diabetes, but he added that as far as he knew, there hadn't been enough time for Peter's condition to escalate and damage key organs, leading to his sudden death.

"This situation seems a little fishy, and we have to keep an open mind for all possibilities." This was the first sentence that Carl uttered, and it wasn't too informative. He was rather fond of describing such scenes as "fishy," as indeed they quite often proved to be.

Carl proceeded with his investigation deftly. First, Bill briefed him on the shoe traces they had found in the hall and bathroom. They stood by their analysis that recently two men and two women had spent some time in the hall, and one of the women as well as Peter and Mike were in the bathroom. In fact, they added, they had also just found traces of a woman's shoes along with Peter's shoes in the kitchen. The woman's shoes didn't match the traces of the ones in the bathroom. After this short report, Carl's first question to Mike was if he had touched anything. He was satisfied with Mike's answer that he only touched the light switch and Peter while attempting to resuscitate him. Then Carl started looking for still extant evidence of intrusion into the house, but he and his aides found none. In the spacious living room that also served as the dining room, the large table was now clean with no apparent traces of food or drink. Generally, the order in the house was in a conspicuously superb state with no misplaced items, as Mike already noted before. There were no dirty dishes or, in fact, any dishes in the kitchen except two clean wineglasses that were lying upside down on a plastic drying plate next to the sink.

Carl asked Mike if he found it strange that Peter had taken a bath the day he was supposed to work.

"Actually," Mike responded, "that could well have happened, because today Peter wasn't working at IBM. I just asked his secretary. He was about to fly to LA tomorrow morning. I think it is entirely possible that Peter just took it easy today, invited a friend over, probably a woman, who left her traces in the kitchen, and had one or two drinks with her. But it is also possible that the friend came the previous night, they had some drinks, and then Peter woke up today in the early afternoon, took a bath, and finally died as he stepped out of the tub. Perhaps the mysterious visitor even had something to do with his trip to LA. But then what happened to Peter between his meeting with the visitor and his death? It seems logical to me that he simply had a long sleep."

"Well, putting everything together, I would agree that the traces of separate women's shoes in the kitchen and the bathroom must mean that Peter had two female visitors," said Carl. Usually, forbearance from drawing conclusions too quickly before considering all available evidence was an important characteristic of Carl's approach in dealing with suspicious deaths; it stemmed from his substantial crime-solving experience. Young crime solvers often swoon over the first bit of 'evidence' that they find and become perfunctory in examining other leads. Then, while they dawdle their time away, those other potential leads rapidly evaporate. In contrast, Carl always looked at his cases from as many directions as possible, trying to first establish a holistic view and using it as a frame into which new findings can be incorporated in a logical way rapidly leading to key evidence before important witnesses disappear. In the eyes of his underlings, Carl really fit the mold of the archetypical homicide detective. This time, though, he didn't know how his conclusion about two female visitors at two different times will play out.

Disregarding whether there was a visitor or not, Carl started to collect items into separate plastic bags, including a sponge and a corkscrew, as well as a large pile of empty Bordeaux wine bottles he found in the garage in a container. Fingerprints were taken all over the place, and during this whole procedure, special care was taken not to move any other items from their original positions.

Then, moving his search into Peter's home office, Carl began looking at Peter's personal items. "Mike or anybody else, did you see Peter's cell phone anywhere?" he asked.

Mike answered him, "He actually had two of them, but I only have one number. I called it not too long ago, but even though it kept ringing, I couldn't locate it anywhere in the house or in his car."

"Is it really so?" Carl murmured. "That is not a good sign. Someone who may have an interest in knowing Peter's connections probably got a hold of the cell phone. Possibly, even as we speak, this person might be continuing to keep track of Peter's callers. I wonder if this person is interested in knowing his business connections or wants to know the identity of his female friend or some other information about him. In any case, if we don't find Peter's cell phone soon, this might be the first important clue pointing to foul play," he added. "In fact, let's hope that the phone company can help us to locate it."

Carl asked Mike if he thought Peter could absent-mindedly leave the cell phone in his office.

"That's quite possible," Mike testified. "He was known to regularly lose cell phones. He was pretty irresponsible in that respect. I believe this was the primary reason he had two phones."

Carl turned to an aide. "Inquire at IBM if Peter left any of his cell phones there."

Now Carl turned his attention to the desktop computer sitting on the desk in Peter's small office. The computer was turned off, and although Carl switched it on, he couldn't proceed without a password. "I'll need to call our computer experts to deal with this," Carl murmured. "After authorization by my department, the service provider must make the password available or create a new one." Mike overheard and informed him that Charter Communication was the internet provider.

Then Carl became interested in Peter's laptop and ordered his men to search for it. For the next half hour, everyone literally searched every nook and cranny in the house, but to no avail. After the missing cell phones, this was yet another thing to puzzle over. The only thing of any significance that one of Carl's associates found in the office was a black suitcase full of IBM contract drafts and legal documents. Apparently, Peter was about to prepare some contracts with LA companies on behalf of IBM. However, it was beyond anyone's imagination that Peter would have left for such an important trip without his laptop.

"First the cell phones and now the laptop. This case is very fishy indeed," Carl summed up the situation again. "But if somebody had a hand in Peter's death, that person wasn't interested in Peter's business dealings. That person was interested in his personal life. But then who could have possibly taken the laptop? If the laptop was indeed taken, or rather stolen, that is already a crime. But without something or somebody disabling Peter first, it's hard to imagine that anybody could have just walked away with his laptop. It looks as though, little by little, we are proceeding," he said with a wan smile. The task to find out the reason for Peter's death appeared to be slightly less daunting.

Carl asked Mike to stay a little bit longer just in case he needed any more help from him. Mike informed him that his friend Doug, who was very knowledgeable in health matters, would arrive soon. The first question from Carl came within a minute. "Can we be sure that Peter's diabetes was not serious?" he asked, glancing toward Mike while still searching for more items. "I can't aver it in either way," answered Mike.

"You know, we didn't discuss our medical situations in detail, so all what my latest information is that Peter was in the early stages of diabetes. Peter was a good football and basketball player back in high school, but then he stopped exercising and perhaps ate too much. During the past few years, Peter gained somewhere around thirty pounds. When you add this to his lifestyle, the stress caused by his marital strife, and his frequent long-distance flying schedule that limited his movement, I'm not surprised that he gained so much weight.

But I think that eating too much junk food was his main problem. Like any number of people who take comfort in an unhealthy activity, Peter found joy in eating."

Suddenly, there was a knock on the door and Doug entered the house. Fortunately, by then Carl's aides had finished taking the footprints inside the house and outside the front door, so the new prints he left didn't confuse them.

"I'm glad to see you here. So you're the health expert I'm told," said Carl, turning to meet Doug.

"Sort of, depending on the issue," he responded nervously.

"Well, I just learned from Mike that Peter had some problems with his blood sugar, and I was wondering how much you know about this. Also, I wouldn't mind if you would please briefly summarize to me what I need to know about this condition. Perhaps that might help me to figure out if it contributed in any way to his death."

"To my best knowledge," began Doug, "Peter was diagnosed with so-called 'insulin resistance' about six months ago. This is an early-stage disease caused by insulin being less effective than it should be in removing surplus glucose from the blood. If this problem isn't taken care of early on, it usually leads to diabetes, meaning that the blood glucose reading remains permanently high unless an effective medication is taken. My friends and I noticed that this mundane change in his health first made him furious. Then came the stage of denial. 'Someone at my age just cannot be diabetic,' he told his doctor over and over again. But the doctor told him that there are even many children who are diabetic and need insulin to survive. Finally, Peter accepted the diagnosis, and from that point on, his only concern was how to keep his health while taking no medication or taking the least amount of medication possible.

"The doctor advised Peter to exercise regularly for at least one hour a day, including weight lifting, and to eat less pasta as well as fatty and sugary foods. Then after eight months, Peter was expected to go back for another checkup to see if his condition had sufficiently improved without taking medication. In many cases, the combination of exercise and weight loss can slow down or even halt the progression of insulin resistance into diabetes, although complete reversal of the process is relatively rare. However, by adhering to the prescribed austere exercise and diet plan, Peter had the opportunity to avoid taking medications for many long years in the future. If he didn't, however, the remorseless process of developing full-blown diabetes would take hold. Thus, for Peter the tug-of-war between an austere but healthy lifestyle, and undisciplined behavior inexorably leading to diabetes began at that point. If you want to know my opinion, there's no way that this early insulin resistance problem could have caused his death unless combined with some other complications."

"Even if this is the case," interrupted Carl, "I am still a little surprised that he wasn't immediately put on some sort of medication."

"You have a point," Doug agreed, "but some experts believe that it may be more beneficial to allow a slightly-higher-than-normal amount of glucose to linger in the blood without medication than to start an aggressive treatment early on. You should also know that Peter was not amenable to advice, and he was neither ready to start any antidiabetic treatment nor willing to give up some elements of his lifestyle. He repudiated the idea of giving up the good things in life just because of a 'little insulin problem.' According to him, this would have been something only health nuts would do. One of these 'good things' that he frequently enjoyed was large cut New York steak. He would drool over the mere thought of it. I heard that he couldn't resist ordering steak whenever he went out for lunch with his IBM colleagues. Unfortunately, he considered this a mere small peccadillo that he could easily make up for.

"Peter also told me once that he had heard that if someone drinks red wine with the steak or a similar fatty food, then that person will be perfectly fine, because the antioxidants in the wine supposedly fend off the otherwise insidiously adverse effects of the fat. That article also explained to him the 'French paradox,' which in a nutshell means that even though French people don't refrain from eating copious fatty foods, they are less likely to die of coronary heart disease because their diet also includes moderate consumption of wine, which protects them. Peter was enthusiastic about this French paradox and saw wine as the best recourse to improve his metabolic health. Once he told me the following: 'If this works for the French, then it should work for me too.' The trouble was that Peter ignored the fact that although moderate amounts of alcohol may indeed help with heart disease, he had a problem with his blood sugar and not his heart. He only concentrated on the idea that in some respects drinking wine may compensate for eating lots of fatty food."

"Then to avert the bad effects of too much fatty food, which Peter was made aware of, he started drinking more wine, perhaps much more than he could handle?" speculated Carl.

Doug agreed. "I assume that more than likely, this may be what happened. Peter was slowly drawn into the idea of the protective properties of wine and began to drink more of it every day. He assumed, wrongly, that the new 'alcohol science' excused him from following the more austere exercise and diet regime prescribed to him and that he can override his doctor's advice not to drink with impunity. To make things worse, Peter also had a sleeping problem. Because of his frequent overseas trips and the substantial time zone differences between Minnesota and the far-off places he flew to, he had difficulties adjusting, and on many days he couldn't have a good night's sleep. And sleeping disorders can contribute to insulin resistance and diabetes. Taking all these risk factors into account and not being serious about exercise and dieting, in my opinion it was only a matter of time before Peter became seriously diabetic, but not even a diabetes specialist could determine the timeline. It might have taken one year or ten."

"Thank you, Doug, this was very informative," remarked Carl. "It seems to me that because there could be some contributing factors, I should not quite yet take Peter's insulin resistance problem off the list of suspects. But now it is time to check with my people and see how far they have gotten." While Carl continued his search for clues, many thoughts crossed his mind, but he didn't share them with the others. "Could it be," he mused, "that on the day of his death, Peter simply ate too much fat or too much sugar for lunch, or maybe even both, without carefully monitoring his blood sugar? This could have led to a sudden surge in his blood glucose, which then caused him to lose consciousness, which he never regained. All this could have been triggered by the sudden change in his blood pressure as he stood to step out of the bathtub. People with high blood sugar often faint and lose consciousness for extended periods of time. Since he was probably alone and without help, he maybe just never regained conscience. But of course, even if he had a too big a portion of food for lunch, why on earth would he go take a bath after that? No, no, eating or drinking couldn't have caused Peter's death. Doug was probably correct. His insulin resistance wasn't bad enough to trigger a comatose state. But what if Peter's diabetes advanced rapidly since his last doctor's visit?"

Rapid progression of diabetes came to Carl's mind because he remembered what had happened to his sister. She had her regular checkup every year, and the last time her blood tests showed only a moderate increase in glucose, which didn't require treatment, only serious adjustments to her diet. However, a month before her next checkup, she suddenly lost consciousness when she stepped out into her flower garden. Fortunately, her retired husband was home, and he called for an ambulance in time. Her blood glucose levels read almost four times the normal value, and she had to be given an insulin injection immediately to avoid a potentially deadly coma. Apparently, in rare cases, diabetes can advance rapidly due to some disorder of the immune system or very serious stress. In connection with his sister's case, Carl learned that sometimes the immune cells receive a miscue from the environment or from within and start destroying the insulin-producing cells in the pancreas rapidly. The damage in Peter's case could be so extensive that there wasn't enough insulin to deal with a massive meal, and consequently he might have fallen into a sudden coma from which he never returned. "I wonder if that's really what happened. I will talk to the medical examiner about this," Carl decided.

After a short pause, Doug resumed his conversation with Carl about the possible causes of Peter's death from a different angle. "Possibly, Peter's diabetes or whatever he had, together with having a hot-water bath, might have also exacerbated some hidden cardiac problem, and all that could have led to a heart attack. Diabetes can trigger heart disease, and hot baths aren't recommended for people with heart problems."

"But, Doug, in all truth, there was no evidence that Peter had a heart condition," said Mike, unexpectedly joining the conversation.

"Believe me, Mike, quite often heart problems remain unnoticed during doctor's visits, no matter how frequent these visits are. Doctors, even the best ones, can never determine the condition of their patients with perfect accuracy. For example, do you remember John Ritter, the gifted actor who was loved by millions of people across the globe? He died recently due to an unrecognized heart problem. In his case, doctors couldn't identify any triggers that could have made his heart condition suddenly fatal. This just goes to show that during our lives, we are constantly switching between the two realities of looking strong from the outside but in fact being a delicate, fragile machine in the inside. This is an incontrovertible truth, and I'm telling you, scientists have made barely any real progress in making our flimsy selves any stronger."

But just when Carl and Mike were ready to accept the possibility that the hot bath had aggravated Peter's diabetes, possibly resulting in his sudden death, Doug cited an even more recent body of research that went against the theory he had just laid out a minute ago. It can be maddening to laymen when scientists can't make up their minds and are able to present two diametrically opposite scenarios, such as fatty food is good for you or not, or eggs are bad for you or not, in the same breath without shame. Doug was one such scientist. "But before we fall in love with the heart attack theory," he began, "I need to mention that another line of research proved beyond doubt that moderately warming the body increases insulin sensitivity of the skeletal muscle and liver, thus easing the diabetic condition."

"How does that work?" asked Carl.

It had never been easy for Doug to communicate science to laypeople, but he tried his best. "Without giving you too many technical details, all human tissues contain so-called stress proteins, which are formed in larger amounts when the human body is confronted with a stressful condition, such as cold or heat. These stress proteins not only help protect the cells and tissues from the harmful effects of stress, but they can also increase the efficacy of insulin. In short, certain stress proteins can overcome, or I should say improve, insulin resistance. Diabetes reduces the production of stress proteins, leading to the worsening of insulin resistance. Thus, while in other people hot baths may indeed have some negative effects on the body or at least no effect at all, they can benefit diabetic patients because the more stress proteins are produced, the more effective insulin becomes at normalizing blood glucose. In fact, stressing the body with cold-water bath would probably have the same beneficial effects on diabetes as the warm bath, but it's unlikely that diabetic patients could be persuaded to volunteer to take extremely cold baths. The demonstrated positive health effects of the cyclical combination of sauna sessions and periodical jumping into cold water are almost certainly due to increased production of stress proteins."

"Do you think that Peter's doctor was actually aware of this new research data and therefore advised Peter to occasionally take hot baths instead of taking diabetes drugs?" asked Carl. "Then Peter could have taken a bath anytime on Tuesday, or maybe he took multiple ones. After all, he was on a vacation of sorts and could afford it. I heard that in a warm bath, some people can think more effectively and can easier make difficult decisions. Some professionals go as far as to claim that taking a warm bath led them to solve seemingly insurmountable problems. For some people, particularly for those who are calm thinkers, a nice warm bath is almost a type of procrastination, allowing them to make decisions later with a rested brain."

"Yeah, I can personally attest to that," interjected Mike. "After having spent some time in the bathtub, my brain is better at coming up with new melodies for the opera piece I'm working on."

It was Doug's turn again to introduce more controversy into the hot-bath story. "Before we fall in love with the idea how hot bath might help diabetes, I should add that some of the same proteins whose production is triggered by hot bath and help with diabetes also promote the growth of certain tumors by stimulating the renewal of tumor-initiating cancer stem cells. I know all these effects of hot bath are confusing, but this is what science have come up with so far."

"I guess we should drop the subject how hot water bath might have caused Peter's death," Carl suggested. "It doesn't seem to lead anywhere."

It was around seven when Carl finally allowed Mike and Doug to leave Peter's house. Carl and his aides decided to stay longer to have a look at the basement, which seemed to function as a separate apartment, with its own bathroom, kitchen, walk-in closet, and a large exercise room in which Peter kept a TV set, a computer, many bookshelves crammed with hundreds of books, and even a king-sized bed. Things were suspiciously orderly there too; obviously Peter had to have hired a maid service recently, Carl deduced. A man living alone, however tenacious he might be, couldn't possibly keep such order without help. He made a note to himself to find out if a maid was hired recently. In the bathroom he found a large cabinet on the wall beside a mirror, and two small cabinets on the opposite wall to store soap, vitamin tablets in three containers, and a plastic bag that contained quite a bit of white powder. Just by smelling it, the drug expert Carl determined that it wasn't a recreational drug; he concluded that Peter probably used it to whiten his teeth. Bill asked if he should collect the containers, but at that point, Carl didn't see it as practical to take any of these items as forensic evidence. "Maybe later," he said.

When Carl returned to the upper floor, he was surprised to see that Mike and Doug were still standing there, both expressing anxiety. "I forgot to tell you something," Mike told Carl. "When I first got here, I clearly saw the marks of the tires of a motorcycle on the driveway."

"That is very interesting indeed. I want to see it immediately!" exclaimed Carl. He took Bill with him, and indeed, in a thin mulch of mud that covered the concrete, they still could clearly see the traces of a motorcycle's tires that must have parked there that afternoon. Carl immediately called the weather service at the TV station to ask when the rain had stopped. They informed him that in that area it had stopped at exactly 2:26 p.m.

This information was very important, because the rain almost certainly would have washed the tracks away if the motorcycle had stopped earlier. "Thus," Carl concluded loudly, "the motorcyclist must have stopped here between 2:26 and when Mike came, which was when, Mike?"

"It was around 5:10," Mike replied, "but by then the driveway was pretty much dry."

One of the aides took pictures of the motorcycle's tracks, but they couldn't find any footprints other than Mike's. This left Carl without any evidence that the cyclist had ever entered the house. "Well, I don't see for the time being how this could help us," said Carl, turning to Mike, "but thank you for noticing this. Perhaps you should be heading this investigation. You have good eyes." He was nice to Mike, but the tone of his voice suggested that he had lost interest in the motorcyclist.

Carl then turned to Bill. "I think as long as we're outside, we should see what else we can find. Would you please check if you see footprints around the door or anywhere else outside? It would be nice to know which direction the two women came from. Now, I want to go back inside and see what else we might be able to come up with." After Carl went back into the house, Bill said to Mike and Doug with a smile, "I'm very sorry for your loss. But you shouldn't worry. Carl will get to the bottom of this. With all his experience from New York, he knows how to catch the perpetrator of any crime and when to pounce on him or her. Luckily, during this last month, we had almost no criminal activity in Rochester. Let's just hope that this will remain the case for a while. That would give Carl more time to deal with Peter's death."

CHAPTER 7

Suspected Causes of Peter's Death

Mike and Doug each went home for a late dinner, but before they separated, they managed to contact Steve and inform him of what had happened. That evening the three friends remained in contact, calling each other for news almost hourly until midnight. Around ten, Doug attempted to get information from the clinic about the autopsy and if it had shown anything interesting so far, but he didn't get much out of them. He even got the feeling that the clinic wanted to stave him off until the morning even though he worked there. Obviously, the person who was authorized to give out information had already gone home.

The next morning Doug went right to the office of John O'Neil, the doctor assigned to Peter's case, who informed him that at the time they first looked at Peter, his body was still in relatively good shape; therefore, his death couldn't have occurred too long before 5:00 p.m. on Tuesday, but he couldn't specify the exact time. Furthermore, he said that the pathologist he was working with had also found that his intestines and stomach were unusually empty, suggesting that on that day, he either didn't eat anything, or he had had serious diarrhea and vomiting. It was almost as if he had taken something to clean his stomach, like people do before gastric examination. The pathologist's assistants, however, were still able to collect some gastric juice, which they would use later along with his blood to determine what he had eaten and whether he was poisoned.

But John cautioned that some poisons and drugs can break down quickly in the circulation, so whoever would do the analysis might have a hard time to find anything at all. "But you must know this better than I do. You're the expert. In fact, if the analysis gets too complicated, you may get a request from us to help us out," John told him. "Anyway, in addition to his empty stomach and some signs of inflammation, Peter was in the early stages of fatty liver disease and some mild cirrhosis. This might be an indication that Peter drank a lot more alcohol than was healthy, and that he was on his way to developing a serious liver problem that could have ended in incurable liver cancer."

Both John and Doug knew the full implication of this diagnosis because the liver is an important organ for the regulation of blood sugar, and the fat deposits there would have made insulin less effective as a regulator of metabolism. Had his doctor examined Peter's liver function regularly, he could have warned Peter of this and other dangers caused by his alcohol abuse. That perhaps would have served as sufficient motivation for Peter to dial his drinking back and start taking real antidiabetic drugs. But Peter's aversion to seeing his doctor knew no bounds; he was too worried about what the doctor might say. It was easier for him to stay snuggled under the cover of his belief that wine would solve all his problems.

That evening, Doug called John once again and asked if he had any more findings. During their conversation, Doug mentioned that on Monday Peter didn't feel well, which coincided with John's hunch that Peter probably had a severe bout of food poisoning. This might have caused him to vomit heavily, perhaps followed by diarrhea that in combination led to serious loss of body fluids, which could be life-threatening. John's assistants were finding more and more evidence for this scenario. To find out for sure though, John ordered microbial testing of samples collected from his stomach, the intestine, and the blood taken from his heart. Doug didn't tell John, but he had the feeling that following this food poisoning idea might be only a sidetrack. But he was pleased to see that John led this stage of investigation in an orderly and not desultory manner.

The data from a complete set of histology samples, together with the microbial tests and other autopsy investigations, was to be ready by Thursday noon. Preparation of tissue samples for staining, which allows them to be examined under a microscope, is a relatively simple but tedious process, and it was already late in the evening. Later that day, Doug found out that John was so burnt out from the great number of other requests he had received that he handed Peter's case over to his boss, Dr. Dean Anderson. Dean was a benevolent man with great scientific credentials, and he happened to be a good friend of Doug's. He was approaching retirement age but was still very agile. He and Doug had a routine, almost daily collaboration, both relying on the other heavily, which made their relationship uniquely friendly. Doug had never worked with John, so he welcomed the change. Dean called Doug up later that day and confirmed that he was taking over the investigation and had already begun to look at the case. He added that the pathologist's and John's first hunch was that Peter died of food poisoning. Apparently, Dean didn't object to this plausible diagnosis, but for Doug this possibility still didn't sound right, and he let Dean know his opinion.

"Well," said Dean with his trademark gentleness in his voice, "I hear you, and you can trust me. We'll try to cut all superfluous sidetracks. But, and I don't want to expatiate on this subject, food poisoning isn't nearly as rare as people think. We get regular briefings on the issue, and the Centers for Disease Control and Prevention—you know, the CDC—recently estimated that in the United States, about seventy-six million illnesses occur because of food poisoning every year, most often resulting from carelessness during food preparation. This results in 325,000 hospitalizations and about 5,000 deaths. Even though the number of deaths might seem small in a populous country like ours, how can we rule out that Peter was one of them? Because of his early-onset diabetes, he would have had a weakened immune system, so he was certainly at a greater-than-average risk of getting infected. However, it may not be easy to identify the infectious bug, even in people who remain alive. In fact, in more than 80 percent of cases, the pernicious microbial agent that caused the food poisoning remains unidentified forever. In dead people this task is even more difficult because of the rapid influx of bacteria into the body soon after death. Nevertheless, since samples from Peter's tissues could be taken relatively fast, we hope to see some detectable outgrowth of the responsible bacteria or virus over the host microorganisms that are always present in great numbers regardless of food contamination. Blood taken from Peter's heart could be particularly useful in determining if he might have succumbed to a serious infection." All in all, despite his earlier concerns, it seemed to Doug that Dean knew what he was doing; Peter's case was in good hands.

By late afternoon on Wednesday, Doug learned from Dean that the analysis of Peter's blood showed that on the day of his death, he had drunk heavily. The first histology data gained from the liver, lung, kidney, and heart were also available; they clearly suggested that Peter's death had occurred sometime between 4:00 and 4:30 p.m. on Tuesday, barely an hour before Mike arrived on the scene. In addition, large areas of these tissues still had not started to degrade, although the pathologist saw many isolated necrotic spots due to widespread cell death in those areas. To expert eyes, these necrotic spots looked different compared to the slight tissue decay that can normally be seen in dead people after several hours.

The pathologist also identified large numbers of infiltrating white blood cells in these tissues, drawn there by the inflammation caused by the broken remnants of dead cells. These cells probably further deteriorated the structure of Peter's organs by releasing agents that caused even more cell destruction. After some deliberation, the pathologist thought that this unusual cytological pattern of necrotic areas could mean that a poisonous substance had triggered widespread cell death. If so, this would rule out heart attack or stroke as the primary cause of death, although at a certain phase of his struggling, both events could have occurred and become dominant. A further possibility was that he was strangled while he was already ill and trying to recover. Since he was apparently both drunk and severely sick, he wouldn't have been able to put up much resistance against an invader.

Considering all these early observations, the pathologist conveyed his feeling to Dean that the real cause of Peter's death seemed to be unrelated to his early onset diabetes, because people with such a condition almost never die from it alone. However, diabetes could still have been an incendiary event that might have made the effects of food poisoning or some other kind of poison much worse, particularly if all this was associated with heavy drinking. The pathologist eschewed from making any more guesses because clearly more work was needed to responsibly make a conclusion.

After learning all this from Dean, Doug placed a conference call to his friends and informed them of what he had learned. They were a little leery of the idea that Peter might have been the victim of food poisoning, but what they really couldn't understand was why he had been drunk so early in the day.

"I think when it came to medical and alimentary issues, Peter simply could not find the means to navigate in the right direction," said Doug. "After all, many people can't decide how to approach drinking alcohol, because there are so many conflicting messages tied to various interests. It's true that excessive alcohol consumption is the underlying reason for many abusive relationships and in fact for some diseases as well, including inflammation of the pancreas, liver problems, certain cancers like breast cancer, abnormal fetal development, and so on. On the other hand, there is a lot of new evidence that moderate drinking of red wine extends life expectancy. I still have the copy of a report, published about ten years ago, which went as far as to show evidence that adult men who drank two glasses of red wine per day lived an average of seven years longer than men who completely abstained from alcohol. Doesn't that sound almost unbelievable? How can you resist drinking wine after reading such article? However, some other studies were far less enthusiastic about the longevity benefits of alcohol, and it has also remained controversial if the actual alcohol itself was beneficial or the substances dissolved in it were."

"And what do the health officials have to say about alcohol? What's their official view?" asked Mike. He usually abstained from alcohol, and all along he had thought that he had been doing the healthy thing.

"Well, I'm not too familiar with this issue," admitted Doug. "But as far as I know, health officials who issue recommendations, and who keep trying to stay objective in the alcohol debate, have so far failed to solve the dilemma of how exactly to inform people of the positive effects of alcohol, without encouraging heavy drinking. Part of the problem these officials still face is that producers of alcohol beverages have a vested interest in advertising the good effects of alcohol, and they put pressure on the media not to demonize alcohol.

"Now, contrast that to the large number of alcohol researchers who, up until very recently, overwhelmingly reported only the negative effects of alcohol, mostly because their livelihoods depended on portraying alcohol exclusively as a dangerous substance. Without publications, they didn't have a chance to get grants, but most of the time their papers were accepted for publication only if they could relate their data to the negative effects of alcohol. For some reason, influential alcohol researchers had a serious bias against alcohol consumption in general. However, this might soon change due to a recent report by scientists at the University of Texas and Stanford University, which may be the most credible report on alcohol thus far. Their study, which included more than 1,800 adults ages fifty-five to sixty-five, confirmed that moderate drinkers, and quite surprisingly even heavy drinkers, were less likely to die than abstainers over a twenty-year span. Moderate drinkers in this study were defined as those who had one or two drinks a day, while the heavy drinkers had three or more drinks. Thus, in my opinion, despite Peter's early-stage liver problems, heavier drinking during a short span was unlikely to cause his death, unless, of course, his blood alcohol exceeded the toxic level. But I do see a possibility in the idea that he couldn't have fought a microbial invader if he was drunk and his prediabetes worsened, and even less so if he was weak because of food poisoning."

"Doug, you're amazing." Mike laughed. "I ask what I think is a simple question, and I get a full-length lecture on alcoholism." And although Mike said this jokingly, Doug thought he sensed some veiled cynicism, and he decided that in the future he would refrain from lecturing his friends.

"So anyway, back to my original question," Mike insisted. "Do you have any idea what might have led Peter to get drunk in the first place?"

"No, I don't have a clue, and maybe that is all I should have said," answered Doug. "Let's discuss these things in a little more detail after dinner. Can we meet at Starbucks tomorrow morning at seven?" They all agreed and said goodbye.

There were several reasons why in Peter's case the medical examiners were more thorough than they would be in a routine autopsy, in trying to get to the bottom of the cause of his death. Peter was still relatively young, and it was difficult to accept by anyone that no foul play was involved. The loss of Peter's phones and laptop computer, which Carl was still unable to trace, seemed only to reinforce this possibility. It certainly did for Carl, who kept pressing the medical examiners for as much data as possible, hoping that it would help shed some light on something that might then put his investigation on the right track. But for the time being, he didn't have much to go on to implicate anyone in any criminal activity. He had no other choice than to cast his net as wide as possible and see if after a tedious process, something or somebody would be caught. This meant that during the entire investigation, even Peter's friends and their wives were on the list of suspects. For the time being, the three friends were completely unaware of this. All they knew was that Carl was certainly giving priority to the case. Although Carl worked at a slower pace than they had hoped for, they all appreciated his deliberation and tenacity. What they didn't see was that Carl had a myriad of ideas that he systematically tabulated in his head, occasionally abandoning some, while adopting others.

After the initial analysis, the opinion of the pathologist and Dean was still that Peter's unexpected death was probably due to food poisoning that had eventually overwhelmed his immune system, causing septic death. But this scenario didn't get much support from the first round of microbial tests that they were able to finish by Wednesday evening. No matter how hard they searched, they couldn't find any of the usual microbial culprits, like *Escherichia coli*, *Salmonella*, *Staphylococcus aureus*, *Clostridium perfringens*, *Listeria*, *Shigella*, or *Campylobacter*, in excess. Most strangely, the total count of bacteria taken from the intestine and heart blood was lower than what was considered the normal range. They also couldn't find signs of any of the most common viruses, like noroviruses and rotavirus, that cause food poisoning. Something that was thus far inscrutable had been going on in Peter's body. Thus, the medical examiners and Carl had to consider other possibilities as well, spurring even more new lines of investigation. This made no one happy; in fact, everybody involved started to feel rather frustrated.

According to the quickly revised plan that Dean and Doug had put together, the medical examiners also looked for the presence of certain agents in Peter's body that might have been produced by bacteria, and they got the results relatively soon after the bacterial counts were in. The most important of these bacterial agents are called endotoxins, which may trigger so much inflammation that the body eventually can't cope with it. Diabetics are particularly vulnerable to these types of toxins and inflammation as well as the ensuing sepsis they may cause. Due to the inflammatory damage of the vascular system, some key organs such as the lung and kidney, suddenly stop functioning, leading to rapid death. Sepsis may kill someone in as short a time as two days. Unfortunately, if the sepsis is in an advanced stage, no known medication can help. Dying of sepsis is not uncommon, even in developed countries. For example, in America an estimated 220,000 people, mostly those undergoing major surgery, contract an infection while in the hospital and die of sepsis every year. These facts, combined with Peter's sudden death, led the examiners to investigate if an endotoxin could be the culprit.

One of these toxins is lipopolysaccharide, or LPS, located in the cell wall of certain bacteria. In the intestine, LPS molecules can be shed from the bacteria's wall, and after escaping into the body's circulation, they languish there for a while. Interestingly, the examiners found suspiciously large amounts of LPS in the blood samples. However, the low bacteria count and the high LPS level just didn't seem to fit together, except in two situations. First, someone could have poisoned Peter with LPS, although it wasn't clear if the amount of LPS they found, roughly ten times above the normal level, was sufficient to kill a grown man. Another possibility was that a cell poison was used, which indiscriminately killed both Peter's bacteria and his cells. Dying bacteria, like *Escherichia coli*, could then release LPS, leading to inflammation and perhaps sepsis. In fact, there was a third possibility as well, which the friends and toxicologists thought of only much later; perhaps Peter had simply inhaled LPS in an environment with unclean air, and in that case, LPS was just an innocent bystander with no role in Peter's death. This inhaled LPS could never have reached the critical amount in Peter's body necessary to trigger an immune reaction. Was this discovery of LPS only complicating the investigation, or was it really the culprit in Peter's death? Nobody knew for sure at that point, mostly because none of them was an immunologist with a deep enough knowledge of this truculent toxin.

On Thursday morning, only Doug and Steve could make it to the meeting before seven in a nearby coffee house to discuss the information they received from Dean Andersen. Doug, still quite downcast, told Steve that he knew Dean very well; he often turned to him for help when a sophisticated chemical or biochemical analysis was needed. To Dean, it seemed that Doug's assistants were sorcerers who were able to analyze any kind of human or chemical samples with almost impossible precision. Doug's lab was equipped with sensitive instruments that could detect minute amounts of metals as well as inorganic and organic compounds.

"Well, Steve, it seems that something very strange is going on. They found quite a bit of LPS in Peter's blood, but Dean wasn't at all sure that the amount was sufficient to cause septic death. The LPS found was certainly enough to trigger some immune response, but this was probably a far cry from bringing Peter's entire immune system down. In Dean's opinion, that would have required at least a hundred times more LPS, although this would be difficult to know for sure because of individual differences. For him, it looks more and more like the LPS route is fizzling out. Instead, he appears to be besotted with the idea that a poison was used that rapidly kills cells."

"Is he equipped to go in that direction?" asked Steve.

"I am afraid that this is something Dean's lab cannot handle," Doug said. "I'll go to Dean and volunteer to perform a very thorough analysis of potential poisons. I hope something will turn up."

"I agree. It was most probably some kind of poison that kills cells indiscriminately. Food poisoning generally wouldn't have this kind of effect. But if a cell poison was deliberately used to kill Peter, it was certainly not one of the poisons, like curare, that blocks neurotransmission in the nervous system. A nerve-blocking poison wouldn't cause cell death in Peter's peripheral tissues and would certainly not raise LPS in the blood. Some other drugs might have damaged his kidney, for example, but they cause death much more slowly. No, we're looking for a fast-acting cell poison, which killed Peter's cells and bacteria alike, leading to this LPS spike. Of course, LPS then might have further enhanced the cell killing effects of that unknown cell poison."

"I tend to agree with you, Steve, but let's not consider the idea that LPS was the primary agent dead just yet. Before we do, would you do me the favor to do a little research about LPS, namely, how much of it is needed to kill a human?"

"I can do that tonight," Steve responded.

"But what I just can't get over with is why Peter didn't call us on Monday evening, and why he didn't respond to our phone calls the next day? He had two cell phones to do that. This is so unlike him. Apparently, he didn't even check the callers. In his position, this is simply unheard of. On Monday evening, something dramatic must have happened to him, which left him so weak physically or mentally that maybe he just didn't have the strength to call. But what could have made a strong guy like Peter suddenly so brittle?"

"I suggest that for the time being we stick to the idea that it was a sort of cell poison, which somehow relates to LPS," said Doug. "Please, don't forget your homework."

Then they shifted their conversation to Peter's apparent visitors who had been in his house shortly before his death. They agreed that one of the visitors could have taken the laptop and cell phones, and Peter was too weak to react; perhaps he was even already dead by that point. And this mystery person took these items because they might have contained some compromising information or personal messages. That person certainly wasn't callous to Peter, they agreed.

"I'm fairly certain," said Doug, "that Carl is dealing with this case deftly and perhaps has already figured all this out and is now working to find this mysterious visitor who might hold the key to solving Peter's death. I wish I had some idea about who that mysterious person is." He sighed. "But for the life of me, I can't come up with a single logical suspect."

"I don't have any ideas either," Steve replied. "Frankly, we're just floundering from one idea to another without actually having a clue. You need to do the analyses we were talking about. Without them, we can't move forward. Of course, the question is, do we really want to deal with this? Or should we just let the professionals take care of this case, and accept the outcome without any remorse?"

"I certainly want to move forward, however daunting this task might prove to be," Doug replied. "I would like to see proof beyond reasonable doubt, either foul play was involved, or that he died of natural causes. Although, considering that two cell phones and a laptop are missing, I have a hard time believing that his death was natural. I would like to see the villain punished for what he did. With Carl on board, I'm confident that we together will get to the bottom of this. Something tells me that if we help Carl, he will be able to tease out the right answer. So are you with me?"

"Yes, I am," Steve replied, sounding very determined. "And once we are in, we are not satisfied with some slap-dash work and conclusions."

It was almost lunchtime when Doug arrived at Dean's office. Although he usually wasn't very good at reading people's minds, when he saw Dean hovering around his desk, occasionally tapping the keyboard of his computer with a serious face, he guessed that Dean might not have a lot of news for him. Without wasting any time with other matters, Dean told Doug all the results he had so far but admitted that for the time being, there were no leads to follow in support of a potential criminal case. "Although the histology suggested that some poison could be in Peter's body, the case for that isn't sufficiently strong. The analysis of the gastric juice doesn't provide a case for food poisoning either, because although Peter clearly had diarrhea and vomited a lot, we couldn't identify any bacterium or virus that might have caused it. In fact, the bacterium counts were found to be lower than usual, and while this finding really flummoxes me, I see no opportunity to expand on it. Besides, I don't see how a food poison could kill cells. The only finding that really stands out for me is LPS, but in my opinion its level was still too low to cause death, at least not without some other additional collaborating health condition. It just doesn't seem like this much LPS could bring Peter down, but I suggest that you talk to an expert if you still want to pursue that idea.

"Listen, Doug," he continued. "Let's admit that presently our information about this whole case is useless, and if you really want to go further, then I'll need your help with the analysis. I need a key observation that can lead us to a firm conclusion. I can already see in your eyes that you want to help. Well then, my hunch is that the same poison that killed Peter might have also killed the bacteria, which could have in turn raised the amount of LPS in his blood. If I'm correct, then this LPS business is just a sideshow, and the real culprit is a cell poison."

"This is amazing." Doug smiled. "Steve and I just came to that very same conclusion a few hours ago!"

"Is that so? Well then, we're on the same page," Dean replied approvingly. "Now, we have to move forward with the poison theory. However, my lab isn't equipped to analyze poisonous substances, but yours is. Can you help us with that?" He didn't even bother to wait for Doug's response before continuing. "You may want to start with the usual suspects, like arsenic, mercury, and perhaps cadmium and the like. They all kill bacteria and can damage most organs as well."

"I agree fully," Doug responded. "In fact, Steve and I already suspected that this kind of analysis would be needed, so my answer to your request is obviously yes. But I need samples of the gastric juice and at least three different organs, and preferably more if you have them. It is a little tedious to clean up the samples for the analysis, but by Friday evening, which is tomorrow I guess, you can expect the results, if we start soon enough. It will not depend on me," he added with determination in his voice. "I really feel that it would be an insult to the memory of Peter if we didn't solve this somehow, or at least try to, as much as humanly possible. I'm confident that following the poison cues, we can fully unshroud this case sooner rather than later. I can't afford to use my assistants' time and our equipment promiscuously, without having a reasonably solid hypothesis, but I know that my friends and I will regain our quietude only after we have done everything we can to solve this case. We all agree that we must try to catch this surreptitious killer."

Doug was about to leave when Carl showed up to check in on the case. "Well, Dean, have you found anything interesting for me yet?" he asked, wasting no time.

"Yes and no, that's the long and short of it," Dean answered dryly. He really hated when people asked him for results before he had gotten the chance to examine the situation from all angles. "By saying yes, I mean that we found a suspicious substance in Peter's body that could have been a poison."

"What is it?" asked Carl quickly, not having the patience to wait for the explanation of no.

"It is LPS, which is an abbreviated name for the bacterial product lipopolysaccharide. It can wreak havoc in the body if there is too much of it."

After Doug's explanation, Carl stood deep in thought for a moment. "But how could LPS possibly get into Peter's body? Could this LPS have just been made inside the body, or did someone try to poison Peter, for example by mixing it with wine?" Carl really tried to fully understand what these scientists had achieved by then.

"You know, Carl, you might be on to something," remarked Doug. "Although, we are starting to think that the LPS is probably just a by-product of the event that killed Peter, not the primary cause. You see, Carl, LPS might be a good lead to something, but the question is if there was enough of it to kill Peter, and right now our answer is no. In any case, I believe you raised an interesting point, and we shouldn't entirely dismiss LPS, particularly if you add Peter's diabetes into the equation. But you know what? It just occurred to me that Steve could be a big help here. He studies cases of medications killing people instead of curing them. Perhaps, just perhaps, Peter took some medication that made him so much more sensitive to LPS that it killed him. But still, there should be an explanation for where the LPS in Peter's body came from in the first place. Was it coming from the dead bacteria already in Peter's body, or did someone maliciously give it to him, or did he pick it up via yet another route?"

"It looks like that you guys still have ways to go," said Carl. "Am I correct?"

"You are," Doug admitted, "but we are moving along without letting our imagination and enthusiasm for finding the culprit to stoop."

That Thursday, Doug skipped lunch in his eagerness to receive the tissue and blood samples Dean had promised him and to make some tangible progress. His desk was strewn with documents he didn't have time to deal with. He had several very close deadlines that he kept pushing back, not only because his work on Peter's death took up a lot of his time, but because he hardly had the presence of mind to concentrate on anything else. Among others, there were three more toxicology cases to review, and he was three days behind in reviewing two cell physiology grant proposals for NIH. He could not stave off any of these reports any longer; it was time to get back to his academic work, and he decided that from tomorrow, that is what he would do.

At the same time, he felt morally obliged to find out what had killed Peter, however slight the chance for that appeared to be at that moment. He made up his mind and made the following promise to himself: "I will not let whoever did this to go unpunished, even if I have to work overtime." Doug had a deep sense of morality and a strong determination to right the wrongs if he felt it was within his power. But he never operated under self-delusion. He had to have the feeling that the task at hand was doable, and in Peter's case, he was confident that with the help of others, they could solve it.

The tissue samples, which included a small amount of gastric juice, arrived by 3:00 p.m., and Doug immediately started a consultation with his two assistants about what to do. "First, Mary will have to take a rough spectral analysis of the most highly suspected metals, and then Joseph will have to refine the measurements that appear to be different from the control values. I hope you will finish it by tomorrow evening." Doug looked at his watch for greater emphasis. "Because I promised the results to Dean by then. If we're lucky, we'll find some answers, and we can take a break over the weekend."

With that, he left to dine with Steve and Mike at Victoria's, a nearby Italian restaurant that was famous for its fine food and wines. They were without their wives, who knew about their meeting and that exonerated them, so they could be there without guilt. A serious discussion lay ahead of them, and they hoped that the fancy restaurant would provide an appropriate background for much needed inspiration.

After all of them ordered the same lasagna with Parmesan cheese and red wine made in Sicily, Doug began the conversation. "Steve, have you learned anything more about LPS? Would it be possible to kill someone with it if mixed with wine or a certain type of food?"

"Well, based on my readings, in principle it might be possible, although LPS isn't well soluble in water and it is usually injected right into experimental animals," said Steve. "Do you remember how much LPS was in Peter's body?"

"Depending on the tissue, it was about five to ten times the normal amount."

"I see. Well, again, according to my readings, this isn't nearly enough to kill someone. To kill a grown man, you would need to raise his LPS level at least a hundred or more times above the level considered normal. But I wonder if the clinic also counted bacteria?"

"Yes, they did, and they found bacteria counts in the gastric system that were lower compared to that of healthy people who died in accidents," Doug replied.

"Then the whole thing is very clear to me," said Steve. "I'll go back to my idea that we already discussed. If Peter was killed at all, he was killed with a drug that killed both his cells and bacteria. LPS molecules then separated from the cell walls of the dead bacteria, made their way into the bloodstream, and spread into Peter's tissues."

"If I understand it correctly, are you saying that LPS was probably not the drug that actually killed Peter?" Mike asked.

Steve turned his head in acquiescence and replied, "Yes, exactly, and I think we should end our preoccupation with LPS and look for other possibilities. Otherwise, we'll just waste precious time. Let's just follow the poison theory. I thought we had already agreed on that." Steve clearly became irritated that someone always keeps bringing up LPS.

After a short silence, Steve turned to Doug and asked if he was looking for arsenic and other poisonous metals in the tissue samples Dean had just given him. "Of course, and by tomorrow evening we'll have the data," replied Doug. "But by the way, could we all meet at my house for dinner tomorrow evening? Sarah will be happy to make some dinner for you and your wives. That way, they won't feel left out of our discussions."

"Great idea!" Mike exclaimed. "Maybe they could even help us come up with even more ideas. After all, this situation concerns them too. They'll be able to view this from a completely different angle, since they're not scientists, and they might help us untangle some of the mysteries surrounding this case. Besides, Nancy could tell us the latest about Helga and her new artistic endeavors. Did you know that her latest ambition is to become a painter and that Nancy introduced her to the local art group?"

"No, I certainly didn't," said Steve, "but in any case, let's meet at seven. I have an important meeting with my future business partner at five, but Liz and I should certainly be able to make it by seven." Steve's face was beaming as he added, "I'm so lucky that I found my business partner. In addition to his business acumen, he also seems to have a deep understanding of science."

When they shook hands and said goodbye, they didn't have the slightest idea if tomorrow night there would be anything important to discuss. But they all thought that a good brainstorming session wasn't a bad idea, particularly in this case. Before their meeting they all were a little downcast over the lack of progress; now they looked forward to a stimulating discussion. Just the thought of their meeting tomorrow cheered them up.

As Steve was about to climb into his car, he heard Doug shouting his name. "Apparently you thought of something important again." Steve laughed as Doug came jogging up to meet him.

"Yes, and this is still about LPS, even though I know that you hate when I even mention the word. But I was just thinking that it might be possible that even a small influx of LPS, no matter how it got into the blood, could have killed Peter if he took the wrong medication. I know I relentlessly keep coming back to this issue, but in my mind, this whole LPS thing still remains a little unsettled."

"It's funny that you mention this, because just as I was walking out, I decided to call Peter's doctor tomorrow and ask if he recently had started to take some sort of medication for his diabetes. Depending on his answer, we'll go from there. At this point, in deference to your continued interest in LPS, I agree that we wait until I call the doctor tomorrow before removing it from the list of suspects. But I still believe that we shouldn't focus on LPS."

"I agree," said Doug, with some relief at the fact that they weren't completely dismissing LPS just yet.

An integral part of Doug's meticulous approach to solving problems was that he refused to discard any possible explanations before all arguments are heard and examined. He preferred to first rack up support either for or against an idea, then exscind the clearly wrong ones, and only after that would he make his decision. This was his immutable doctrine when he dealt with science, and it had served him well so far. Additionally, after investing time and energy in something that produced some minimal results, he didn't easily abandon it. And finding LPS was such a result. Thus, whether the LPS theory would eventually fizzle out or not, he simply couldn't treat this odious substance as a completely lost cause, not yet. The fate of LPS as a likely suspect really hinged on what the ongoing analysis would turn up. Until then, he couldn't settle on a conclusion in either direction.

As Doug was walking off, it was Steve's turn to suddenly call him back. "We also have to find out if Peter had gone somewhere on the day of his death. Maybe he picked up a bug somewhere. Did you hear about what happened a year ago at the Hormel plant down in Austin?"

"No, I didn't," answered Doug, who knew surprisingly little about the small town just about forty miles south of Rochester.

"Well, as you probably know," Steve began, "Austin is home to the Hormel Food Company, one of the biggest food producers in the US. Hormel has a slaughterhouse in its pork plant where, years ago, during just one day, twenty-one workers came down with a mysterious neurological disorder that appeared to be a kind of autoimmune disease. Then at our neurology department, someone determined that these workers had inhaled very small particles derived from pig blood and brain tissue that had been sprayed into the air by high-pressure air hoses. Basically, the immune systems of these workers reacted badly to the proteins that they inhaled with the carrier particles. The proteins spread into their blood and activated their immune system, and the immune cells began destroying the affected person's own nerve cells, causing permanent disabilities in some cases."

"Do you know how many people worked there that day?" asked Doug.

"That's exactly the point I wanted to make. The truly mysterious part of this incident is that why only twenty-one of the workers became sick, while hundreds of others who had been similarly exposed to the same proteins remained healthy. It's almost a travesty in my eyes that we still don't know the reason for these differences among people. I think the next big advancement in medical research will have to be to somehow slash through this elusive barrier in our knowledge. But for now, the point is that this and many similar examples teach that humans can react to almost any drug very differently. Take for example the allergens that sicken certain people but leave the rest of us alone. Peter could be particularly sensitive to an environmental hazard that would have no effect on the rest of us. This, then, could have been worsened by the alcohol or his early diabetes, or who knows what else."

"Well, thank you for complicating the matter even further. Now, we really should go and get a good night's sleep," Doug told him. "We'll need the full power of our brains tomorrow to figure out how to proceed. Besides, tomorrow I'll have all kinds of other drudgeries to deal with as well. I feel depressed just thinking about them."

While Doug was driving home, he kept thinking about LPS, and his intellect came up with yet another possibility. The floor in the major areas of Peter's home, including the basement, was covered with wall-to-wall carpeting. House cleaning was exclusively Helga's job each weekend, and since Helga had left him, Peter hadn't had the energy to vacuum the floor. His house didn't yet look quite derelict, but every day it was getting closer to that point. Corners here and there were already becoming overrun with dust. Although some of his visitors certainly noticed the deteriorating order in his house, nobody had the heart to bring up the issue.

Doug saw it all come together in his head. The absence of regular vacuum cleaning usually creates an ideal condition for various kinds of dust mites to thrive in the carpet. Almost certainly, Peter's floor carpet was teeming with these tiny bugs. Dust mites, once they got into the air, can cause serious allergic reactions in many people. Whenever Peter walked on the carpet, a veritable cloud of dust mites, hitchhiking on dust particles, would have been launched into the air, after which Peter would inhale them. Moreover, Peter recently bragged that he had added jumping rope to his limited exercise program at home to help control his weight. Jumping in that environment would greatly increase the number of airborne dust mites. This would also have increased his breathing rate resulting in even more dust mites entering Peter's air duct and lungs.

The major allergen associated with dust mites happened to be LPS, which could then easily be transferred to the bloodstream. This could certainly explain why Peter's body had more LPS than average people. And maybe Peter could have been allergic to LPS, which isn't unusual. After all, many people are allergic to pollens and other common particles. Thinking of that possibility, Doug recalled that Peter often appeared to have flu-like symptoms during winter months, which seemed strange because most people have this kind of allergy during the pollen seasons in the summer. Could dust mites have caused this? During winter, ventilation in most American homes is generally limited due to superb insulation. While this brings down the heating costs, it also creates the perfect environment for mites to thrive. He instantly decided to ask Carl for his permission to take a sample from the carpet in Peter's exercise room in the basement to analyze it for LPS.

CHAPTER 8

ZINC: A NEW SUSPECTED KILLER SUBSTANCE

On Friday, Doug went to his office very early in a sullen mood; he had too much on his plate for the day. He decided not to wait for Steve and called Peter's doctor to ask if Peter had received any medication. The doctor's short answer, "none," once again appeared to kill the idea of LPS playing a role in Peter's death, no matter how it got into his body. Although he should have felt good about finally being able to throw out one suspect, he was a little disappointed, as any scientist would, to see that his elaborate theory about the role of LPS was ruined. In part this was because there was nothing else to replace it, but he had also begun to become attached to the idea. But this is the process for achieving a discovery or finding a scientific solution to a problem. While trying to solve problems, scientists often formulate several theories that might seem ludicrous, particularly in the retrospective, until they find a new one that may prove to be closer to the truth. Of course, in the present case, Peter still could have taken something without his doctor's knowledge that could have affected his LPS tolerance; this was something still to keep in mind.

Right after his conversation with the doctor, Doug received an email from the NIH informing him that he needed to write reviews of two grant proposals that had been sent to him several weeks ago, or inform them immediately if he couldn't do it, in which case he would be swiftly replaced by another reviewer. Doug was compelled to deal with this request straightaway, because the last thing he wanted to do was to impede the review process NIH depended on so much. He was always very conscientious in these matters and now felt guilty that he hadn't responded earlier.

Like most investigators who are involved in medical research, Doug had a grant from the NIH. The amount he received was modest, but it still paid for 30 percent of his salary. It also helped him maintain a research lab with two assistants to carry on with a small research project, in addition to providing services to the clinic. However, to do any really serious research, Doug would have needed another grant to hire PhD scientists who would develop the proposed research programs under his mentorship. He had submitted a grant that was rejected in the first round, which now he had to revise and resubmit. The grant dealt with the role of a special kind of substance in increasing the survival of cancer cells. This was a neglected area, with only Doug and a research group in Spain making any progress. The reviewers of his grant proposal required him to perform experiments with animal tumors to prove his concept. The crux of the problem was that he never had sufficient research money to do that. To solve this conundrum, Doug convinced the Mayo Clinic to provide him with a small grant that allowed him to carry out a major confirmatory experiment.

Based on the new and very promising data, Doug resubmitted a revised proposal and was very hopeful that this time he would get his proposal approved. If he failed, he was quite aware that the Spanish group would shortly advance to the point where it would be impossible to catch up with them and they would receive all the recognition and glory as well as the further financial support that he was hoping for. In a way, the fate of Doug's entire future scientific career rested upon that one grant. Although, he explained this very carefully when he forwarded his revised grant application, he was also aware that this fact alone, even if combined with quality science, may not be enough to move the reviewers and the head of the study section. There were just too many grant applicants for the money available, and this harsh economic reality was all too clear when grant awards were decided upon. Doug's creed of conviction used to be that a grant proposal is either good or bad, and that is all what matters. But in recent years, his belief in the fairness of the grant system had slowly eroded. Although he saw some notable exceptions, he was now one of the many scientists who believed that good science alone wasn't enough to get grants. Since serious financial restrictions made funding of all good research impossible, it was almost inevitable that personal biases were filtered into the distribution of funds, and the NIH was no exception, at least in his opinion.

While previously Doug didn't have the skill to develop personal relationships with potential reviewers, he was slowly learning the art. Recently he managed to meet an influential scientist who worked at the Fox Chase Cancer Center in Philadelphia, and he was one of those who showed the accumulation of the same compound Doug was working on in human tumors using a uniquely fitting instrument. He let Doug know that he believed in and had a high opinion of the work that Doug was doing, and he even invited Doug to give a seminar sometime early the next year. After talking some more about their common interest, he proposed to Doug that they set up collaboration between their research groups. Doug's work could shed some light on how this compound accumulates in the tumors. Thus, with this new research in hand, Doug had much better prospects of being awarded the resubmitted grant.

Doug started his review with the grant application that a young scientist from a small and obscure institute had submitted. The proposal was for a series of experiments to prove that when a hormone activates a protein in the outside of a cell membrane, this leads to a domino effect resulting in the activation of additional proteins, most of them located inside the cell. As Doug dug deeper into the proposal, he became increasingly irritated because the applicant had obviously just stitched together some ideas from other publications, without even coming up with a new hypothesis. Even though he tried to look at the application without bias, after two hours he had to conclude that the proposal lacked the required originality and strength, and overall it was a paltry piece of work. It took him only another hour to write a devastating review, which probably determined the future career of this young scientist. He briefly thought about this, but soon he was at peace with his decision. In his view, only good scientists, like himself of course, deserved grants.

The second proposal came from an older investigator whose salary was entirely dependent on this grant. She was nearly above the age limit of employment, and the only way to extend her research career was to get this grant. The real problem that stood out for Doug was that this lady had already completed about 80 percent of the proposed work, and he was simply appalled at such kind of "high security, low risk" grant applications. This investigator was very unlucky that her proposal had ended up on Doug's computer. Since even the remaining proposed experiments seemed far from being innovative to Doug, he quickly produced another withering critique before 4:00 p.m., and it took him just another hour to forward everything to the grant administrator. "Job well done," he congratulated himself.

In the meantime, Doug's lab assistants weren't slacking either. With painstaking precision, they analyzed the mass spectrum peaks corresponding to each of the suspected poisonous metals. Except for one peak, they just couldn't find any significant differences between the samples derived from Peter's tissues and the control samples, taken from a young man who had been healthy before but died in a car accident. Surprisingly, this peak coincided with none of the suspected metals, but with zinc. Whichever tissue they looked at, the zinc level was about three to five times higher in Peter's body than in the control. The two assistants had just agreed on that conclusion when Doug rushed into the room and asked how far they had proceeded with the analysis.

"We're pretty much done," said Mary, the younger but more authoritative of the pair.

"We saw no increase in any of the suspicious metals in the gastric juice or the tissues," Joseph added. "But we did find something interesting. Do you see this peak?" Joseph moved the printed spectrum toward Doug. "I'm sure that you can't even guess what that might be. It's zinc."

It took a while for Doug to respond. "Why on earth anybody should have that much zinc in his body? Where exactly did you find all this zinc?"

"It was everywhere," Mary answered, "in the blood, the lung, the liver, the heart, the gastric juices—just everywhere."

"Now this might be a game-changing observation," Doug murmured. "Although frankly, I have no idea what it might mean. But in our present situation, we cannot afford to be too fastidious, so we must start working with the data we have and go from there. But I certainly have a lot of things to discuss with my friends tonight. In any case, it's time for you two to go home. On Monday we will discuss this and the other three remaining cases. Please don't forget them. Have a good night's sleep. It seems to me that we'll need all the fortitude we have to continue this investigation. I see some more studies ahead of us until we can ferret the meaning of this surplus zinc out."

Although research is a great deal like detective work, this was the first time in Doug's life that he felt like a detective. Up to now he had been performing this investigation without a clear lead; all he had were some vague ideas, mostly revolving around LPS, not knowing if it was even worthwhile to pursue any of them. "But now this zinc discovery might change all that. Zinc might be a better lead than even LPS was," he thought to himself. "But what can I make of it?" Even though at that point he had no idea of the implication of high zinc, he felt less pathetic than at the beginning of the investigation.

Mary and Joseph were already walking out the door when Doug asked them one last question. "Did you also look at the levels of iron, copper, and calcium?" His train of thought was that if Peter took many vitamin pills, which usually contained all these elements, then their amounts should also be high, similarly to zinc.

"Yes, we did," Mary replied, "and they all were normal except for copper, which was about 50 percent below the normal level."

"Well, that makes sense, because zinc inhibits the absorption of copper. It also seems to me that we can conclude that the extra zinc didn't come from vitamin pills."

On his way back to his house, Doug kept thinking about this sudden, rather puzzling, turn of events. Why exactly zinc? It is certainly a very important element, but otherwise it is normally innocuous. Would this prove again to be a sidetrack just like LPS? He refused to believe that; his instincts suggested that this was going to be a stronger lead than LPS had been, but he still had no idea how. He had never heard of a case of when excessive zinc consumption had caused death, let alone used in a murder. But then again, how could he be sure that it was a murder case? Could it even have been suicide for some reason? But where did zinc fit into all this? His befuddlement caused by this new data was only growing, and he was relieved when he got home, and for a little while he could forget about the whole day. He was going to help Sarah make dinner, because his friends were coming soon, and time was running out to prepare the food. This allowed him to hold his theories in limbo for a while. And he hoped again that perhaps serendipity might yield some clues in the meantime.

Doug was about to add the meat to his favorite soup when the phone rang. It was Carl, wanting to know if he would have time tomorrow, even though it was a Saturday, for a short meeting in Doug's office. Carl was usually quite cautious with his requests for other people's time, particularly at weekends, but now he almost implored Doug to meet him. Doug assured him that they could meet toward the end of the day. He hoped to finish three remaining toxicology reports for the hospital and to have enough ideas to supply the detective with some leads for his investigation by the time they met. He often had to be in the clinic on Saturdays, sometimes even on Sundays, to keep up with his duties and occasionally think about his cosmological theories. Sarah wasn't happy about it, but she seemed to understand that Doug had a difficult time providing services to the clinic, while at the same time maintaining a research program with two research assistants, not to mention this recent flurry of work related to Peter's death, which really took away precious family time. Doug also regretted that recently he hadn't been able to devote enough time to deeper thinking about his theories. He had it worked out in a very coarse form, but to work on the many subtleties of it would require much greater effort. And even after it was complete in his mind, it might still turn out to be false like so many efforts by others before him. Even with the best of his efforts, with the time available to him, he might produce nothing more than an interesting fantasy that would be forgotten as quickly as it had been conceived.

CHAPTER 9

ZINC AS DOUBLE-EDGED SWORD AND THE POSSIBLE CASE FOR AN ACCOMPLICE SUBSTANCE

Sarah and Doug found themselves wishing that time was malleable; they were slightly behind schedule with their dinner preparations. Fortunately, the guests were late too. Mike, with Nancy in arm, arrived first, well after seven, and ten minutes later, Steven arrived with Elizabeth. After a little exercise in the fine art of small talk, they wasted no time sitting down at the dinner table to begin the first course, a delicious meat soup with handmade curled pasta. After that Sarah brought out a large plate filled with steaks, baked potatoes, stuffed cabbage, and a selection of cheeses including a bleu cheese from France. On a separate plate she brought a nicely decorated black forest cake. At any other times, such an excellent menu would have provided the background for a relaxed conversation. But now they all knew that the dinner, which almost evoked a burial feast, was just a pretext for them to discuss the latest developments relating to Peter's death, and everyone was very much aware that it was only a matter of time before someone would bring it up. Indeed, in the middle of the first course, Doug somehow shifted the general conversation about weather, Elizabeth's pregnancy, and each other's health to the topic of Peter's death.

Doug started the discussion by saying that he and Steve thought they had reason to believe that a poisonous kind of toxic substance had caused Peter's death, although they were still far from getting to the bottom of this case, as they felt compelled to do. Then he started to explain that the incipient forensic investigations had so far only resulted in vague results; in short, they had revealed an increase in zinc and a bacterial product called LPS in Peter's body. He added that neither LPS nor zinc could really be considered lethal on their own, or even together, at least not at the amounts they were found in Peter's tissues. The zinc was particularly troubling to Doug, because he knew that even the extreme fivefold increase in blood zinc they had found was far from enough to kill a man. During his nutrition studies, he had encountered the issues associated with zinc deficiency and overload. In fact, zinc had fascinated him so much that he had read many research papers on the subject, although he was far from understanding this subject at the level of an expert. Still, due to these excursions into zinc research, he had accumulated enough knowledge of this metal to know that, in his view, it was greatly underrated by both the public and by nutritionists.

Mike was the first to break the ensuing silence. "Listen, Doug, like most people not trained in the health sciences, I know practically nothing about zinc, but in my humble opinion, maybe too much zinc could cause trouble if combined with something else, like a disease or drug. Shouldn't this possibility be considered? Why do you think that zinc alone couldn't have killed Peter? What if zinc and LPS together did the damage? By the way, I'm just playing devil's advocate here," he added. "You and Steve are the experts."

"Well, I can tell you some basic facts about zinc off the top of my head. Our bodies contain two to three grams of zinc, 70 to 90 percent of which is in the skeletal muscle and bones. The reason why zinc is so important for life is that it binds to and affects the functions of several thousand proteins, and there is solid proof that it seriously affects the functions of over 300 of them. Considering that a human body may contain up to 100,000 different proteins, the dependence of 5 to 10 percent of them on zinc means that deficiency in this element can wreak havoc if that problem isn't dealt with in time. Zinc deficiency is something that rarely, if ever, goes unnoticed by the body, and one can be sure of reprisal if that happens. Thus, to summarize, zinc quite literally regulates virtually all aspects of our body's vital functions. From conception through pregnancy, childhood, and adolescence, intake of sufficient zinc is essential. Otherwise, many vital bodily functions will be stymied."

Doug seemed to be in full lecture mode as he proceeded. "Unfortunately, even today, zinc deficiency is rampant in most developing countries, which can have dangerous consequences. For example, zinc deficiency is associated with growth retardation, perhaps related to reduced activities of some growth factors, depressed immune function, reduced release of insulin from the pancreas, delayed wound healing, numerous skin diseases, impaired sex drive and reduced sperm production as well as reproductive capacity, anorexia, diarrhea, increased incidence of infections, increased risk for miscarriages, alopecia, mental lethargy, and probably many other health problems that I forgot. If, for example, someone at your place of work bleats too often or gets irritated easily or has become less enthusiastic in performing his or her duties, there is a chance that zinc deficiency might be the underlying reason, although almost no one would even consider it as the cause. It's really a pity that the role this rare element plays in our lives hasn't yet permeated public knowledge. I think most people aren't aware that if they want to stave off diseases, one of the most important things to do is to make sure they consume enough zinc every single day. I'm telling you, zinc is one of my most revered trace elements in food."

"Is zinc deficiency common in America?" asked Elizabeth.

"It absolutely is. In fact, it is estimated to currently affect up to 70 percent of the US population. For different reasons, this primarily affects the young and the elderly. As I just said, people should be more aware of this and pay more attention to their daily zinc requirements, because zinc deficiency is one of the most damaging, and quite insidious, deficiencies that can lead to a slew of major health problems. Although to be fair, taking just the right amount of zinc is easier said than done. Professional nutritionists try to advise people on this, but their opinions are often incredibly divergent. They also advise their readers to keep track of so many different nutrients and their calorie contents, that if we were to follow all advice, we would hardly have any time left for anything else. Doctors tend to concentrate on elements like iron because of its role in the hemoglobin and oxygen supplies, and they usually pay less attention to zinc. I suspect that many doctors are still fairly in the dark when it comes to the nutritional value of rare elements such as zinc, but that's only my personal view."

"Is this zinc deficiency a new thing? Which kind of foods have zinc? How much of it do we actually need?" asked Elizabeth in concern. "I am actually starting to get a little worried. What if my baby isn't getting enough of it?"

"Well, I will try to remember everything I can, but there is also a lot more information out there that you can look up," said Doug, while confronted with the decision of whether he should take one or two gorgeous-looking pieces of stuffed cabbage. After he decided on two, he continued. "As far as I can recall, the recommended daily dietary intake of zinc for adults is between eleven and fifteen milligrams, but just as with other nutrients, the correct amount that we need depends on our age, lifestyle, body weight, and health conditions. For example, infants need less zinc, while pregnant women like you, Liz, require more than others. So you are right to want to start eating more zinc. Also, the elderly, alcoholics, and vegetarians usually have low zinc intakes. I also read that all kinds of tea can also reduce zinc intake, and so can coffee. I drink a lot of coffee and tea, and sometimes I come to think that perhaps I should be better off by drinking less of both just because of zinc."

He immediately regretted mentioning tea, because Sarah was just about to serve some. He quickly tried to get their minds off that fact, adding, "Generally, one multivitamin tablet contains twelve milligrams of zinc, which is pretty close to the daily requirement, although zinc is not absorbed as well from a tablet as it is from a meal. If Americans were to take one such tablet every day, and they would eat just one serving of a meal rich in zinc, I highly doubt that we would have a problem. And this is far from being hard to do, because zinc is found in a wide variety of foods. I don't recommend eating a lot of red meat, but poultry and particularly oysters are rich in zinc, and they are healthy for many other reasons too. Again, just off the top of my head, I would recommend milk, cheese, yogurt, shredded wheat, beans, chickpeas, crab, and shrimp, and there are many more of them, but I can't remember them all. Fruits and vegetables, except beans and peas, contain negligible amounts of zinc. That's why vegetarians must pay extra attention to their zinc intake and take regularly complete vitamin pills supplemented with zinc."

"I can believe that zinc deficiency causes these problems, and everything you just said is very useful to know," Steven cut in, with some irritation in his voice. "But Peter's case is different. He had too much zinc in his body, not too little. And, Doug, you even agreed that zinc alone might not have been the cause of death. Then, the way I see it, if our conclusion is true that excess zinc itself wasn't the cause of Peter's death, then there had to be something else, like a drug, that helped zinc kill the cells in his body. For the time being, let us name this mysterious substance the *accomplice substance*. Doug, do you know how cells take in and remove zinc? Can cells take in unlimited amounts of zinc, or is this process normally limited? What I'm really getting at is that perhaps this theoretical accomplice substance in Peter's body caused his cells to bring in too much zinc, which eventually led to their demise. Are you aware of anything that helps zinc to enter cells in abnormal amounts?"

Doug sat in silence for a moment. "I can't think of anything just now."

Steve explained his thoughts further. "You know, I've seen so many cases in which a drug that used to work just fine suddenly acted violently because the patient started taking an additional drug without notifying the doctor or he or she changed diet drastically. I can give you an example. As a prophylactic drug, lithium is used to treat bipolar affective disorder. If used properly, it decreases the frequency and severity of manic and depressive attacks. However, extreme care is needed to make sure that lithium isn't used together with certain other drugs, for example ibuprofen, which can increase serum lithium levels, resulting in side effects like vomiting, diarrhea, seizures, and even comas. And there are hundreds of other known examples of drug interactions than can cause very bad side effects.

When you buy a prescription drug, the drug maker in most cases provides a list of other drugs that should be taken cautiously, if at all."

"You might be on to something," exclaimed Doug. "Peter might have taken an accomplice drug that transformed zinc into a perilous substance. Of course! Why didn't I think of that? You're coming up with all these scenarios with such ease. I envy you for that," he continued with a smile. "But here's the thing concerning how cells take in zinc. During evolution, all living organisms, humans especially, developed a very efficient balancing mechanism to keep the levels of free and protein-bound zinc molecules relatively constant inside a cell. This balance is the result of several fine-tuned mechanisms working in concert, starting with the absorption of zinc into the intestines. The transportation of zinc in and out of the cells, involving many different transport proteins, is a similarly fine-tuned response to the needs of cells.

"Once inside a cell, most of the zinc is bound to so-called storage proteins, and from there it is then transferred to the many other intracellular proteins in a highly regulated manner. Because of the efficiency of these regulatory mechanisms, excess ingested zinc generally doesn't result in serious toxic effects, as I already mentioned, although it can commit misdemeanors like nausea, vomiting, abdominal cramps, and diarrhea when humans are exposed to zinc oxide fumes or orally consume very high concentrations of zinc salts. One reason why excess zinc is not highly toxic is that its uptake by cells through the surrounding membrane is very limited. Also, if more zinc is taken up by cells than is needed, that triggers more synthesis of the storage proteins, which will in turn eventually reduce the amount of the harmful free zinc. Then the surplus zinc is excreted from the cells and eventually from the body. This is a beautiful example of how finely regulated our metabolism is. And this is just one substance. Think of the many thousands of other compounds in our body that are similarly regulated.

"Now picture something else. An accomplice substance in the space outside the cells may somehow facilitate the entrance of zinc into the cells, as Steve's sage remark suggests. Imagine a revolving door at the entrance of a very popular art show set up at a relatively small space. Depending on its speed, it will let in fewer or more people in any given time regardless of how many people are waiting outside in line. If the door's speed is set too high, it will let in too many people into the waiting area, overwhelming the box offices. Similarly, too fast an influx of too much zinc can throw the cells out of balance because they may not have enough zinc storage proteins, and free zinc through a chain of oxidation-reduction reactions can cause enough damage to the cells to kill them. In fact, there are diseases in which the accomplice isn't a substance but the pathological environment, which generates enough free zinc to damage the cells. For example, this can occur in the brain during periods of ischemia, seizures, or trauma. In these conditions, zinc is released from restricted areas of neurons, called synapses, and then taken up by neighboring neurons into less specific cellular areas, causing their death.

"Now, how might all this relate to Peter's diabetes? Well, zinc also accumulates in the pancreatic islets that contain the specific insulin-producing cells. Unfortunately, if the pancreas is inflamed, zinc can kill these cells as well. This can lead to diabetes but not death, at least not with the speed with which Peter died. Besides, the examiners haven't found any problems with Peter's pancreas. However, if in Peter's body a mischievous substance was present that helped the content of zinc to rapidly rise in all tissues, it would explain his sudden death. Frankly, I'm just hedging my bets here, trying to follow Steve's lead. This might turn out to be a preposterous idea, but at the very least, we now have a reasonable hypothesis to follow."

The initially languorous discussion had suddenly become an exciting one, though you couldn't tell from Doug's unfailingly ponderous tone. But at one point, even Doug had gotten so excited that he couldn't help but stand up and begin pacing next to his chair as he gathered his thoughts. The idea of this accomplice substance called for the opening of yet another bottle of wine, a Riesling from Australia this time, which matched the upcoming dessert very well.

"But if we now consider zinc the killer, then there is another reason why an accomplice, like a drug or disease, would be necessary," Doug continued. "You see, if a person consumes a lot of zinc, albumin takes a large portion of it out of action. Albumin is a major protein in the blood that avidly binds zinc but always releases some of it to supply the needs of cells and tissues. Obviously, the more zinc is consumed, the more of it is available for cells, but this is not nearly enough to kill them under normal circumstances. Thus, the accomplice must be a substance that is able to grab zinc from albumin and then carry it into cells in large amounts. Once inside, the zinc must then be able to leave this carrier substance because it can damage the cells only in free form. If this reasoning is correct, then the accomplice really is a kind of chemical substance, not a disease."

Steve seemed to pick up this line of reasoning. "I believe that the salient point is this: If what I suggested and what Doug just said are true, we are indeed looking for an *accomplice* substance that carries zinc into cells, which then initiates destructive oxidative events that the cells can't handle. After a battle between the oxidants and antioxidants, the oxidant zinc wins, and the cell perishes."

"Exactly," agreed Doug. "I am going to try to get permission to take more of the frozen tissue samples from Peter's body and see if a new compound pops up in them."

Doug took some time to sketch out the next steps on the pad he always carried, and then he turned to Steven. "In the meantime, could you search the internet again to see if any small molecule has already been discovered which binds zinc and is perhaps even used to kill cells? After all, this accomplice was *your* idea."

"Yeah, I'm on it. I'll also try to think of scenarios to explain how that much zinc could have gotten into Peter's cells." Almost everyone seemed pleasantly surprised with this turn of events, especially Doug and Steve; this fertile discussion had spurred them into action, with a real prospect of finally solving the case. This was indeed an astounding development, considering that just hours ago they felt despondent and completely in the dark. None of them said it loud, but each of them thought that this was a sign that God hadn't forsaken them on their arduous journey to find the truth. However, it wasn't time for celebrating just yet; it was clear that a lot of work remained. Doug was aware that he and his assistants would have to bear the brunt of the work, which, after his initial euphoria, dampened his enthusiasm a little bit.

While all this was unfolding, the guests didn't seem to notice that Sarah stayed exceptionally silent. She served all the food and made small remarks here and there mostly in response to praise for her sumptuous food, but she didn't seem to have any desire to involve herself in the discussion about zinc as related to Peter's death. Her withdrawal might have seemed understandable considering that, particularly toward the end of the evening, the conversation had taken a scientific turn that she probably couldn't follow; but it also seemed like she just didn't have an interest in sharing her opinion at all. Perhaps, her reticence reflected her suspicion that this whole debate was only held for Doug's benefit, so he could put his intellect on display. At one point during the conversation, Mike asked for more ice, and when Doug stepped into the kitchen to get some, he found Sarah sitting at the kitchen table, keenly bending toward the door in a listening position. Sarah's aloofness surprised Doug. "Sarah, why don't you come in and sit with us?" he asked.

"I'm just too tired. Besides, if you really want to know my opinion, I think all this talk about zinc and some accomplice substance is so far-fetched that to me it's getting almost ridiculous," she said in a derisive tone that stunned Doug.

Sarah was a ravishing woman who had always been supportive of Doug. But the way she talked now displayed none of the vivacious imagination that Doug liked so much about her. Something clearly made her peevish, and she couldn't hide it. After Doug looked at her for a brief period of shocked silence, she continued.

"I can't participate in your discussions unalloyed. You see, sometimes people just die. Plenty of people get sick from things not even their doctors know about. I think you and your friends are just complicating this whole thing and wasting your time." She ended with a dismissive shrug; she clearly scorned their discussion about zinc and anything that related to Peter's death. "You're looking for a villain, but I'm afraid there isn't one. But I also don't want to force my opinion on you or discourage you from exploring all possibilities. I realize that you can't leave any stone unturned, and that you're just trying to heal the wounds that Peter's death left behind. But still, I can't contribute to this discussion."

Doug appreciated that she un-bosomed herself to him; he tried to show his understanding side.

"Maybe you are correct and we're looking for something where there is nothing," said Doug while filling the ice bucket. "But you don't have to hold back from telling us what you think. We're at a point when every opinion helps. And if you think that this evening I hurt you with something, I wish to propitiate you. Incidentally, I believe that Steve's accomplice idea is a good one, and perhaps we should give it a shot." Although there could be truth in Sarah's blisteringly negative comment, Doug was shocked when he heard it; she had essentially said he and the others were stupid for pursuing the accomplice theory. This incident left Doug perplexed for a while.

It was already almost eleven, and while Sarah still was in the kitchen keeping aloof from the guests, they decided that enough progress had been made for the day, although once more they couldn't be sure if any of their idea had any merit. Soon, they would know. But it was clear that they were still forlornly lacking ideas of what had really happened to Peter, so they couldn't afford to forfeit any new leads. With the zinc finding and perhaps with LPS as well, they had only skimmed the surface, but one or more key elements were missing from the puzzle; they clearly needed more data for their ideas to hold any water. So the task for Doug and Steve was to dive much deeper to crack this case, perhaps by finding the highly hypothetical accomplice substance. It was the best hope they had, so the thought of not following the idea didn't even occur to them for a second.

In the meantime, Sarah had begun loading the dirty dishes into the dishwasher and was happy not to hear any more about zinc, the accomplice, or anything else related to Peter's death. This also conspicuously gave the final signal to break up their meeting. Although the guests also noticed that Sarah had been more reserved than usual, nobody thought that this was because she had lost her zeal for the case, or that further dealings with Peter's death were outside her interest. Nobody suspected that she had any good reason to avoid discussing the case.

But Sarah soon realized that she had probably been too dismissive of the guests' discussion. After the guests left, she told Doug that she was sorry for retorting to him in a way that even she admitted was harsh. She assured Doug that she hadn't meant to come across as hostile to their ideas; what she was really trying to say was that all this extra work for Doug kept him even more away from their family life. Doug happily accepted her explanation and promised that he would try to reach a conclusion as soon as possible, and after that he would start spending a lot more time at home. At that moment, it didn't appear that anything might be looming over the solid foundation of their marriage, which for Doug was the single most important thing in his life. He hated anything with the potential to complicate his marital life; he saw through the examples of several of his acquaintances that marriage can be ruined all too easily.

CHAPTER 10

A BUSY SATURDAY FOR DOUG

On Saturday morning, Doug arrived at his office early, and after some hesitation, he started the day with a phone call to Carl, asking him to come by on Monday afternoon. Despite having lost faith in the LPS theory, he also asked Carl for his permission to take a small sample of the carpet in the basement of Peter's house to determine how many dust mites were in it. After a long explanation of why this would be helpful to the investigation, Carl finally gave him permission to take a sample that very same day at 6:00 p.m., though in the presence of his assistant.

Doug had a lot of assignments in front of him, and after the phone call, he quickly settled down to work. He started by examining three toxicology cases that he had to report to an internal committee. Files containing all related records and inquiries were already sitting on his desk. Dean and his assistants had really done a good job; all he had to do now was only rake over the data piece by piece and give his recommendations. It was 1:00 p.m. when he finished the last report, which he immediately dropped off at the main office. Then he set out to review two manuscripts, which he finished a little after five just in time to go to Peter's house, where Carl's assistant was waiting for him. Doug cut a small carpet sample from the exercise room in the basement and put it into a sterile container. He dodged the question from the assistant about why he needed the sample, saying that he was running late for dinner at home, which was true. Then after dinner, he took his sons to a nearby playground, as he had earlier promised them. The next morning, which was Sunday, Doug and Sarah decided to skip the church service to stay in bed a little longer. Sarah appeared to be genuinely sorry about how she had spoken to Doug during the dinner on Friday night and wanted to make sure that he felt placated. Doug was initially a little wary of Sarah's conciliatory mood because he thought that they had already made amends, but in the end, he was happy to do it again. The children had grown into the age when they needed no immediate attention in the mornings. On weekends they woke up at around seven and then played or read in their rooms until their parents decided to get out of bed. No reprehension was needed; they accepted without complaint their parents' wish of not being disturbed until they had exited their room of their own free will.

As it turned out later, their friends also skipped church that Sunday. Elizabeth and Steve were deeply religious, and if it hadn't been for Elizabeth not feeling well, they probably would have gone to the mass at the nearby Catholic Church. Nancy was also a strong believer, but she always caved in when Mike didn't feel like going, which happened a few times every year. That Sunday was one of those times. What none of them wanted to admit was that they all also felt the need for more rest before attending the funeral service, which was set at 10:00 a.m. next morning. The anguish waiting for them there would require a rested mind.

Somewhat unexpectedly for Doug, after lunch Sarah decided to visit her brother, Jim Fletcher, who was an artist living in a small town near Little Falls Northwest of Minneapolis. Because Little Falls was at least three hours away, they only had a quick lunch, so she could leave at 1:00 p.m. In a rather languid voice, she asked Doug and the children if they wanted to accompany her. Doug chose to stay home to catch up on reading some important articles and writing some letters, because on Monday, the funeral would take away a good chunk of his day. Besides, he was to give one of the eulogies at the funeral, and he wanted to have some quiet time to think about what he was going to say. The children, on the other hand, were simply not in the mood to travel for hours and particularly not to that place. They made the trip once a few months earlier, and all they remembered was that they had had nothing to do there.

Jim was a fit man, still slender, although he was around fifty. He had never been married and had no experience with children. He had many pictures in a large studio and several statues outside in the garden he had molded from steel. Last time the children had visited him, he was holding an opening ceremony for a new exhibition, which was attended by few dozen guests who mingled with each other, and nobody was left to watch the children. They ventured outside to play, where they came across a strange roundish object they thought was some sort of art piece. They tapped on it out of curiosity, and a swarm of wasps immediately flew out and swarmed them. They both received several excruciating stings before they could rush back into the house. Sarah was horrified but tried to console them by saying that at least they now knew that neither of them were allergic to bees. This didn't help much. The whole debacle was probably the major reason the children weren't overly eager to return to that place where they experienced only boredom and danger.

As Doug stood close to the window, he saw Sarah carrying her large computer case and a black plastic bag to her car. Sarah often took her spacious laptop case that could hold two laptops and additional accessories. She regularly used to drop off used clothes and other items in similar plastic bags at the Salvation Army. That's what Doug assumed Sarah had in the bag then too, so at that time, he didn't attach much significance to this sight, although for some reason the image remained indelibly imprinted in his memory, which he time to time conjured up for no apparent reason. Such ephemeral observations, which seem insignificant at the time, are often stored away in our brain, and they can resurface in a flash in the proper context. These memory flashes can, in fact, clarify uncanny situations and actions that occurred in the distant past. Nature created our brain with astonishingly prescient insight, ensuring storage of information that comes to the forefront whenever it becomes useful for our analysis and decision-making. Our brain is suffused with memories, and brain researchers are hard at work to find out how can we evoke them at the precise moment they are needed.

CHAPTER 11

THE FUNERAL AND A NEW TWIST IN THE MYSTERY OF PETER'S DEATH

Starting the week with the funeral service was especially hard for Peter's friends to bear. They hadn't only lost a dear friend, but they were also unable to forgive themselves for their inaction that had lasted almost a full day. In retrospect, that was the most critical period in Peter's life, and they knew it. The contrition they all felt just didn't want to go away.

By the time Doug arrived at St. John's Catholic Church, all the bereaved who could make it were already there, gathered in small groups. Everybody remained mostly silent, only saying hello to the newcomers. The environment of sorrow brought out once more all the emotions Doug had been trying to suppress. He was so overcome with grief that he hardly paid attention to the eulogies or any of the other events preceded his own eulogy, which he gave on behalf of Peter's high school friends. When it was his turn to stand at the podium and address the bereaved, he began by recalling all the fun he and Peter had had together, particularly after their high school football team defeated both their archrivals, Winona and Mankato. He and Peter had also considered circumnavigating the globe once they could afford it. He remembered how happy he and his friends were when Peter returned to Rochester to complete the old high school gang, and soon after that beginning their weekly poker parties. He also mentioned that they held very similar views on religion and the other aspects of life that really mattered, but he didn't go into details. He held back his tears for most of his speech, but finally broke down into sobs when toward the end he asked the question that was on everybody's mind: "Peter, why did you have to go so early?" he cried out.

As he had been writing the eulogy, he had planned to hint that he was suspicious of foul play, but at the last minute, he decided that such hint might be inappropriate and prove untenable, and so he remained silent on that subject. He finished his speech by offering solace to Peter's friends and relatives, adding that he was prepared to do everything in his power to help the Mayo Clinic's examiners find the medical reason for Peter's premature death. He stressed the word *medical* so as not to give any reason for anyone to suspect that something malevolent might have been involved.

In some ways, the funeral was as strange as Peter's death. For one thing, Peter's first wife, Anna, wasn't there, and even her children didn't show up, although later she claimed that she hadn't received the official notification of the funeral in time. Sarah was also absent; she called Doug the night before with the news that she had come down with a nasty case of the flu and she was unable to drive. Whether she was indeed sick or just faking it, Doug would never know, but to question Sarah's sincerity was something that never occurred to him. Although, if he had been more observant, this might have been another hint that something was off between her and him, but at this point of his life, he didn't pay heed to warning signs.

Helga was there, dressed in black, but she left soon after the ceremony, saying that for the remaining part of the day she hadn't been able to find a replacement at the hospital she worked in. This was a reasonable excuse, although nobody really believed it, but deep down everyone happily consented to it. Nothing is more inconvenient than trying to console a volatile person like Helga, who had divorced the deceased just one week ago because of her own capriciousness as it appeared to some people. People around Peter were aware that Helga was the one who had pushed for the divorce and that she didn't stand beside Peter when he would have needed her the most. After a tragic event like this, people tend to forget to consider both sides of the story. Perhaps well before his death, seeing that no matter what he did wouldn't change Helga's attitude, Peter had already had enough of this unrequited love affair? Perhaps he had decided he didn't need Helga anymore and had already found somebody to replace her?

Several of Peter's coworkers at IBM also showed up for the funeral. Doug briefly noticed a young woman in all black who sobbed during most of the ceremony. She held a black handkerchief with white trimming in front of her face, which prevented Doug from recognizing her. After the ceremony she quickly disappeared.

Peter's parents made all the preparations for the funeral, but they were so devastated by the loss of their son that they didn't have the strength to organize a funeral reception, either before or after the ceremony. They were both in their late seventies, and incidentally both had undergone major surgeries in recent months. It was hard enough for them to drive from their house in St. Cloud to deal with the many mental, organizational, and financial, issues this sad event required of them. With Carl's permission, they planned to stay in Peter's house for at least a couple of days to sort out with the family's lawyer how to deal with the property of the deceased, although in Helga's case, this had already been settled and purportedly she won't fight any remaining decisions. The friends understood that it would be unseemly to stay longer with the family, so around noon they too left to see after their duties that this miserable but otherwise normal workday held for them. As it later turned out, they all skipped lunch that day for one reason or another. For Doug, all he felt was nausea, and he knew his stomach would accept absolutely nothing for a while.

Doug also needed the time that lunch would have taken from him. In addition to a report he had postponed from Friday, he still had several short reports to prepare, each taking him a little more time to complete than he had thought. He also had a short conversation with Joseph about what to look for in the carpet sample.

"We will be obliged to toil away without payoff for some time," he told Joseph. "Peter was a very good friend of mine, and I really want to figure out what caused his death. I owe him much more, but at least that much. Are you and Mary with me on this, even if it means sacrificing some of your free time?"

"We're with you 100 percent," Joseph replied in a firm voice. "You must know by now that we actually like challenges, and this case seems to provide ample opportunity for that." Neither Mary nor Joseph had ever left Doug in the lurch when exceptional effort was required. Sometimes they went home dog-tired after a hard day's work, but they always did beyond reproach everything Doug required of them in this sleuthing project and others.

Doug barely completed the reports by 4:00 p.m., which was when he expected Carl to visit him. He was just about to use his remaining time to give further instructions to Mary and Joseph in other matters when Carl knocked on his door.

"I'm sorry I came a little earlier than we agreed," he said while his eyes flicked around the office, "but my wife expects me to meet her at 5:30. It's her birthday, and we already made reservation at a restaurant."

"Well, frankly, I don't have too much to report, so we should certainly finish by then," answered Doug.

"So," Carl started slowly, "how far did you get with your chemical analysis? Did you find anything of interest that might give us some clue of what might have happened?"

"Oh yes, we keep finding interesting things, but we're still missing one or more important pieces of the puzzle."

"What are they?" Carl demanded specifics with intense curiosity in his voice.

"Peter might have been poisoned. The evidence for this is that in all his organs that the pathologist examined, he found cells dying in larger numbers and in different ways than one would normally expect to see in a dead person, considering the relatively short time period between his death and the collection of the tissue samples for analysis. Most cells in a dead body are of course functionally dead after several hours, but they look different when someone dies of an accident, a natural cause, or a disease like cancer, as if someone accidentally or intentionally ingests a poisonous substance that destroys cells rapidly. By the way, many poisons actually kill slowly."

"So, in Peter's case, what seems to be the killer?" asked Carl.

"That," Doug responded, "is the problem we are facing now. The only strange thing we have seen so far, other than the LPS, is abnormally large amounts of zinc in his body. But the rub is that this couldn't kill Peter, unless he also ingested another substance that helps zinc enter the cells and kill them. You see, Carl, even if somebody has a huge amount of zinc in his or her blood, cells normally ingest only a tiny portion of it. However, if there was an 'accomplice substance,' as we call it, that helped to carry the zinc into Peter's cells, this could have made the zinc deadly. And from what I have heard so far from other scientists, zinc really can conflagrate and kill cells very quickly, in a kind of implosion taking only about ten minutes, if enough of it can get inside. Cells are very fragile structures, and the combination of zinc and some accomplice substance could easily damage them beyond repair. Cells are equipped to deal with moderate increases in zinc influx, but they are helpless when there is a zinc deluge. If too many cells die in a tissue, the result is chaos and dysfunction, and this eventually leads to the death of the respective organs and then the whole organism."

"Do you have any idea what this accomplice substance could be?" asked Carl.

"I have to admit, this is well beyond my expertise, and I have asked Steve to help me out with this. He promised to thoroughly search the internet for answers, and hopefully he'll come up with a potential accomplice substance that has already been discovered to help zinc kill cells. I also set up an appointment with an endocrinologist from the University of Minnesota, George Carrick, who will see me tomorrow. He knows far more about zinc than I do. He got several NIH grants to study the effects of zinc on cell proliferation and metabolism. It was from him that I heard first that 'zinc is a double-edged sword,' meaning that zinc is needed for the normal functioning of cells, but it can also kill them. But as a cautionary note, I want to stress that in addition to the cell poison, the concatenation of other factors might also have played some role in Peter's death."

At that, Carl pointedly looked at his watch. "I think I understood everything what you said, and many thanks for that," Carl said.

"I hope this man can help us come up with some answers. Tell me if you can. What are the next steps?"

"Tomorrow we will start analyzing some of the tissue samples that we still have, which will hopefully help us find this hypothetical accomplice substance. I really have strong hopes that there is one. Fortunately, we still have enough frozen tissue samples left. But we'll need some luck, because the kind of molecules I'm thinking of could be quite unstable, even in frozen tissues.

"So tonight I will meet Steve to see if he has a potential suspect for the accomplice substance, and then we can focus our analysis around it. This was his homework for Friday, but during Peter's funeral we obviously didn't talk about this," he added. "A good hint, if he finds one, could tremendously help our search, and I really hope that he has. If he couldn't ferret anything out, frankly I don't know how to continue. We can't just randomly scrutinize thousands and thousands of potential compounds. We have neither the time nor the manpower to spend on this project. If in a few days we will still be bogged down by paltry data, we will have to seriously reanalyze the situation without predilection and perhaps stop our work altogether. Believe it or not, last Friday my wife suggested just that. She hinted that the data we have obtained so far is indeed very poor, and therefore we won't be able to avoid the conclusion that Peter simply died of an illness, nothing more. Looking at the case objectively, she might have a point."

Then it was Doug's turn to interrogate Carl. "Before you go, would you tell me how things are developing on your end? Could you figure out who was the last person Peter was with?"

"Of course, I've spent some time investigating this," Carl responded, "but the only information I have that I'm sure about is that Helga wasn't with him, neither on Tuesday nor Monday evening. Nor could she provide any clues about who could have been Peter's last visitor. On this last point, I can't be sure that she was completely truthful. Sometimes I get the feeling that she wants to extricate herself from this case, but other times she leads me to believe she wants to help. At this moment, I have no way of knowing which one is the case. One thing I can say about Helga is that Peter's death didn't seem to be a particularly traumatizing experience for her, unlike for you and your friends. I get the sense that she was barely affected by it. Doug, you seem to be dismissive of the idea that Helga might be innocent. You should know, I have decided to gradually ratchet up the pressure on her, and then we might be able to extract out of her whether she is hiding something. In any case, I'll start several entirely different lines of investigation," he conceded with some disappointment in his voice. "And that's what I'm going to do, starting tomorrow."

"What do you think the odds are that it was indeed a murder?" asked Doug.

"It more and more seems that we are indeed dealing with a homicide. I don't know if the crime was deliberate or if it was just accidental. But the person who did this to Peter had to know him well and perhaps, only perhaps, had to be familiar with his house as well. We might even know this person. He or she may be among us. And you know what else? I called Peter's phone today, and it was still on, but the signal was so weak, perhaps coming from a rural area that the phone company couldn't locate it. But it's certain that whoever stole the phone also took the charger with it. Most of us keep our phone charger in a special place, which gives one more reason for me to believe that this person had some knowledge of Peter's house. The cobra is at once audacious and circumspect, depending on the situation. But he or she will slip up eventually, I'm sure about this. The important thing is that we should try to remember unusual events, however insignificant they might seem, that caught our attention around the time of Peter's death."

Doug couldn't explain why, but as Carl emphasized the importance of minor unexplained events, for a moment he once again saw Sarah carrying her large laptop case and plastic bag the day before. This was one of those occasions when in the retrospective, Sarah was imponderable to Doug; this is perhaps the reason why this specific event came back to his mind with such frequency, although he didn't necessarily tie his observation to anything. After being in a quandary for a few seconds, he decided not to mention his observation to Carl. On what right should he bring his loving wife into the investigation?

Then Carl suddenly touched his head as if he had just remembered something. "Oh, yes, and I still don't know whether this is significant, but I found out from Helga where Peter used to buy his wine. I went to this Apollo liquor store beside Best Buy, and I asked the storekeeper to tell me about Peter's drinking habits. He told me that about two months ago, he began to buy red Bordeaux wine, usually twelve but sometimes even twenty bottles every week. Any storekeeper would remember a zealous wine customer like that. Each week he probably made quite a bit of profit on Peter alone. Then I asked Peter's doctor if he knew about Peter's drinking habit. The doctor said that since Peter didn't like the idea of starting antidiabetic medication, and in fact became bellicose when this possibility was discussed, he advised him to first exercise every day, then drink one or at most two glasses of red wine during dinner, and finally take a warm bath. His recommendation, according to him, was based on very new research, which this doctor amazingly kept track of, that showed that red wine and a warm bath can help people with insulin resistance and early diabetes. I remember that in Peter's house, you already told me about how warm water can help people with diabetes. But the mystery I can't put a finger on remains: why he had the propensity to drink exactly the Bordeaux brand, never trying out other brands?

"The doctor also assured me," Carl continued, "that by advising Peter to combine exercise, red wine, and hot baths, he wasn't doing anything that was against the latest trends in science, although it might be rather unconventional in medical practice. He also confirmed that at this stage, he hadn't yet prescribed any drugs for Peter, but he meant to do so in a couple of months. The doctor let me know that he denounced Peter's approach of not wanting to take medication, but he could do nothing more to convince him. He added that in hindsight, Peter probably forfeited the crucial advantage of starting medication early on, and thus avoiding potential complications related to his insulin resistance. But again, I couldn't do more, the doctor claimed, to assuage his objections toward drugs."

From all this, the first important thing that stood out for Doug was that unless Peter got some drug from somewhere else, he was officially not on any sort of medication. The doctor was also clearly struggling with his conscience for allowing Peter to stay off diabetes medication for so long, but it was certainly only for months not years.

Doug shared with Carl his second most important conclusion about the doctor: "I am simply astounded that the doctor's knowledge about science is so up to date. I do nothing else but science, yet I simply cannot digest the gargantuan amount of papers coming out in hundreds of life science journals every month. It must have been difficult for the doctor to accept that even with all his knowledge, he still couldn't break Peter's strong aversion toward manufactured drugs. Frankly, Peter's decision to not take medication bordered on the mindless."

"In some ways, I understand Peter's aversion toward drugs," Carl reflected. "Years ago, before my cancer, I had some problems with my cholesterol, and my doctor prescribed a pill for it. The warning on the box said that in a small number of cases, people who take the pill might experience muscle cramps or more serious muscle problems. I had been taking it for three days when I already started to notice some cramping in my leg and some minor tingling in my toes, reminiscent of some sort of neuropathy. I still really don't know whether this was real or only my imagination, but I stopped taking the pills, and the symptoms went away in less than a week. Since then I have been exercising regularly, which might not be obvious to you, and trying to control my diet to keep my cholesterol in check. In my view, doctors prescribe medications too freehandedly. I don't know why. Are they in the pockets of drug companies? Or they just don't want to take the risk that a patient might sue them for not giving them drug treatment?"

"Most probably, it is a combination of several things," Doug replied. "But it is true that there are too many frivolous lawsuits, and I know several doctors who were driven out of business because of that. It's no wonder that many doctors are afraid of lawsuits, and often unnecessarily prescribe medications and expensive tests, under the pretense that they are necessary for proper medical care. This practice may never change unless Congress does something about limiting these lawsuits. For years there has been a lot of talk about it, but no action."

"Just out of my curiosity, did Peter ever mention to you his fascination with warm baths and wine?" asked Carl.

"No, Peter never explained or even mentioned to me that he was taking warm baths to improve his condition, although he usually didn't divulge these kinds of things. He was quite aloof in that respect. But I knew that Peter occasionally drank red wine because he thought that it helped eliminate fat after a larger portion of New York steak. But Peter's doctor advising him to drink wine on a regular basis is completely new information for me."

"What component of the red wine do you think could have helped Peter with his diabetes?" asked Carl.

"I have to admit that I'm not up to date with the research on this topic. But I promise you that I will ask George Carrick tomorrow if he knows any more about this subject. I occasionally read papers about the good effects of antioxidants present in red wine and various fruits, but my knowledge is really insufficient to give you a complete answer."

"Do you think that this antioxidant thing might just be hype?" asked Carl.

"I am starting to suspect that in some cases this might be the case, although I believe that in general, adding certain antioxidants to the diet supports human health with some exceptions." Doug chose his words very carefully, because medical sciences and surrogate food industry didn't like any doubt raised about the exclusive beneficial health effects of antioxidants. "To give you some examples why the antioxidant issue is so complex," Doug continued, "let's start with wine. It is becoming clear that even a gallon of wine doesn't have enough antioxidants to have an impact on common health risks like heart disease, diabetes, and cancer. Thus, antioxidants may not be the substances in the wine that allegedly increase life expectancy, although it could be that they work only when they are dissolved in alcohol. Since wine contains many different antioxidant molecules, maybe they work better together than in isolation. This is another possibility. In the few studies I have read, a single antioxidant was always used, which might not be the right approach. But then, not all antioxidants necessarily have good effects either. For example, during the '90s, vitamin E was touted as an antioxidant that might reduce the risk of cancer. Unfortunately, subsequent studies later revealed that ingestion of high amounts of vitamin E unexpectedly increased the risk of certain cancers, and the situation is very similar in the case of vitamin A. Scientists probably still don't really know why. Perhaps the Food and Drug Administration should discourage natural food stores from selling these vitamin tablets without warning signs. In any case, my prediction is that in the future, the allure of taking antioxidants to solve already existing health problems will diminish and they rather will be used for prevention."

It was nearing 5:30, so Carl had to cut the conversation short. He gathered his coat and shook Doug's hand. Before he reached the door, he stopped and looked back at Doug. "To be honest, I'm nearly clueless," he admitted. "I'm going to need any help I can get to develop a good case."

"I'll do all I can. Have a fun evening with your wife," said Doug.

"I most certainly will," Carl answered with a beaming face and walked out. This evening was something he had been eagerly looking forward to for weeks. He had let his colleagues know well in advance that, for anything short of alien invasion of Rochester, he would not be available that evening.

CHAPTER 12

STEVE MIGHT HAVE FOUND THE ACCOMPLICE SUBSTANCE

Friday had been a productive day for Steve. By noon he had finished his clinical work, and thus he had the full afternoon to search for the potential accomplice substance. The case appeared to be simpler than he had anticipated. After typing in the words 'zinc-induced cell death,' the first thing that appeared in his search was an article describing that in cell cultures, large doses of zinc indeed caused cell death, even if it was presented alone to the cells. Leukemia and lymphoma cells were among those that succumbed relatively easily to zinc. But this never led to any useful therapy because certain types of healthy blood cells were also very vulnerable to high levels of zinc. Thus, using high doses of zinc to cure cancer would kill the person along with the tumor. Besides, almost all the papers he found accentuated the difference between *in vitro* cell experiments and real clinical use, mainly because zinc avidly binds to some blood proteins like albumin, and thus it would be hard to maintain free zinc at a sufficiently high level to kill cancer cells.

Then, Steve found several articles claiming that much lower amounts of zinc can also rapidly kill cells in combination with some other compounds. The organic compound most mentioned was PDC, belonging to the pyrrolidine type compounds, which has a chemical zinc binding group. PDC can carry large amounts of zinc into cells, bypassing the normal and highly restricted gateway in the cell membrane through which zinc can gain entrance into the cell interior. Steven also learned that the combination of PDC and zinc kills normal cells as well, although certain cancer cells were somewhat more sensitive; so, at least in principle, these two substances in combination could be used for cancer therapy, although only with great caution. Then, Steve found a US patent that described how to avoid toxic side effects by directly injecting PDC and zinc into a tumor. Steve couldn't find any information on whether this patent had ever found its way into clinical practice. But Steve found the method of directly injecting PDC and zinc into tumors to be fascinating. The patent provided evidence that within thirty minutes of injecting these agents into the tumor, 80 to 90 percent of cancer cells around the injection site were found to be dead, and if a sufficient number of injection sites were used, about 70 percent of all tumor cells were killed. The patent owner argued that the method could be used to reduce the tumor's burden before other cancer treatment methods were employed, particularly at anatomical locations where surgery wasn't otherwise possible. This method could also be used together with chemotherapy and radiation treatment.

Steven wondered why no one was working on this technology, although he had enough experience not to be too surprised. Small companies generally aren't exactly flush with money; they rarely have enough of it to push a procedure or product like that through human clinical trials. And most pharmaceutical companies would be reluctant to invest in something so unprecedented.

He also knew that this is also the fate of many other good ideas that haven't yet been tested on humans. Big companies very rarely license procedures without first seeing positive effects from a well-executed human clinical trial. They don't want to find out, after investing many millions of dollars, that the procedure is an abject failure; in fact, they usually expect the originator of the idea to provide the money for an initial trial. As a result, less than 1 percent of prospective anticancer agents make it to human clinical trials, meaning that the cancer scientists who developed the other 99 percent wasted substantial amounts of research money. It's really anyone's guess how many potentially groundbreaking ideas to treat cancer have remained unexplored because of the ignorance or financial prudence of drug companies.

After some more thinking, Steve considered a scenario that even if PDC isn't being used to treat patients, someone still might keep it for further research, or maybe even some malicious purpose. Someone might have even changed its structure and created an entirely new compound. In any case, it would be interesting to see if Doug could find PDC or anything similar in any of Peter's organs. Steve took it for granted that Doug would be compelled to try this route. He was convinced that this lead, at long last, had a very good chance in helping them untangle the knot of confusion they were in; at least, this time he was convinced they weren't just grasping at straws.

That Monday evening, Steve showed up a few minutes before Doug at their favorite downtown Starbucks coffee shop. Steve was anxious to share the information he had gathered on Friday. But he had to wait until Doug recounted in his usual punctilious manner his conversation with Carl. Steve was dumbfounded to hear that Peter had apparently become an alcoholic after grossly misinterpreting his doctor's advice.

"You know," Steve said after digesting the information, "it's quite an interesting coincidence that some wines contain a lot of zinc, depending on the soil where the grapes are grown. If you meet your alcohol expert tomorrow, try to figure out if he knows how much zinc is in Bordeaux wines. Otherwise, we can just do another internet search. Hopefully some chemists somewhere have done this kind of work. The reason I think we should keep this wine thing in our minds is that if every day Peter drank one bottle of a wine rich in zinc, then this could have been enough to kill him if someone also gave him PDC or something similar around the time of drinking. But I still can't get over the idea that Peter drank so much wine with his doctor's knowledge. I wonder why he didn't consider that Peter might develop alcoholic liver disease, which in fact he was close to. He should not have skirted this alcohol issue, and in fact I don't know of any other doctor who would encourage a patient to drink any alcohol of any quantity. I still feel queer not paying enough attention to Peter lately."

"No, no," protested Doug, "you don't understand something. What happened was that the doctor advised Peter to drink moderately, something like two glasses of wine a day, and reluctantly agreed to postpone putting him on drug treatment. This would have been appropriate advice if Peter had combined it, as his doctor strongly advised him, with an appropriate diet and a rigorous exercise program. This was a give-and-take relationship, and the doctor thought that by allowing moderate wine drinking, Peter would adhere to the other, more important recommendations. This doctor very much knew what he was talking about however unorthodox his approach was. He certainly based his opinion on the many recent studies in which alcohol experts reached a consensus that the effects of alcohol can be best described with a bell-shaped curve, which is slightly different for men and women. Only moderate drinking, meaning two glasses of wine a day for men and one for women, provides the benefit of increased longevity. If someone drinks wine at a level far above or below that amount, he or she starts to lose this benefit rapidly. If Peter could have just understood this science-supported truth, he could have lived a healthy lifestyle while still enjoying wine. But he clearly didn't listen to the doctor's recommendation, and who knows, this might somehow relate to his premature death. By the way, it appeared to me that his doctor is a very kind, soft-spoken man, probably close to retirement. Perhaps it would have been better if he occasionally berated Peter for his habits. But it looks like the doctor tried to handle Peter's resistance to taking drugs within the margins of medical error."

Just as he finished his long monologue, the term *PDC* finally reached the conscious part of Doug's brain. "PDC," he repeated the word again. "What is PDC?"

Steve immediately realized that his friend couldn't possibly know about this drug. "I'm sorry, Doug, I'm just tired because of the funeral and everything else. I forgot I haven't told you about PDC yet. Well, on Friday I did the search as promised and found many studies all saying that PDC is used together with zinc to kill cells."

"But what is it?" asked Doug, slightly annoyed over getting his desired information piecemeal.

"Okay, I'll start over again from the beginning," said Steve. "PDC is a complex compound that is put together by chemists from two smaller compounds. One of them can bind to zinc, while the other one helps the entire molecule enter the cells in large amounts, via nonconventional routes. When zinc is bound to PDC, much more of it can enter the cells. If there is enough PDC in the blood, it can really swamp the cells with zinc. As you said at dinner the other night, since the cells aren't prepared to deal with this massive influx of zinc, after a short struggle they die. But what you may not know is that if somehow, a tiny amount of PDC and zinc could be targeted to the tumor, this complex would eclipse all known cancer drugs in its effectiveness. My feeling is that someone is working very hard on this even as we speak. But I'm not sure that person will succeed. Medical chemists are still ill-equipped to selectively and effectively target cancer drugs to the tumor tissue. The trouble is that if PDC and zinc aren't targeted specifically at the tumor, this combination can engender a mess in normal organs. But in another patent that I read on Friday, the authors describe experiments wherein PDC and zinc were directly injected into experimental tumors, and they wiped out most of the cancer cells in a very short time frame, as little as thirty minutes. I've become kind of fascinated with this PDC and zinc stuff. Just think about the incredible advantage this method might offer to stunt tumor growth when surgery is no longer an option, or to use it to replace chemotherapy that weakens the body and causes your hair fall out. But I still believe that the key is efficient targeting of the PDC zinc complex to the tumor. Without this, they may offer only a beguiling promise."

Doug listened to Steve's account intently, and after sitting there ruminating for a short while, his imagination kicked into full gear. "Just think this through, Steve. Medical robot technology is already available for drug delivery. Someone could insert a sort of computer-aided long catheter or hollow wire into the appropriate vein or artery and use a minipump to push this anticancer stuff with great precision inside the tumor in the target tissue. Then, from the outside, one could inject a small volume of PDC and zinc into the catheter and apply pressure to deliver it into the tumor. This could be a noninvasive procedure to rapidly reduce the size of the tumor, and that would make additional treatments a lot more effective! Isn't that brilliant?"

Steve hated to put the brakes on Doug's vivid imagination, but being in the medical profession himself, he felt he had to.

"At first, I was also very enthusiastic about this approach, but the more I thought about it, the more doubts I began to develop. For one thing, many tumors have irregular shapes, and that makes it very hard to determine where to inject these drugs. Furthermore, tumors generally have a high internal pressure. This means that if you were to try to inject one with a drug solution, a large part of it would be ejected back from the tumor and might cause collateral damage to the surrounding normal tissues. Maybe these are among the reasons that big companies aren't all rushing to this technology. However, there is some new research suggesting that certain drugs may reverse the high internal pressure of tumors, so the inventor and we can still hope that this method isn't totally doomed."

Doug looked disappointed seeing his theory so easily repudiated, but Steve quickly tried to refocus the conversation before they could get sidetracked again. "Anyway, I did some research and found that PDC is commercially available. This means that it could very well be the accomplice substance that we've been looking for. Doug, I believe you should start looking for traces of PDC or a similar compound in Peter's organs, although I know this will be a new imposition on your and your assistants' time. PDC should give you a very characteristic peak on the mass spectrogram because it does not occur naturally."

"Yes, my assistants probably won't be too happy, but I will get it done," said Doug. "And tomorrow we will begin looking into obtaining PDC from somewhere. You know what? I just thought of somebody who might have it. I think I'll send an email message to my friend, George Carrick, telling him to bring some PDC tomorrow. He worked with zinc. He may well have some PDC left if he ever used it."

Doug knew the behavior of scientists very well since he was one of them. Most would hoard a large number, sometime hundreds, of chemicals over their careers, usually in amounts more than they really need. In addition, the trajectory of their scientific interests may change, and the drug they ordered yesterday might no longer be needed tomorrow, at which point there is usually quite a lot left. This is one of the better-hid realities of the science world, which will probably remain unchangeable forever. No one has ever assessed, and probably never will, how costly this waste is to the government. This is one of those touchy topics in science funding that are hardly ever even acknowledged by anyone. Besides, in practice, it would be almost impossible to even start reducing chemical waste. In any case, Doug hoped to use this waste to his advantage, and prayed that George was one of these drug hoarders.

In the afternoon Sarah suddenly felt better and returned home with the children long before Doug arrived. "Well, I see you no longer felt the need to escape?" asked Doug teasingly.

Sarah didn't get Doug's joke and left it unanswered. "We had some cold cut for dinner," she announced instead. "I didn't have time to prepare anything warm. I left everything on the table for you."

"That's fine with me," Doug replied. "I need to send an email first, anyway." In fact, he would have liked to eat something warm; during his drive to home, he had longingly imagined a hot soup and some grilled meat, but he thought it was pointless to admit that.

Amazingly, George answered his email almost immediately. It turned out that just as Doug had predicted, George had indeed worked with PDC years ago, and he was pretty sure that it was still sitting somewhere in his industrial freezer. He promised Doug he would bring him around fifty milligrams of the compound, which had suddenly become so valuable. So Doug at least wasn't thwarted by a lack of this precious compound. He was still very guarded in using the expression "accomplice substance"; whenever he mentioned it, he liked to add, "If there was one." He knew the importance of paying heed to any potential biases, like taking as fact the existence of the accomplice substance, which may divert his attention away from other possibilities. In that respect, he was like Carl.

CHAPTER 13

Hunting for the Accomplice

Tuesday didn't start off very well for Doug. Heavy rain and high winds lashed the city, and he found himself wondering if George could make it. The weather forecast had even placed a tornado watch in effect, and not far from Rochester, there was an actual tornado touchdown. Sarah complained that once again she didn't feel well, but she gave no specifics about her condition. It could have been a second bout of stomach flu, or maybe she was just tired. Their older son, Roy, also said that he was too sick to go to school. Their younger son, Brian, wasn't yet old enough for school, and Sarah still took care of him full-time. That day, however, Sarah asked Doug to take the two boys with him in the morning.

Doug was little reluctant to take on this extra duty, but it was clear to him that Sarah brooked no argument and he had no choice. Usually they took the children to the home of a retired teacher, Martha Warwick. Martha knew Sarah, Doug, and their friends from the high school where, back in the day, she had taught American history to them. Her husband had died of liver cancer three years ago, and after retiring two years later, she didn't have much to do. She loved to help parents out by taking care of their kids, because in some way she saw it as a continuation of her teaching career. Her new business dovetailed very well with the needs of young parents who had no other place to turn to when due to an unexpected situation they had to leave their children somewhere.

DUALITIES IN HEAVEN AND EARTH

The kids liked Martha very much, because she told them many interesting stories from American history, especially about the Civil War and its aftermath. In her realm, kids were always happy; they didn't even know what boredom was. She also had a stupendous stamp collection, with stamps from about 140 countries, which she had started to collect in her early childhood; she still added to it occasionally. Her favorite way to pass time with the kids was to search for the countries the stamps were from on a large map of the world that hung on the wall in her living room. When she was in a good mood, she would give small gifts to the kids if they found a small country in Asia or Africa. This was often tricky because many countries had changed their names since the issuance of the stamps, which even Martha had a hard time keeping up with. Fortunately, she also had maps from earlier times, dating back to the period between the First and Second World Wars when most of the colonies were still in place, and it was even more fun to compare these with the newer one.

Martha always taught the kids, Martha's kids as she called them, some very basic history and geography lessons, providing some background behind the name changes. Occasionally she would even let them watch a movie related to major historic events. Once she taught them about the history of India while they watched the movie *Gandhi*, which provided a very good illustration of what had been going on in that country before it gained independence. Another time she talked about the French Revolution while they watched *Maria Antoinette*. Certainly, Martha's kids, unlike those taught by teachers less focused on geography and world history, had no problem distinguishing similarly sounding European capitals like Budapest and Bucharest, or locating Ethiopia on a map. In short, Martha was the paragon of a devoted, masterful teacher who could translate her lore into a language that even small children could understand.

Unfortunately for Doug, on that day Martha could take the kids only for the afternoon, but not the morning. Occasionally, particularly during the weekend, Doug would bring the boys to his office in a remote wing of St. Mary's Hospital, at a safe distance from patients and laboratories using radioactive material. According to the Mayo's policy, parents weren't allowed to take their children to work with them, but this time Doug really had no other choice. The guard at the main door probably assumed that the kids were going in for a doctor's visit and didn't bother to ask questions. Some time ago, Doug had given into tech-savvy Roy's pleading with him that he and his younger brother allowed playing games on his computer provided they didn't make too much noise. This is how the day started for them that Tuesday morning as well. Fortunately, Doug had two computers in his office; he kept the older one only as a backup for himself and for the children to play on. Usually, as on that day, after two or three hours, his "rules of engagement" no longer held any sway over the kids, but Sarah would turn up by then and take them home.

After the children had settled into the office, Doug went over to the laboratory to see if Mary and Joseph had already started preparations for the PDC analysis. "We still have some samples left of the liver, heart, kidney, gastric juice, and blood that will probably be enough for one analysis or maybe two if we really stretch them," Mary informed Doug.

"We also have a problem," interjected Joseph, "and we wanted to wait to discuss it with you before taking the tissues out from the freezer. Since we don't have PDC as a control, we can't identify it in any of the samples. We need to spike one sample with PDC, and then see if we find a peak from the other sample coinciding with it." Joseph drew a breath to explain the situation in more detail, but Doug interrupted him.

"Settle down, Joseph. It's okay. I know all that. It is fine that the specimens are still in the freezer. I just got an email from a person at the university who has PDC, and he will bring it here around eleven. Now, you can start thawing the tissue samples. By the time you have prepared them for the analysis, we should have the PDC. After that we'll still have a plethora of problems to solve, so let's try to get past this one quickly. But I'm very happy to see that you were cautious with the sample preparation." Doug had the special ability to simultaneously praise and impel people.

His children were engrossed in a computer game, and Doug still had half an hour to do some thinking before George arrived. "I wonder," he murmured to himself, "if Peter had a specific reason to drink the Bordeaux brand wine, or if it was a random choice? Could it be that Peter chose Bordeaux based on the belief that it may have something in it that could help his diabetes? It seems unlikely that he just made a random decision to select and stay with the Bordeaux."

Then he remembered his conversation with Steve, when Steve had said that some kinds of wine may contain more zinc than others. In fact, from his nutritional studies he knew, as he once already explained to his friends, that the soil in certain regions can be very rich or very poor in various trace elements, which is then reflected in the metal composition of ground water and plants that are grown in that area. For example, he remembered that in some US states, the soil, and thus the vegetable grown on this soil, was very poor in selenium, which is also an essential element like zinc.

On a whim, he called Carl and left a message telling him to call back immediately. Within a minute, Carl indeed called him back, expecting some great news.

"Carl, do you still have those wine bottles you took from Peter's garage? Have you done anything with them?" asked Doug.

"We did fingerprinting and confirmed that Peter had touched each bottle," Carl answered. "We did nothing else and intend to do nothing else for the time being."

"Did you by any chance see some wine remaining in the bottles?"

"Quite a bit actually. If we were to combine the wine from all the bottles, it would probably fill about half a wine glass."

"That's fantastic!" Doug exclaimed. "Carl, I just got an idea. I need to analyze the wine in the bottles, separately. If you can arrange it, would you please bring all the bottles to my office on Wednesday morning? Don't try to pour the wine into a glass or anything. We have automatic pipettes that are suitable to do that without losing the slightest trace."

"I can do that," said Carl. "By 9:00 a.m. tomorrow I'll be in your office with the bottles, except for one that I have to withhold as evidence. I'll withhold the one that has the least amount of wine in it, I promise. But I'm telling you in advance that you will have to take out the remaining wine with gloves on your hands. Then I will return all the bottles to a sealed plastic bag and store them together with the other items I took from Peter's house."

Doug was so caught up in his own thoughts that he didn't even hear Carl's last sentence. *What if,* he thought to himself, *the Bordeaux wine had a lot of zinc in it, and what if it was also spiked with PDC?*

"By the way," continued Carl, regaining Doug's attention, "I never mentioned this to you, but last Wednesday I found one of Peter's cell phones in a garbage can in front of his house, waiting to be disposed of. It had a number different from the one you gave me. I already gleaned some information from voice mails, and I identified a few callers, but you and your friends weren't among them. You obviously only used the other number, but that phone is still missing. It finally stopped working, so now we certainly can't trace it. In any case, most of the callers were apparently from IBM and other businesses. But three calls that might relate to this matter stood out for me. One was early in the afternoon last Monday, and another one late in the evening the same day, from the same guy from Stewartville. We figured out that his name is Ulrich Siebert, and since then I've learned that for a long time he was Helga's boyfriend. I'm still trying to find out what he's doing for a living now. However, I know that he, just like Helga, is from Germany and that he worked at the Mayo Clinic from 2004 to 2008. If my information is correct, Helga used to work in your laboratory for several years."

"That is all correct," said Doug with an irresolute expression. He was very much surprised to learn that Ulrich and Peter knew each other. After he had gotten over this new information, he said, "I know them both very well, and indeed Helga worked for me for a while. So far, your findings dovetail very much with my information. And I also found Ulrich to be a nice and open guy, and even though he is an atheist and thinks that I've fallen victim to the power of false ideology, we still agree on many other things. It would really be a shame if he got mixed up in all this. But I wonder who the third caller was?"

"It was Helga, but she called the next day, around 3:00 p.m. It's obvious that she knew nothing of what had happened to Peter at the time of her call."

"Is this how you know for sure that Helga had nothing to do with Peter's death?" asked Doug.

"No, you misunderstand me. All I said was that Helga didn't kill him. Whether she was part of a scheme, abetting someone who arranged Peter's death, I cannot say yet."

"Well, these phone calls should lead us somewhere," Doug murmured. "Have you tried to talk to Ulrich?"

"Yes, I tried, but… you won't believe this. It sounds like something out of a sitcom. Around noon on Wednesday, I called Helga to ask where Ulrich was, and she informed me that a common friend had told her that the very same morning he had left for Switzerland on a business trip he had been planning for a long time."

"That's really an egregious turn of events," said Doug with a deep sigh. "Any idea of how long he will be in Switzerland?"

"Helga didn't know. Her guess was that not even Ulrich could be sure of how much time was needed for him to conclude his business deal. We actually got information from Delta Airlines that he had indeed bought a one-way ticket to Switzerland, which would fit Helga's story."

"Well, I hope you will learn more about what Ulrich did in Stewartville, where he lived, and things like that. There must be other people who knew him in Stewartville."

"Of course. I was planning to do all this anyway," said Carl. "But tomorrow Helga will fly to Washington, DC, to visit some art museums. Apparently, there are several major modern art exhibitions going on there, and she decided that she can't make progress with her art without seeing all the most recent trends. She'll return only after the weekend."

"Well then, it looks like we are stuck for a while on that front too," Doug concluded. "Both Germans seem to be imponderable to the point that it's almost suspicious. This whole situation really has all the elements of a farce, although for me it is very serious. In any case, in the meantime, please, at least try to find some people in Stewartville who knew Ulrich," Doug begged the inspector. "The way I see it, we have some vague connections, but nothing firm yet. Maybe this is the point where our analysis will help connect the dots that are starting to show up."

Then Doug touched his head. "Carl, I still have one more question. Do you remember if someone by the name Agnes Lambert called Peter on Monday or Tuesday?"

"Yes, I actually do remember that someone by the name of Agnes Lambert called Peter just after 10:00 a.m. on Tuesday. Why, who is she?" asked Carl, hopeful for still another lead.

"She was Peter's secretary, and Mike called her right after he found Peter dead. She didn't seem to know anything what had happened to Peter, and Mike assumed that she didn't call Peter on his off days, although Agnes didn't say anything about that."

"I guess I need to put one more person on my radar," Carl concluded.

"Carl, I need to tell you one more thing before you hang up. Agnes was Peter's next-door neighbor, on the right side."

"His secretary and his neighbor—what a coincidence," Carl murmured. "Doug, let me be honest with you. I'm close to retirement, but it seems that Rochester is forcing me through one final test of my abilities, perhaps my most difficult one ever. I have never encountered a case as strange as this one. We keep making suspicious observations and finding circumstantial evidence here and there, but never anything solid." The dearth of useful information in the case clearly infuriated Carl, and he didn't even try to hide it. Doug noticed his exasperation and assured him that he strongly believed that through their analyses, a pattern would at last emerge in this chaos. Carl put down the phone in a slightly better mood.

Later in his life, Doug occasionally thought back to this conversation. While he suspected that Agnes might be a source of useful information, which is why he brought her up with Carl, he never even considered that she and Peter might have formed a relationship or that she might have had anything to do with Peter's death.

George arrived ten minutes later, but Doug didn't mind it. While he waited, he finished another telephone conversation with a member of the Mayo's advisory panel about several cancer cases from last week. Then Sarah called; she was at the main entrance, ready to pick up the kids. At least this exigency was solved for the day.

No more than two minutes after Doug returned to his office, George knocked on the door. He too had had other businesses to take care of at the Mayo that had taken a little longer than expected. George was a well-built man with a scraggly beard and a sizeable wart under his left eye. Despite being married for almost five years, in his friends' imaginations, he was still a hard-drinking womanizer, which he fervently denied.

"What's up, Doug? How are things going?" he asked jovially. "But first thing's first, here's the PDC. I wrapped it in aluminum foil because it might be light sensitive."

Doug took it and immediately called Joseph, who rushed away with the precious gift. "Well, it has been quite a while since we had a good chat," said Doug.

"Yeah, it's been what, four years? I guess when we met in Paris at the International Congress of Clinical Oncologists."

"I remember it well," Doug mused for a moment about the good times they had in Paris. "You took a room in the George V Hotel, close to the Champs-Elysees. I always wondered how you could afford that."

"I got a 30 percent discount because it was off-season. Remember, it was early February and by Paris standards, it was quite cold. I guess only scientists are crazy enough to go to Paris in the winter."

"Now I understand." Doug smiled. "Well, in any case, let's get down to business. The reason I wanted to meet you is that my friend Peter unexpectedly died, and nobody can seem to figure out what killed him. If I can be honest with you, this harrowing event has been absolutely devastating for me and my friends. If a crime was committed that we aren't aware of yet, my friends and I don't want the scoundrel to live his or her life in freedom. But so far, looking into everything that our instruments permitted, we could only come up with some hints that Peter ingested something that killed many of his cells. This something might be zinc, which was found in high amounts in all his tissues. But we also think that these levels of zinc alone couldn't have killed Peter. We think there must have been some other substance present, which helped zinc enter Peter's cells. At this stage, I am looking to you as the expert, and you know that I am always amenable to good advice. To be a little more informative, PDC or a similar compound is what we are looking for as the second substance. If we find something like PDC in his system, it might explain how the zinc could have killed Peter. However silly this idea might sound, this is the only reasonable hypothesis we have so far."

"Well, silly or not, you can test it now," said George. "By the way, keep in mind that you need to store the PDC at minus four degrees Celsius or colder. I wrote it on the container."

"Very good." Doug nodded. "My assistants have already prepared the samples for analysis. After lunch they can get to work. Joseph thinks they should be done by 4:00 p.m. He's a very bright and hardworking young man. With the instruments we have, he can ferret out almost any information we need. Maybe someday we can even help you out with your analyses."

"You're very lucky, and I'm jealous." George chuckled. "I have a lazy postdoc who puts off work as long as he can. He can languish in the lab for days without producing anything useful. First, I tried to dangle incentives in front of him, like sending him to a conference on cancer, but to no avail. I can't take much more of it. Eventually, I'll have to fire him, but I can't do that for another two months. The worst part of it is that he's already started to act like a victim and spread all kinds of false allegations about me. The poor fellow doesn't know yet that without a recommendation letter from me, nobody will hire him."

They made small talk like this for a while, and then George commented that he was hungry.

"Would you like to have lunch?" Doug suggested. "There is a Greek restaurant just across from Barnes & Noble. They cook well, their gyros are excellent, and we won't have to wait for long."

Gyros sounded good to George. The place was only about ten minutes from St. Mary's Hospital, and they found a quiet corner seat right away. Doug then proposed that they have a glass of Bordeaux with their gyros.

"I am getting fascinated with Bordeaux wine," Doug revealed to George, as they began to eat. "I always knew that it's a good, delectable wine, but I still have no idea exactly why Peter preferred it, so much so that he refused to try any other wine. This inexorably drives me to the conclusion that this wine has some peculiar property that Peter was aware of. After all, in my experience, you can find equally good or even better wines from France and Italy in the same price category. A big question for me now is if red Bordeaux has any health advantages over other wines, as long as it's not abused of course."

"Knowing your interest in wine, antioxidants, and zinc, I expected a question like this, and I brought three related articles with me," said George, setting them on the table. "As you can see, the first one is an older paper published back in 1992. It shows that red Bordeaux wines far outperform other wines in their resveratrol content."

"I don't want to annoy you, but what is resveratrol?" asked Doug. "I am a little behind with reading the literature."

"You're not only behind on the literature, but you must have been living on another planet or in a cave recently. Everyone's talking about resveratrol nowadays. Nobody has appraised you about it?" exclaimed George. "Resveratrol is a natural compound found in the skin of grapes, peanuts, pines, legumes, and berries. Extracts from these plants have been used in traditional Chinese and Japanese medicine for thousands of years, without anyone actually knowing that the active substance in them was resveratrol." George sounded eerily like Doug when he got into lecture mode. "Some scientists claim that it even helps people suffering from cancer, diabetes, aging, and God knows what else. Other scientists, however, warn that it's not clear yet if its effects are always positive, or if regular wine consumers can even get enough of it without getting dead drunk. Alcoholic people may get enough, but at the expense of liver disease, or their driver's license.

"At this moment, scientists are somewhat divided on the health benefits of resveratrol. In any case, because it may be good for diabetes, in my opinion this could be the reason why Peter began to exclusively drink Bordeaux wine. I read in several studies that if enough resveratrol is injected into diabetic mice, initially it helps control blood glucose. Unfortunately, at least one study also showed that resveratrol can kill the cells that produce insulin, which in the long term could make diabetes worse or can even cause diabetes in a previously healthy person. I also read an article somewhere claiming that if someone consumes 500 milligrams of resveratrol a day in a tablet, which indeed some people do, it will inhibit renewal of neurons in the hippocampus. Thus, if someone advised Peter to drink Bordeaux wine to alleviate his insulin resistance problem and avoid full-blown diabetes, in my view that was probably not the best advice, particularly because alcohol and resveratrol might have damaged his pancreas, liver, and brain functions. At least that's what I gather from the literature. If those authors are wrong, I'm wrong as well. I'm not in the position to either promote or repudiate the health benefits of resveratrol. But back to the wine, chances are that even Bordeaux brand doesn't have enough resveratrol to make a dent either way for an average wine drinker. But evidently, if Peter drank one bottle of wine or more a day as you said, he drank way too much. He wouldn't have benefited from the good effects of the wine but could have suffered from the negative ones."

"That's what I thought," Doug agreed. "I believe that, after health experts condemning alcohol for decades, there is now a new renaissance of wine drinking, and some enthusiasts might feel exonerated even if they drink too much. Perhaps we shouldn't be so complacent about wine drinking, and certainly the public should be better educated to make sure that alcoholics can't make excuses for their alcohol abuse, and at the same time moderate drinkers don't feel guilty. What I mean is that responsible drinkers shouldn't be chastened just because of the irresponsible ones."

"I totally agree," George confirmed, "but this is something much easier said than done. But going back to Peter, it is also known that people with insulin resistance and diabetes have a relatively high rate of mortality from coronary artery disease. You see, the second paper I brought gives evidence that antioxidant flavonoids in red wines reduce this disease. If this was Peter's goal, then perhaps red Bordeaux wasn't the best choice. For example, in this second paper, the red Châteauneuf-du-Pape is cited as a high flavonoid wine, much higher than the Bordeaux brand. Maybe other winegrowers have belied this claim, but I'm not aware of it if they have."

Doug was growing impatient, so he tried to redirect the conversation once again. "Well, what I really wanted to know is if you have any information on how much zinc is in the Bordeaux brand, especially compared to other wines?"

"That's why I brought the third paper," said George with an ostentatiously satisfied face. "Now, we seem to have something here. Different appellations of the Bordeaux contain anywhere from three to ten times more zinc than other red wines. In my view, if Peter drank an entire bottle of Bordeaux wine in a short time, say a few hours, it could explain the high zinc you found in his organs."

"But would that be enough to kill someone?" Doug asked.

"Certainly not, unless of course there was something else that prevented Peter's body from expunging zinc, thus making it harmful. And that something could be PDC, which can smuggle a lot of zinc into cells. Once I worked with a combination of PDC and zinc, and I found that together they killed high numbers of cells very rapidly, namely a half hour or less. PDC and zinc in combination are vicious, remorseless cell killers."

"That's what the cytological samples showed us," Doug exclaimed. "Cells apparently died in such large numbers that it would have killed Peter sooner or later. Perhaps the missing link in Peter's death is PDC after all."

George had become so interested in the case that he decided that, after some more visits with Mayo doctors, he would return to Doug's office by 4:00 p.m. They agreed that it was pretty much pointless to talk any more about PDC and zinc at that point; they had to wait for the results.

But before they left the restaurant, Doug suddenly remembered that a couple of years ago he had indeed heard about resveratrol. "George, wait. You said that resveratrol has something to do with aging? I heard somewhere that a certain substance extended the life of worms and some kinds of yeast by 30 to 50 percent. Could that be resveratrol?"

"It certainly could be," said George.

"Well, not long ago, on the public TV, I saw Charlie Rose interview a scientist. I believe his name was Dr. Sinclair, who is leading the research on this substance, which as you say is probably resveratrol."

"If he was the Dr. Sinclair I'm thinking of, then the substance he was talking about was indeed resveratrol," said George. "In fact, he founded a company to produce and study resveratrol and some other compounds that structurally resemble it and are biologically even more active. He's so confident that resveratrol prolongs life without any side effects that he claims to take it every day, and a growing number of people do the same."

"Based on what you said before about the side effects, this guy must be very brave," said Doug with admiration in his voice. "Don't you think experimenting on yourself seems a little irresponsible?"

At first, George wasn't sure where he stood on the issue, then he said, "As you probably know better than I do, any drug can have unpredictable effects. I have only a little bit of toxicology experience. But what I do know is that he isn't the only one to experiment with untested drugs, and I'm sure that you have also heard many examples of this. Case in point, I recently met a biotech guy who is using a certain substance to prevent further loss of his hair. He thinks that because in general his substance makes stem cells more viable, it also will stimulate stem cells in the scalp to regenerate hair. As you can probably see, I also have a hair loss problem. There was a time when it was falling out in handfuls, and it really depressed me for a while. But the guy gave some of his funny white cream, and I am telling you, it really works! It completely stopped my hair loss, although I don't see much new hair growing in yet. My attitude is that it cannot be gainsaid that self-experimentation can be sometimes valuable provided the experimenter has a clear idea about the possible side effects."

"Do you have any explanation for why you don't see new hair growing?" asked Doug.

"I really don't," George replied hesitantly, "although the reason might actually be very simple. As far as I know, after losing hair, stem cells are needed for regrowth. But they remain viable only for a certain time. If after several years they don't get appropriate hormonal stimulus, then they will die out and nothing will replace them. In my case, this is exactly what could be happening, so at the spots where my hair disappeared about three years ago, I may not have a chance to regrow it, because the hair follicles are probably irreversibly atrophied. But wherever I still have viable stem cells, this guy's cream can still prevent hair loss and in fact stimulate new hair growth. However, a recent paper claims that even many years after losing hair, the dormant stem cells are still there and can be resuscitated if they get the right mixture and quantity of hormonal stimuli. I don't know whom to believe. All I know is that I'm confused about whether I'm too old for hair regrowth, or if I still have some hope. I'm still hopeful about regaining more hair, and I keep scouting the relevant literature for positive news."

"Do you see an opportunity to replace lost stem cells with viable ones?" Doug asked.

"Absolutely I do. I believe that eventually this is going to be the answer for hair loss," replied George. "The problem, though, is that before implantation, we first have to teach the stem cells to make hair and not become cancer cells. As you know very well, in case of embryonic stem cells, the problem is, the ethical dilemmas aside, is that they have such a great potential to become any cell type that they can very easily become cancer cells as well unless genetically modified prior to use."

"Yes, I'm a little surprised that I very rarely hear stem cell scientists talk about stem cell cancer risks in public," said Doug.

"Oh yes, they are afraid of losing the public's trust and with that their funding," George explained. "I don't talk to the public about cancer risks either. And the reason for that is that we all hope that by kicking out certain genes, we'll be able to prevent embryonic stem cells from becoming cancer cells. Of course, the trick is to do this in such a way that they retain the ability to become any other kind of normal cells, like heart or liver. Thus, we hope to solve this problem first, and as quickly as possible, before the cancer problem becomes a serious health issue.

"And don't forget that we can use adult stem cells as well, which don't seem to carry any significant cancer risk," added George. "I truly believe that the mesenchymal stem cells extracted from bone marrow or fat tissue can be used very safely to regenerate certain diseased tissues. And now even skin fibroblasts can be genetically tweaked to become stem cells. If we can get these adult stem cells to work, then using embryonic stem cells will become a non-issue, both from scientific and ethical viewpoints. But to achieve this, will still require from specialists a lot more tenacity and hard work and less brazen smugness."

"I am with you on that," said Doug smilingly, recalling his own experiences with smug grant reviewers.

George continued, "I believe so much in adult stem cell therapy that I would gladly participate in a human trial using these cells to try to grow hair. The cells are there. All it would take is a surgical procedure to implant them and then provide the right hormonal stimulus that would put them on a differentiation path to produce hair. I guess this is again something easier said than done, but I have high hopes. In fact, I heard something about someone in Florida starting to do this." At this point George happened to glance at his watch, and this made him concerned. "I'm sorry, Doug, but I have to go see a physician I'm collaborating with. Thanks for the lunch, and I'll see you around four."

Joseph and Mary finished the analysis by around 3:30, and the three of them immediately started studying the digitalized peaks that the mass spectrometer had produced. Getting the results is always very exciting for scientists at any level. Would the data rebuff or confirm the hypothesis, or at least demonstrate that it is suitable to pave the way for further meaningful experiments? The control PDC clearly showed up at the correct expected location on the spectrum when compared to the database, but they couldn't find an identical peak in any of the tissue samples. Strangely, in Peter's liver, heart, and lung, but not in his blood or gastric juice, there was a distinct peak of an organic compound, very close to the location of PDC, which was not present in any of the control tissues.

"Whatever it is, it is not PDC," commented Doug, and the assistants agreed. "This new peak is certainly flabbergasting, but at the same time, I find it promising, because once we find out what it is, we will be a whole lot smarter."

At the very moment that Doug drew this conclusion, George showed up. "Have you found PDC or not?" he asked.

"No," said Doug, "but there is something else here instead, and you might be able to help us figure out what it is."

George looked at the control PDC peak and then at the new peak from Peter's sample, noting that the two were actually very close to each other in the spectrum. "You know what," he suggested, "it would be worthwhile to check if this other strange peak coming out of Peter's samples has the signature of a zinc-binding chemical group. If so, I would be ready to bet that somebody synthesized a compound that is structurally sufficiently close to PDC to carry zinc into cells. PDC can be altered to have a different compound with practically the same zinc-binding property."

"It's interesting that you say that. My friend Steve also explained to me earlier that the structure of PDC can be modified in a way that would help seek out tumors easier and more selectively."

"Why would someone want to do that?" interjected Joseph.

George was quite familiar with the subject, and he attempted to explain the *why*. "Biotech companies do this kind of trick quite often. Since the effects of PDC and zinc were published earlier, these publications count as prior art, and therefore no one can patent this combination as a new composition. To overcome this hurdle, someone probably tried to make a new molecule, structurally close to PDC but sufficiently different to be acceptable for the patent office. For example, that person could attach yet another chemical group or an antibody to the compound, allowing its specific targeting to a tumor. That would be a significant structural improvement, and the patent office may not have a problem accepting it as a new compound, although it would all depend on the examiner reviewing the application. But if this is the case, then we're clearly looking for a seasoned professional, who probably left his or her trace somewhere," said George.

"That doesn't sound too far-fetched," Doug mused. "Someone who works for a biotech company created this new compound for essentially the same purpose the PDC would have been used for. George, I'm loath to admit but most probably you are correct," said Doug, smiling.

"Yes," George agreed, "my theories generally have a dismal track record, but this time something's telling me that I'm on the right track. Doug, if I were you, tomorrow I would try to identify this new peak to see if it has the molecular signature for binding zinc. If it does, then I suggest you take a close look at biotech companies in and around Rochester to see if somebody used that compound for any purpose recently. Most probably it has nothing to do with this case, but I know a research group at the university that works on this subject and claims that supplementing drinking water with zinc can help people with diabetes."

George's idea to look for a nearby biotech company would almost certainly yield results, because most of them leave some traces behind, like a patent or a publication in a professional journal. After a short silence, George expanded on the idea. "Published patent applications or issued US patents are the best places to start the search. This person was also probably in contact with Peter and was either trying to help him with his diabetes, or intentionally tried to kill him for some reason. If I had to guess, I would say it's more likely that the person who gave Peter this hypothetical PDC substitute might really have just been trying to help him, and he might not have been aware of the potential repercussions. But even if that person wasn't a deliberate murderer, he or she was certainly incompetent. But in court, giving someone a drug out of incompetence but with purpose to help and giving it with the intention to kill are two very different cases."

"We have been working closely with a detective from New York on this case. Would you allow me to present this suggestion to him, as the opinion of an expert?" asked Doug.

"Yes, of course you can, but now I really must go," said George. "I hope I was able to help you a little bit and not just complicate things further. This seems to be an extraordinary case, but I believe that so far you have been playing your cards right. There must be a reason this unnatural compound showed up. Such compounds aren't just left lying around, so it's ridiculous to imagine that Peter took it by accident, without knowing what it was for. But if your detective really got his training in the New York Police Department as you say he did, you and he together will be able to solve this puzzle."

George was finally about to leave when he stopped in his tracks. "Doug, next Monday at three, we're going to have an interesting seminar in our department. I guess you already know that it's located in the new Molecular Biology building. The speaker is from Northwestern University, and he's going to talk about how staying up at night affects our health."

"That topic sounds very interesting," Doug replied. "Thanks, I will be there. Perhaps after that we can have dinner before I head back to Rochester. Until then, take care and thanks again for everything."

George said goodbye once more and at last walked out the door.

Later that Tuesday evening, Doug arranged a conference call with Steve and Mike. He explained to them what he had learned from Carl on Monday, and he couldn't hide his slight disappointment with the detective. "I don't want to be too negative about him and look captious, but the most important people in this case are just flying away at their convenience. It's getting quite exasperating. Couldn't there have been a way to restrict their travels until their roles in this case were clarified, one way or the other?"

"Calm down, Doug," Steve protested. "Unfortunately, the level of suspicion in both cases was probably so low that Carl had no way to get a warrant, and the judge couldn't have issued one even if he wanted to." Steve had seen quite a few murder cases before, and he had a good sense of what could and couldn't be done. "I think Carl probably hasn't even asked for any warrants, because he knows better. He can't just drum up some false evidence. At this point he doesn't have any easy answers in this case, and he obviously wants to avoid any malfeasance that later could derail the investigation. The only option he has is to gradually ratchet up the pressure on them and watch closely to see if at some point one of them makes a blunder. He obviously decided not to steamroll them, however, as other investigators might have. Carl is following a meticulous process. Besides, he might be telling us only a fraction of what he has been doing, and this might not include some very crucial information. Just yesterday

I concluded that he has probably been investigating us and our wives as well, which would actually be a routine practice for a homicide detective."

Steve's conclusion that they were all suspects didn't surprise Doug and Mike; in fact, it sounded very logical. "Undoubtedly," Steve continued, "if Carl wants to do a good job, he should not allow anyone to unduly sway his investigation. This explains his aloofness, which I agree can be irritating at times. But the best we can do is to not expect him to share everything with us while furnishing him with all the information we have. Trust me, Carl is the kind of guy who knows exactly when to pounce on his suspects and swoop in with arrests."

"All right, then let's see what else we can do," said Doug. More important than Carl were the experiment with PDC and his consultation with George, which Doug also informed his friends about. After his story, he asked Steve if he could yet again search the internet to figure out if Ulrich had been involved in any kind of research that involved PDC or a similar compound. "Steve, the other day you mentioned that there was a patent on intratumor injection of PDC and zinc. Was Ulrich involved in this? This might just be the key to solving the case. And, Mike, would you check the list of Minnesota companies to see if Ulrich's name is mentioned in connection with any of them or any PDC-like compound, including PDC itself? The key questions are now the following: Does this new peak have anything to do with PDC? If PDC is used as an anticancer agent, why would Peter have taken it for his prediabetic condition? And can we tie Peter's use of PDC or a similar compound to Ulrich with certainty?"

Both Steve and Mike promised they would do what Doug asked of them, although Mike murmured that perhaps all this should be Carl's job. Before they hung up, Doug brought up the issue of Peter's peculiar wine drinking habit. "George dug up an article showing that Bordeaux wine is very rich in zinc, in fact much richer than any other wines chemists have analyzed so far. The soil in the Bordeaux area contains a lot of zinc, and this shows up in the wine. After all, plants are like humans. They are what they eat, or in the case of plants, what they absorb. I highly suspect that it was Bordeaux's high zinc content that led Peter to drink it exclusively. He must have gotten this information from someone he knew, who is knowledgeable about wines."

Tomorrow, yet another difficult day was ahead of them, so they wrapped up their conversation. But now they were at least following a solid lead, not a wild guess. Maybe even Sarah would change her mind and agree that this investigation was getting more than their collective phantasy.

On Wednesday morning, Doug was in his office by seven. Fortunately, Sarah felt better and could run all the errands she was behind on and deal with the children at home. Her mood also changed for the better; the sullenness she had displayed during the previous few days had disappeared, and she appeared to be back to her old self.

Once again, a substantial amount of work waited for Doug at the office, and he tried to get through as much of it as he could before Carl's visit. These days were anything but dull for him. On the positive side, it now looked as if this time the NIH reviewers might have had a more favorable take on his grant proposal; as a good omen, yesterday an administrator had sent him a message asking him to provide some clarifications on the financials. To know if he had any reason for optimism, he called the NIH officer in charge. The administrator refused to give Doug a yes or no answer when he asked about the likelihood of funding. Even this apparent hesitation in making a final decision was sufficient for him to retain some guarded optimism, but he wished he was better at getting clear answers.

Doug was wallowing in this confused emotional state between dread and hope when Carl knocked on the door. He had brought six wine bottles from Peter's house that he judged to contain enough wine to be worth analyzing. "I don't know if it matters," he told Doug, "but two bottles have a different Bordeaux label than the others. I don't want to sway from your normal procedure, but as you suggested, I took the bottles separately. Because of the differences in the labeling, I think it would be a good idea to sample the wine from each bottle separately. Some difference might show up in their composition that could turn out to be important in our investigation. I already inquired at the liquor store Peter frequented, and they have never carried the Bordeaux appellation that was in the two distinct bottles."

Doug asked Joseph, who was working in the laboratory, to put on gloves and with their electric pipette place wine samples into separate sterile glass containers and clearly label them so they could identify the bottles later. Joseph emptied all the bottles, in the presence of Carl and Doug, and then immediately started working with Mary to fulfill Doug's behest.

"Do you have anything else to tell me?" Doug asked.

"Not really," Carl answered hesitantly, like someone who isn't exactly sure how much he should give away. "Soon I'll have to be on my way to Stewartville. Ulrich lives there, and I have the judge's permit to enter his apartment. I really hope something important will turn up. I'm bringing a computer expert with me who's going to help me access information in his computer. I know that this might look like an invasion of Ulrich's privacy, but we have to do it."

"If you can, could you please look for the words *zinc* and *PDC* in his documents?" Doug asked him. To make sure that Carl wouldn't forget those words, he wrote them down on a piece of paper, which Carl put away carefully.

"I will certainly look for them," he promised with a smile. Carl always liked to get free tips for his investigations.

In the meantime, Joseph had finished measuring the collected wine and told Doug with a big smile that he had been able to extract four to five milliliters from each bottle. "Fortunately, Peter wasted a lot of wine," he remarked. "Now each sample should be sufficient for two analyses. There won't be any doubt about the data. Should we only look for zinc, or should we also look for this PDC-like stuff?"

"You should definitely look for both," responded Doug. "We need information on both. I would hate to miss something. Thus, one sample will be used for the zinc, and the other for the mysterious compound. In fact, both PDC and zinc could be delivered in wine or in a tablet, but so far Carl hasn't found any suspicious tablets, or have you?" He turned to Carl.

"No, not yet, but this is a good point. Maybe we should search the house more thoroughly," the detective answered. "I will certainly make a note to do that."

Then Doug turned once again to Joseph. "During your lunch break, please buy at least six different brands of red wines to see how their zinc contents compare to Bordeaux. I will write a letter that will authorize you to bring them in for research purposes. By the way, how have you guys proceeded with the sample preparations for the next round?"

"Just fine," said Joseph "We can do the actual analysis this afternoon. A chemist has promised to take a hard look at the unknown peak to see if it has anything to do with PDC, but so far he hasn't been able to give me a firm date for delivering the data."

"Can you also do the wine analysis this afternoon?" asked Doug.

"Yes," answered Joseph after a short pause; "but only if we stay a little longer."

"Just one more thing. Have you done something with the carpet sample?" Doug didn't seem to understand that Joseph would have to stay overtime to do all these things.

"No, we still haven't gotten there either," Joseph replied, "but I'll do it tomorrow, unless before then you give me yet another urgent task."

Doug wasn't sure if Joseph was just kidding or if he was genuinely irritated and running out of patience. He chose to ignore his answer. In the meantime, he almost forgot about Carl, who had already packed all the bottles into his sterile plastic bag and was about to leave.

"I wish you good luck, and let's keep in touch," said Carl. "This whole thing is still fishy and getting fishier every day," he murmured while leaving. "My god, what did I do to deserve this infernal case?" He kept brooding over this impasse they still appeared to be at, despite the occasional progress they had made. Although his demeanor and remarks often indicated otherwise, Carl had never yet felt completely hopeless in his work. Occasionally, some unexpected turn of events temporarily discouraged him, but he always regained his composure quickly and refocused on the tasks ahead of him.

CHAPTER 14

Were Helga and Ulrich Meant for Each Other After All?

It was only the middle of the spring of 2011, but Helga thought it was time to prepare for the summer and renew her wardrobe with some practical but attractive clothes. She also had plans to travel to the Appalachian Trail in July, which she had wanted to do since she was a little girl, even though she lived in Germany at that time. She also hoped that just being in the beautiful environment would help her to find directions for her life and leave behind the scruples she still fought. For the trip she needed a new pair of hiking shoes and some other mountain gear. So one hot Saturday afternoon, she made a trip to the Apache Mall, which was her favorite place to shop in Rochester. There, at the entrance to Sears, she literally bumped into Ulrich, who was walking out. This was an awkward situation, which temporarily confused them. They wanted to talk to each other, but it took some time to figure out where to start. They hadn't sought each other's company in a long time, but their chance meeting made all the difference.

Ulrich was still the slender handsome boy Helga had wanted to marry, but with an even more mature and attractive face. He didn't fail to notice that Helga was rather scantily clad, wearing a low-cut top that he couldn't help admiring. Helga's appearance sorely reminded him how much he missed the flagrant sexuality she always radiated. But it was much more than Helga's beauty that Ulrich yearned for;

he terribly missed the close emotional connection and solicitude he and Helga had shared during the earlier phases of their relationship. There was no way around it; he still had strong feelings for Helga, despite the many things that had happened since their separation.

Although the emotional fallout from their separation had resonated in Ulrich for quite a while, now all the squabbles he had had with Helga no longer seemed to matter. Seeing this vivacious young lady again, in a split second he went through a whole gamut of emotions that blotted out all the bad memories of her from his brain. He even forgot how unstable and rancorous Helga had been during the last few months of their union before they finally broke up. Momentarily he also forgot that it was Helga who had not kept her fealty to him and was married to someone else.

Ulrich couldn't have explained why, but deep inside him he had a mysterious feeling that their chance meeting might be a signal for a new beginning. Was their meeting prearranged by a higher authority? Despite being an atheist, in the heat of the moment, he came close to believing that. He gave no consideration to the fact that Helga was married or that a renewed relationship, even if Helga was interested, could be a repeat of what they had gone through before. But after his initial confusion, his brain once again started to dominate his emotions, and despite being happy to see Helga again, he tried to stay cool.

After chatting hesitantly for a while, they decided to go somewhere for a coffee. While sipping their drinks in the Barnes & Noble cafeteria, they began to tell each other what they had been up to lately, as if they had only been separated for a few weeks. Ulrich explained that he was consumed with trying to turn his company in a new direction. He told her about the PDC he had been developing against diabetes and how everything had been looking so promising, until the patent office discovered that someone else had already published similar data about the same compound before him. This development was horrific for the company, which now bordered on bankruptcy. Now the onus was on him to avert complete disaster and reorganize the laboratory in Stewartville, which initially meant layoffs.

"Isn't there any way around this?" asked Helga. She was truly moved by Ulrich's troubles. She had a good sense of what this turn of events might entail. She imagined Ulrich's company as a beleaguered fortress, under heavy siege.

"I thought there was," replied Ulrich. "We made another compound that was slightly different from PDC, and it works even better."

"I don't want to pry, but if it's not a secret, in what way is it better?" asked Helga.

"It is better than PDC to bind and carry zinc to cancer cells and thereby kill them. Also, if it is used alone, it can steal zinc from a certain protein more aggressively than PDC does, thereby inactivating it. This protein is a key regulator of inflammation, and by subduing this process with the new compound, we can also halt the progression of diabetes."

"What happened after you developed it?"

"The patent office responded that the structure of this new molecule was still too close to PDC, and they once again refused to allow it to become a patent," said Ulrich. "It is sometimes difficult to grasp the logic behind the patent office's decisions, and I certainly didn't understand it in this case. Anyway, in the meantime, my doctor discovered that I have early-stage diabetes. You can imagine how this situation made me feel. I work tireless on a treatment for diabetes, and during this process, I become diabetic. I felt like all this was the travesty of a lifetime. I am now only 175 pounds and still exercise every second day. Yet, for some reason, I'm losing the cells that make insulin in the pancreas. This onerous process must be due to so-called low-grade inflammation, which my compound works against. This gave me the idea to begin to test this compound on my diabetes. I made tablets from it and several other ingredients, and now I take one every day. I don't want to come across as someone full of conceit, but in about six weeks, my blood sugar went from something like 40 percent higher than normal, down to practically normal. Now I feel amazingly well."

"That's incredible. If somebody else would tell me all this, I wouldn't believe him," said Helga. "I'm so happy for you."

"There is just one little disadvantage to this drug," continued Ulrich. "It is more soluble in alcohol than in water. Therefore, to make it more effective, I often dissolve it in Bordeaux wine."

"And what's your doctor's opinion about all this? Wouldn't he say that what you're doing carries huge risks?" asked Helga, looking shocked.

"Of course, he told me that. Doctors abhor people like me and want to stifle the patient's opinion. Their worst nightmare is a patient, however knowledgeable he or she might be, suggesting a new, untested treatment. As a doctor myself, I understand fully what a jungle the world would become if everybody would just make their own treatments. If I practiced the medical profession, I would hate to see patients like me. So, obviously, my doctor predictably said that he was completely against using such rogue medication, and in principle, his advice was absolutely the right one. Even if it might take years until the serious side effects show up, in most patients they would eventually surface. In fact, medications often cause unforeseen side effects, even if they have been widely tested and certified prior to clinical use."

"So how could you convince your doctor to let you use this new drug for your treatment?"

"I eventually persuaded him without antagonizing about the worth of trying my drug. Seeing the results, he admitted that my drug seems to work faster and more effectively, at least in the short term, than other noninsulin drugs. To start with, my doctor isn't a confrontational person. He didn't want to strongly argue with me the case. He probably realized that this drug might not be just another scientific shot in the dark, particularly after I explained to him how it works. Also, you would have to see that for him it was difficult to argue with another doctor who is also a scientist. Eventually, we agreed that he would check up on me every week for two months, and after that once every month, to see if I'm developing anything that might be considered an acute side effect, which then might become chronic. He periodically will check the functions of my heart as well as my liver and kidney, and if he finds even a slight deviation from the norm, I will stop taking the drug and switch to an approved medication. Tomorrow I'm scheduled for another checkup, but I examine myself every day too, although obviously my instruments aren't as sophisticated as his. In any case, I hope that we won't find anything to debunk my belief that this drug works without serious side effects. Helga, you also should see that I'm not trying to kid myself. This drug almost certainly has some side effects, just like any other drug, but I just hope they are not serious, and the benefits outweigh the risks."

"I don't want to play devil's advocate with you," said Helga after a long silence, "but why are you going through all this trouble, when there are several thoroughly tested and effective drugs available out there?"

"Sometimes I ask myself the same question," responded Ulrich with a roguish smile. "I think I have a dual nature, like almost everybody else I know. One side of me tells me to play it safe and take an FDA-approved medication. The other side of me, however, tells me to trust myself and take the risk to prove my drug's worth. For now, my inner risk taker seems to be winning. But there's more to it, Helga. You may not know this, but practically all antidiabetic drugs increase either insulin production or insulin sensitivity, but none do a good job in protecting the very cells that produce insulin.

"In contrast, my drug does all three very well. That is why I think it's better than any of the prescription drugs, perhaps except insulin. But my drug, unlike insulin, can be used without the risk of hypoglycemia—that is, without reducing blood sugar to a dangerously low level, which can be even more life-threatening than high blood sugar. So it's not like that I have a kink that I want to be taken seriously, but I truly believe that my drug will be a very useful addition to the arsenal of antidiabetic drugs."

"Now I finally get it," Helga proclaimed. "You don't want to forfeit the crucial advantage of your drug over the others just because of some hypothetical safety issues, is that it?"

"That's exactly right," answered Ulrich. He was happy that he had got his point across, although he wasn't sure if the tone in Helga's voice was the one of understanding or taunting.

Helga was clearly still digesting all this news. "And what's going to happen to your company? You made it sound like it's in its final hours."

"No, not yet," Ulrich protested vehemently. "Even after these few successive snags, we still have some hope, in fact a great deal of it. We have patented this new compound in combination with two other anticancer substances, which together kill tumors rather effectively. We contacted venture capital firms all over the world and have already gotten two positive responses, both from Switzerland. Unlike PDC, this new compound can be targeted directly to the tumors, causing only very small side effects, and this is what drew attention from venture capitalists. This targeted molecule, when applied together with the two cancer drugs, can kill all the cells in even a large tumor, in only a matter of days."

"Interesting, then what are your next steps?" After this new development, Helga was more interested than ever.

"Sometime in early September I'm going to fly to Basel and then Zurich to give presentations to these two venture capital firms. I really hope this will be a turning point for my company, but I need to handle this whole thing very delicately, and I'm still learning how to do that. Once we have earned some serious income from selling this new cancer treatment method, I will tweak the structure of PDC even more to create a molecule effective against diabetes, which will finally be acceptable to the patent office. I'm not going down without a fight. I'm a guy who thrives on crisis."

"Maybe it's time for you to start believing in God and implore Him for help. It seems to me you could use some," quipped Helga. "I'm just kidding, of course. You don't seem to be that desperate."

"No, I'm not at all desperate, not as yet," agreed Ulrich with a smile.

"Then are we still on the same page concerning God?" asked Helga.

"It seems that we are," said Ulrich. "Although, just a little while ago, a strange thing crossed my mind. I had the thought that maybe a higher power arranged this chance meeting between us."

"Are you serious?" Helga frowned.

"I don't really know myself. I'm not sure how to explain it, but whenever I'm in trouble, something unexpected always comes seemingly from nowhere and saves me. It's almost as if someone were watching over me from above. This is partly what takes my worries away. But let's talk about this some other time."

Helga was ready to change the topic as well. Earlier in their conversation, she pricked up her ears when Ulrich specifically mentioned Bordeaux wine. During one of her recent visits to Peter, she noticed a bottle of Bordeaux on the kitchen table, and she was curious to know if there was something special about this brand, so she off-handedly asked Ulrich about it.

"Well, I don't think there is anything necessarily special about it." Ulrich tried to remember if he had heard anything particularly important about this wine that might impress Helga, but he had to acknowledge that he had no specific information on hand. "Like other red wines, it certainly contains antioxidants, selenium, zinc, and other good stuff, but I'm not aware of any specifics. But frankly, I drink it only because I like it, and some Bordeaux brands are relatively affordable."

"I hope you didn't turn into a drunk in the meantime," Helga joked.

"I couldn't afford to even if I wanted to. Besides, I still have a company to run, which is difficult even for someone who never drinks. But we have talked about me far too much. How are things going with you, Helga? Are you still with Peter?"

"Well, sort of…" Helga was clearly unwilling to pursue the subject any further.

"Okay, I can see that things with Peter aren't perfect, but if you don't want to talk about it, that's fine. What did you come here for to buy?"

"Well, I definitely want to buy a pair of good hiking shoes and perhaps some other mountaineering supplies. I'm planning to go hiking on the Appalachian Trail in July. I want to experience real mountains, not just the little bluffs we have here. Also, it's a nice place to meditate, which I haven't done for a long time, and I feel a great need for doing it."

"Is Peter going with you?" asked Ulrich. Without even waiting for an answer, he blurted out, "If not, I would be happy to accompany you. I'm sure you remember that I love mountains too. Besides, I don't see why we can't still be friends," said Ulrich, looking up at Helga entreatingly. It crossed Ulrich's mind that he might have made a gaffe with this rather forward offer, but at the same time, he sensed that Helga might not oppose the idea after all. It was 100 percent Helga who had decided to break off their relationship in search of more safety, but it was all too clear to Ulrich that, for whatever reason, her marriage to Peter wasn't working out either. Looking at the situation from that direction, going on a vacation with Helga wouldn't be the same thing as flirting with a happily married woman. For the time being, Helga chose to ignore Ulrich's offer and instead talked about family and friends in Stuttgart.

The more they talked, the more Ulrich realized that he was indeed still in love with Helga. He supposed he had always known that, but now the difference was that he didn't consider a reunion with Helga as hopeless as he had before. In some ways, Ulrich and Peter were in similar situations, only a few years apart. In both cases, from the man's perspective, the breakup was not based on anything tangible or explainable; the reasons for Helga's decision were somewhat mysterious, although both did make some avoidable mistakes as well. In fact, there was nothing mysterious about it from Helga's point of view. In Ulrich she missed security to compensate for her own feelings of insecurity, and in Peter she missed true love. These two things just didn't come together in either man. The reason for her apparent coldness wasn't that she was emotionally unstable, or that she enjoyed cruel games with men's hearts, nothing like that. It simply took her some time to discover what she missed in Peter and Ulrich. Although with both men there were certainly early signs that foretold what was to come, during the first few years her attraction to them overcame the misgivings that came to the surface only later. She didn't think she had acted unreasonably in either relationship, and she hadn't. She moved out only when she saw no other solution. At the same time, as she gradually became more and more aware of her own values, which came with the feeling of increased security and self-respect, she also began to look at both men differently.

The direct and simple manner with which Ulrich handled their happenstance meeting both surprised and impressed Helga. But the entire time, she purposefully remained rather cool and said nothing consequential, and she didn't even try to hide this. In a way she and Ulrich treated their meeting similarly, cautiously exploring where the other stood, although Ulrich was still in love with her, while she still was figuring out all her emotions.

"You know," Helga broke the short silence, "all this came out of the blue for us. Let's just continue with our business today, and then we'll see about the trip. Here is my card with my phone number on it."

"Oh, you have your own company now?" asked Ulrich. "Your card says 'Helga Schmidt Portraits.' Does that mean that you've become a painter?"

"Yes," said Helga, "or at least that's something I strive for. But believe it or not, I also have a full-time nursing position in the Stewartville hospital. It's strange that we have never met in that small town. Can I have your card too?"

Ulrich pulled one out; it read, "Ulrich Siebert, PhD, Founder and CEO of Diabetic Solutions LLC, Stewartville, MN."

"Then we're close. My studio is few miles from the airport at the edge of the town," said Helga. For a while she hesitated, wondering if she should mention the fact that the studio was in her house where she lived alone, but then she decided that for the time being, it would be too much information. Instead she said, "Maybe I'll visit you in your office once. Where is Diabetic Solutions located?"

"Just two blocks away from the hospital." Ulrich laughed. "This is really extraordinary. And my apartment is about five blocks south of the hospital. Helga, I would be happy to see you anytime, either at my home or in my office."

Then something changed in Ulrich's face. He suddenly became very serious and looked at Helga dead in the eyes. "Listen, Helga. I know that after your arrival to Rochester I didn't do a lot of things as I should have done, and in the retrospect, I understand the anger and frustration you had with me. But if you would just give me another chance, you would again be in the center of my life, just as you were in Germany, and I would work on whatever it is I did wrong. Our chance meeting today might be a sign of something better to come."

"Ulrich, I'm still technically married," Helga reminded him, but she didn't add that her divorce was imminent. Ulrich tried to gently grab her hand, but she withdrew it. She wanted more time, much more time, to think this thing through. She didn't want to make a sudden decision in the heat of the moment and then regret it later; Ulrich was moving too fast for her. "But let's keep in touch. I'll talk to you later, Ulrich," she said finally, quickly shaking Ulrich's hand, and on a second thought, she gave him a peck of kiss on his face. As Helga walked determinedly away, Ulrich watched her retreating figure desirously with langur and wafted a kiss in her direction. He still felt the sensation caused by her touch for a long time.

Helga was moved by Ulrich's eagerness to express his repentance, and she knew that he was being completely earnest. Ulrich had never admitted to making any mistakes in their relationship, and this was a factor in her decision to separate from him. While she headed toward Macy's, in her mind she toyed with the idea that maybe it was a good thing after all that she and Ulrich had parted ways years ago; with regard to their future, this might turn out to be a blessing in disguise. In the intervening time since their separation, both had grown internally, and now they looked at each other differently, in a less confrontational and more mature way. She also liked that Ulrich appeared to be less stubborn and introspective than he had been before.

In retrospect, life with Ulrich hadn't been dull at all. Besides, in truth, she had made a lot of mistakes as well. She never really attempted to capitalize on the richness of what the frontline research that Ulrich was involved in could intellectually provide her with. She never really asked Ulrich how his work was going or even what it was about. Maybe Ulrich's smoldering intellect that she had once worshipped as intellectual superiority was too intimidating? She had to admit that the only thing she really cared about at that time was that Ulrich spent a lot of extra time in the laboratory instead of spending it with her. She suddenly realized that at the time they didn't recognize that they were wasting the enormous gift they have received from life; they could have pondered together, if they so wished, some of the most exciting insights into the workings of the human body. "I wonder," she said to herself, "if we will, or we should, ever get a second chance to do that without contrition. Restarting a relationship while being contrite about the past will lead to failure again. And we should feel solicitous for each other."

During the last few years, Helga had learned that any person with high ambitions can achieve a lot more with a strong, unwavering, loving supporter beside him or her. A splenetic unsupportive partner isn't really a partner but a hindrance, and as she analyzed her life with Ulrich in more depth, she had to acknowledge that she had indeed been more of the latter. As she looked back, she now began to understand that without her standing by his side, Ulrich had only one wing, and with that he couldn't fly high enough. They could have and should have flown together. Perhaps it still wasn't too late? That thought kept reverberating in her brain. "After all, we're still both young, but now we have enough experience to really value each other. What we lost can't be brought back, but we may still have time to start a new life. But what if I'm just dreaming and reality will soon hit me again? Restarting a relationship with somebody you loved before is always risky, but then again, isn't starting a new relationship with someone you have never known even riskier? Oh well, these heavy thoughts don't really belong in a mall." Helga sighed. She smothered her qualms about Ulrich for the time being and began to concentrate on the shopping she had come for.

She found a pair of very sturdy but feminine and comfortable hiking shoes and two beautiful blouses that she simply couldn't resist buying. She added a pair of jeans and a sweatshirt to her cart, and with that she was done with her shopping. When she got home, she was still thinking about Ulrich; she was still under the influence of the pleasant experience she had just encountered at the mall. She kept going over everything she had just learned about Ulrich; he had matured a lot and was even more handsome than before. He also seemed more balanced and in control of himself than before. Even though his business didn't seem to be flourishing yet, the confidence in his voice impressed her. She had never heard him speak like that before.

During the last few months of their relationship, she had considered Ulrich a complete loser. Now that he radiated his newfound confidence, however, Helga began to feel the beginnings of an attraction toward him. She wasn't sure if she could ever get as close to Ulrich as she had been before, but she also didn't see any harm in fantasizing about it. The more she thought about it over the next few days, revamping her life with Ulrich seemed more and more like a real possibility. She realized her heart was inching back to Ulrich, but so slowly that she could stop it any time if she wanted. At that moment, her emotions were still oscillating wildly, and nothing seemed certain anymore; all she knew was that she wasn't quite ready to drop everything that instant and return to Ulrich. All that she was ready for was to keep her heart and mind open and wait, without a rush, to see where this meeting would take her—one more thing to meditate about during her trip to the mountains.

As the days passed swiftly and she began to line up motel accommodations and other arrangements for her hiking tour in July, she found herself wishing that Ulrich would call her. She didn't really know what she would say, but she just wanted to hear Ulrich's soft voice in her ear. The silence began to suffocate her; she felt trapped. Often, she was on the verge of giving in and calling Ulrich to accept his proposal to accompany her on the trip, but for days she was skittish to do that. Suddenly life seemed insipid without Ulrich. At times she wished she could fling her arms around his neck and just stay like that forever. A few more days passed, when one evening the phone finally rang, but at the other end of the line was Joseph. Neither of them had contacted the other for a long time, but they had kept each other's phone numbers. Joseph seemed to be in an exuberant mood. With great excitement in his voice, he informed Helga that the Minnesota Art Institute had selected twelve photographs from his collection for an exhibition with nine other contemporary photo artists.

"That is really something. Congratulations," said Helga, but she barely could hide her disappointment in her voice at the fact that the caller wasn't Ulrich. But she didn't want to discourage Joseph's enthusiasm. "When is the exhibition scheduled?" she asked, pretending she was interested.

"I don't have the exact date yet. But big art centers like to schedule their shows years ahead of time, so mine will probably be sometime in 2013 or 2014. I'll send you an invitation for the opening well ahead of time," said Joseph, still hardly able to control his excitement. "But enough about me. How is the art business going on your end?"

There were two questions that irritated Helga horribly, that she was literally afraid of being asked. The first was, "Do you have any shows lined up?" This question usually came from her artist friends who had just gotten an invitation to display their art in a gallery, and they wanted to share their good news. Often, these artists weren't really interested in whether Helga had a similar opportunity or not, but they asked the question anyway. The second implacable question was, "Have you sold any pictures lately?" It seemed to Helga that people with no true knowledge of art measured an artist's success solely based on the number of pictures they sold. Although for a while Helga did sell quite a few pictures and had nothing to be ashamed of, in the last three months she hadn't sold any. The country was still not out of the recession, and in difficult times like that, luxuries like art are usually the first things people cut back on.

Now Joseph had managed to imply both questions in one sentence. Helga was loath to admit the truth, but after some hesitation and skirting around the issue, she answered in a prickly voice that she didn't have any exhibitions lined up, nor had she sold any paintings recently. Then, as soon as she could, she said good night and hung up the phone. She hoped that she hadn't been openly hostile toward Joseph, but she couldn't help being annoyed at him. She suddenly found herself harboring something close to hatred toward him, for no apparent reason other than she had expected the phone call to be from Ulrich.

After heaving a deep sigh, she finally called Ulrich with the clear purpose of asking if he still wanted to join her on her trip. But for quite a while she was unable to articulate this question; instead she kept asking Ulrich about how he was doing. Ulrich was somewhat surprised but admitted that he had also been contemplating calling Helga; he had never gotten around to it, because he was unable to find a good excuse. He added that after they separated in the mall, he wasn't sure if she ever wanted to see or talk to him again. "I couldn't decide about the meaning of your peck of kiss," he said.

Helga appreciated Ulrich's honesty, but she was faced with the problem of how to handle this emotional event that she didn't yet fully understand. Her mind swirled with questions with no answers. If she got back together with Ulrich, would she feel the same old attraction for his seemingly new personality? Would she ultimately experience the same letdown in the end? Would it be recidivism or progress in her life? And finally, how would her own erratic heart, which she was fully aware of, behave under their rekindled relationship? She wished she could rely on the prescience that sometimes came to her aid, but she felt none. The thought that she and Ulrich might fail once again sent shudders through her body and put a lump in her throat.

After making some meaningless small talk, Helga finally stopped beating around the bush and asked Ulrich if he still meant what he had said about wanting to join her on the trip.

"Of course, I meant it, but what about your husband?" he asked.

At this point Helga felt that Ulrich deserved an honest answer, and she chose to stop skirting the issue. "Well, the truth of the matter is," she began, "that we are in the process of divorcing. For me, keeping this marriage afloat has become hopeless. I admit that the onus is partly on me, and I still feel occasional twinges of guilt. But I just couldn't stay in this marriage any longer. I just can't help it, nor can I fully explain it. There was certainly something missing from this marriage, although I don't know exactly what. I suppose it was passion that I was missing most in Peter. Slowly but surely, it made the marriage unbearable. But there are no hard feelings between us. Far from it. At this point, Peter probably wouldn't even mind staying together. I believe that up until the moment of filing the divorce papers, he hoped that I would turn around. He doesn't seem to understand that I no longer love him."

Ulrich was a little surprised by Helga's admission, even though he had sensed that something wasn't right between her and Peter. "Don't take me wrong, but are you sure your decision is final?" he asked.

"Yes, my decision is irreversible. I knew this when I made it, but I'm afraid Peter still might not get it. In any case, we already live separately, and we expect to receive the divorce decree and adjudication within a week, although a captious administrator might still delay the process. So I'm going on this trip as a divorced woman with no obligations to anyone, free as a bird. It's an interesting feeling. I came to cherish independence so much that I'm not sure if I could ever give it up again. Overall, it can't be said that my marriage to Peter was a good experience, but I've learned a lot about how not to do it next time, if there ever will be a next time."

"I understand you perfectly," said Ulrich. "I have no obligations to anyone either. Maybe this is the recipe for a perfect trip together. But just for the record, Helga, do you remember how I always became taciturn when you mentioned the word *marriage*? I also valued a certain level of independence, although at that time there were also other reasons. I always said that you could truly love someone without being married. At the time you didn't seem to understand the value of independence, but now it sounds like you came around quite a bit."

"I did," Helga acknowledged. "But what's the situation with you now? Do you have anyone serious in your life?"

"I have had a few on-and-off relationships, but right now I have nobody serious… except perhaps you," he added archly. Helga seemed hesitant to respond, which emboldened him to continue. "You were the only one I ever truly loved, I swear. It devastated me when you left. Right before you did, I was in a celebratory mood because my experiments on diabetes had finally started to yield results. When I suddenly realized that I was all alone, I was depressed for weeks. Nothing brought me joy anymore. Nothing gave me motivation to work. But you see, at least I survived. In retrospect, I believe, in fact I'm quite sure, that we made a mistake in separating."

"Well, that remains to be seen," Helga replied. "Right now, I think we should focus on the trip." Helga sensed that the conversation was spiraling out of control, and she wanted to get a grip on it; she wanted to avoid yet another emotional roller coaster, of which in recent years she had had her fair share. She didn't want to talk about the past. She simply wanted to see if they could get a fresh start without looking back to past regrets, without any could-haves, should-haves. She was afraid that talking about the past might poison any chances they had of starting anew; she wanted to talk about the future, but not just then. So she turned the conversation to a more practical matter, the trip, which they had to go over anyway.

During the following weeks they had several similar conversations. Now that Ulrich felt more secure about Helga, he called her almost every day. Ulrich admitted over and over again that he had missed her very much during all those years of separation and was heartbroken when he heard about her marriage to Peter. Although she still didn't want to talk about the past, she patiently tried to explain again and again from different angles, always trying to keep her cool, why she had left him and what she had seen in Peter. "To have security and commitment from your partner is very important for a woman, and back then I saw neither in you. Just being the casual partners that we were wasn't enough for me in the long term. I was simply at the end of my tether when I left you. Now I know that during our many intimate moments you gave me something, which is impossible to describe, that Peter couldn't, however hard he tried. But I want to be fair to him. He really did try very hard to satisfy my needs. He was good to me, but now I know that I was just missing that essential element of a good marriage—potent, unconditional true love. You see, love, commitment, and security all need to be there for a woman to be happy in a stable relationship, and neither you nor Peter could simultaneously provide all these to me. That's why I don't regret ending my relationships with either you or Peter. On the other hand, I don't feel either that I just trifled away my time with you or Peter because I had learned a lot from both relationships."

"And now do you feel a little bit different about me?" asked Ulrich.

"Yes, I admit that now that I'm talking to you again, I feel differently, and I believe this different feeling can become anything. Some evenings I feel that in a way I still love you, but the next morning when my brain is rested and thinking more logically, I convince myself to avoid getting too serious with you again at least for a while. Frankly, I still worry about your commitment and priorities. I care less about security, since I have it now. Besides, as I already told you, I might not be able to bring myself to give up my independence ever again."

"That's perfectly all right," said Ulrich. "I believe that we can remain friends for now and just wait to see what the future holds for us. I believe that I don't have to remind you once again that I'm not the kind of guy who would rush into marriage." Helga could picture the wry smile on Ulrich's face when he said that. But during one of their conversations, Ulrich admitted that he warmed up to the idea of marriage a little. "I came to realize that in a marriage you can still have a lot of independence with a like-minded partner."

On one or two occasions Helga couldn't resist caving in to Ulrich's desire to talk about the past, and she admitted that she perhaps had something like a dual personality. "I came to see myself," she was chagrined to admit, "almost as two different people living symbiotically within each other. One is caring and kind, and the other is extremely individualistic. On the one hand, I strive for stability in my life. You see, with you I yearned for more stability. But when with Peter I finally had it, I found it boring, and I felt I needed to move on once again, not thinking of the consequences. The two sides of me somehow inform each other of their excesses, and in the long term, they keep me in balance. But in the short term, one or the other excess may dominate temporarily, which probably drove both you and Peter crazy. Maybe you were right all along. Maybe marriage just isn't for me after all, although I'm not ready to solemnly announce it."

One evening they also talked about Peter's diabetes problem and the fact that prior to the diagnosis, he had rapidly gained weight.

"I'm not surprised," said Ulrich. "It sounds like he was and perhaps still is very much in love with you, and men in his situation often can't help but eat and sometimes drink too much. I was no exception to this almost axiomatic truth. In the first months after our separation, the same happened to me as well, and I had to work hard to regain my normal weight. I would add that he travels frequently with no opportunity to exercise much, and this doesn't help with weight control either. Even worse, his food options during these trips are probably limited to junk food, and I haven't even mentioned sleep deprivation, which is yet another risk factor for obesity and diabetes. But of all the risk factors that can cause diabetes, chronic stress is the most important. Stress at work, stress at home… we all endure a lot of stress in our lives. I'm sure that stress at work was the major factor in my diabetes."

Mentioning of stress made Helga to think for a while. Then she asked, "Do you think I also could have contributed to your and Peter's diabetes? I probably caused a lot of stress to both of you."

"Oh, let's not go into the subject of who caused stress to whom," suggested Ulrich. "Besides, I think work was really the main source of my stress."

Ulrich wanted to deflect the conversation to another topic; he tried at all costs to avoid thinking of Helga as a person who caused stress to her partners. "Helga, is Peter taking any medication? Because if he is taking some antidiabetic drugs, that could also lead to weight gain. Insulin itself and most other drugs that increase insulin secretion or efficacy could do that. This can become a vicious cycle. You take medication for diabetes, which leads to weight gain, which could make diabetes worse."

"No, to my best knowledge, he doesn't take anything," answered Helga. "He is very much against prescription drugs. In this respect, he is like you. In fact, sometimes he can make very derisive comments about people, particularly younger people, who take medication. In his opinion, it's the person's own fault if he gets sick, and the reason he is angry with himself is because he knows that he now might need some medical help, and for that he blames himself. He's a stubborn person, really difficult to deal with sometimes."

Ulrich had a sudden idea. "Helga, do you think Peter would be willing to experiment with my drug?" he asked. "I've kept taking it since our meeting in the mall, and a week or so ago, my doctor still couldn't find anything wrong with me. My blood sugar has nearly dropped down to normal level, and I have kept my normal body weight."

"It's so nice of you to offer your drug to Peter, but knowing him, he wouldn't take any drug, particularly an unapproved one, unless perhaps it came from a natural source." Helga sounded firm in her summary of Peter's attitude toward drugs, and they dropped the matter for the time being. However, as far as Helga knew, Peter wasn't technically diabetic yet, and he seemed to be able to control his insulin resistance. But, after she put down the phone, she kept thinking about Ulrich's offer, and the more she thought of it, the more she liked the idea. She knew it wouldn't be easy to convince Peter. So she would have to find a way to allay his concerns by depicting Ulrich's pill as an innocuous drug like some of the natural compounds that we eat with plants or in medical supplements. Amazingly, Helga had no second thoughts whatsoever about the safety of Ulrich's drug; she took it for granted. The only uncertainty in her mind was whether she could convince Peter to take it.

Since Helga had made up her mind to divorce Peter, she was much more at peace with herself and with everybody else. So much so, in fact, that she even started to care about Peter. She slowly came to realize that late in their marriage, she saw Peter through a biased lens that gave her an unfairly negative image of him. But recently she had gone through some subtle changes in her approach to Peter; even though she had irreversibly lost her love for him, she respected him more now than ever before. During the months leading up to the divorce, she was completely oblivious to the positive sides of Peter's character, of which there were many. Now that they were separated, she again began to appreciate him; her caring self began to dominate, and her antagonism receded. At the same time, all this change in her attitude was without feeling any remorse about her decision.

Although Helga didn't have a lot of spare time during her nursing duties, every month she very carefully read the scientific journal *Diabetes Care* that was in the doctor's office she was assigned to. This journal is published monthly by the American Diabetes Society, mostly for laymen who are not directly involved in research but are interested to know how to prevent diabetes or how to help their diabetic loved ones. Almost every issue contains low-sugar and low-fat recipes, along with the newest methods for treating and monitoring diabetes. Helga used to copy the most interesting articles and file them under "Diabetes" in her filing cabinet. There were three other diabetes-related journals in the office. One of these journals, simply titled *Diabetes*, was for doctors and scientists, and Helga found the text too complicated to read. But the two others were also geared toward laymen, and she occasionally browsed them as well.

Not long ago she had found an interesting article in one of them; it made a case for the positive health effects of moderate red wine consumption. The author of the article wrote that red wine is full of antioxidants that stave off inflammatory diseases. Helga learned that coronary heart disease, diabetes, and cancer are all considered to be inflammatory diseases. The article went on to say that wine may even help with Alzheimer disease, although no human studies had yet been conducted to prove it. In addition, according to the article, certain substances in red wine, such as resveratrol and zinc, can also improve the diabetic condition, and Bordeaux wine was cited as containing both in fairly large amounts. The article concluded that moderate consumption of wine increases life expectancy by several years, and resveratrol is probably responsible for this specific effect. Helga was so impressed by all this that she made up her mind to pass it on to Peter. So the next time they met to discuss some details of their divorce, she brought up the issue. This wasn't the same sullen Helga he had come to know during the recent months. She not only told Peter about the Bordeaux wine, but she even brought a bottle with her.

Peter wasn't surprised to hear all this about red wine, because his doctor had already told him about it, albeit with stern warnings of alcohol abuse and its consequences. But the doctor hadn't specified Bordeaux, and Peter never passed up on free education, particularly from Helga. After opening the bottle and tasting the wine, he decided that he would rather drink a couple glasses of this wine every evening than take medication. It helped that the price seemed to be reasonable as well. Exercise was also on his list of things to do, but he was lax about it even when he wasn't traveling. The only thing he tried to do with some regularity was jump rope, but even that he did sparsely. He had lost his desire to exercise regularly, even though he had played all kinds of sports at a younger age. Once he discussed this with his doctor: "I know people who played sports when they were young, like I did, and they feel compelled to keep at it in their later years as well. But in my case, somehow exercise doesn't come naturally. Apparently, it's not ingrained in my brain or something like that. I have difficulties to getting myself motivated enough to start up again."

"I don't know the exact reason for these differences," the doctor admitted. "I only know that you are not exceptional at all. Most middle-aged men and women belong to the nonexercising category, often hiding behind the excuse that after work and family programs, no time is left for regular exercise. But the reality is that for most people, exercise just doesn't have the same allure as TV or other forms of entertainment, even though exercise can be combined with watching TV. That's what I do every evening. You just have to find a way to do it jauntily, and once you get in the mood, you can keep it up."

Recently, Peter had also changed his diet. He didn't bother to see a dietician; instead he searched the internet for useful tips, and based on what he found, he completely stopped eating white rice and potato and reduced his usual intake of pasta by half. With that, he hoped to reduce his carbohydrate intake, which was important for keeping his blood sugar in check. In compensation, he began to eat more beans, legumes, and egg whites for protein. He also partially replaced red meat and pork with turkey, chicken, and fish, although he couldn't give up his customary half-size New York steak whenever he had the chance to go to a restaurant. Health experts made him aware of the problems with red meat, and he knew that he should eat more fish, which is rich in healthy omega-3 fatty acids. But he had heard all sorts of rumors about mercury in salmon, most probably blown out of proportion, and he was afraid of eating fish. He also occasionally bought Hormel's famous luncheon meat, Spam, which some people disparage as cheap, low-quality food. Peter was always surprised to hear this because he found Spam quite tasty, particularly the turkey brand. He hated airline food, and sometimes he would pack a turkey Spam sandwich with him to tide him over until he had the chance to eat in a restaurant.

Once, Peter took two cans of Spam, one turkey and one regular, to a weekly poker meeting and asked Doug if he could analyze them for fatty acids. "I read somewhere that turkey has more good fatty acids than beef or pork. I wonder if this is also true for turkey Spam."

"No problem." Doug chuckled. "Next Monday I will have the results for you. We have a wonderful gas chromatograph that will tell us which type of Spam is healthier."

The next Monday, Doug presented Peter with a chart with all kinds of peaks, each representing a fatty acid. He led Peter through the charts in full lecture mode. "As you can see, Peter, the turkey Spam has less unhealthy saturated fatty acids, but it has more of the healthy unsaturated ones. We scientists would say that the turkey Spam has a 'healthier fatty acid profile' than the other kind of Spam, which is made from pork. This must be true for any meat product, so try to stay with turkey if you want to stay healthy," Doug concluded with a smile.

Since that conversation, one of Peter's preferred foods was turkey in any form, including turkey Spam. He was sure that his doctor would praise him for these dietary changes, and during his last visit, he proudly talked about his new approach to food. To his surprise, the doctor hesitantly said, "I hear from some of my friends who do nutrition research that the dietary recommendations concerning saturated fat and egg yolk are profoundly misguided because they are based on poor research. These nutritionists tell me that there is no connection between fat consumption and obesity, or the number of eggs we eat and blood cholesterol. However, those recommendations are so ingrained in the community of scientists and health policy makers that scientists who have the opposing research data have difficulties to get their message through. Because of these differences in dietary recommendations, I'm a little confused what advice should I give to you. Should I tell you the official line, which probably will change soon, avoid saturated fat and egg yolk as much as possible? Or should I tell you that you can eat saturated fat and egg yolk the same way as you would eat fish and poultry? Or perhaps you can eat anything you want as far as you keep your diet and body weight in check? The sometimes widely different opinions of nutritionists and the apparent lack of a common ground just makes my work as a doctor sometimes really difficult."

Peter appreciated the honesty of the doctor, even though he was completely confused by it. He was surprised that Doug was among the scientists who weren't aware of the latest nutritional findings and adhered to the official recommendations. Not knowing whom to believe, he decided on a middle-of-the-road approach, and after his talk with the doctor, he again occasionally included red meat and scrambled egg in his diet without feeling guilty.

When Helga presented him with the Bordeaux, he was quite pleased. "At least moderate wine drinking isn't controversial," he thought. "Thank you very much, Helga. Red wine will be great. I love the idea," he told her warmly. "I'm very weak when it comes to wine." Admittedly, he currently had the habit of drinking half a glass of whisky or vodka before going to bed. Although his doctor explained to him that he shouldn't do that, citing higher risks of all kinds of cancer and less sleeping time, he could not break off the habit. "But now here is the perfect solution," he thought to himself. "I'll drink wine instead, and it will even be good for me. If a doctor and Helga both tell me that it's healthy, and it's even confirmed by science, it certainly must be true."

"Do you think I should stick to the Bordeaux?" he asked Helga. "Or would any kind of red wine do the trick?"

"I have an article at home. I regret not bringing it. But from that article, it's clear to me that you are better off drinking the Bordeaux than any other kinds of wine. This wine has a lot of resveratrol, which should be good for your insulin problem, and it is also rich in zinc, which is good for many other things. There has been an ongoing fungal invasion in the Bordeaux region, and some experts believe that this is where resveratrol comes from."

After settling the wine issue, they spent the rest of the evening sorting out who would get what. They both agreed that they wanted to leave as little up to the court as possible and that they didn't need a lawyer. They kept their discussion very friendly and agreed about the distribution of their wealth in less than three hours. Helga would receive significant cash compensation, while Peter would retain the house with all its belongings, except for Helga's personal items and furniture that she had bought to satisfy her personal taste. A couple of days later, Helga returned to take her belongings. When the last items were placed onto the truck, Helga said to Peter jokingly, "I hope I just sent you into bankruptcy." Peter just smiled, because he thought that he had gotten the better end of the deal.

One Saturday morning, a few days before her planned trip to the Appalachian Trail, Helga received a call from her mother, who still lived in Stuttgart. "I have some bad news," she said gently in German. "Your father has been diagnosed with pancreatic cancer, and he is now in the local cancer center. He had some problems with digestion and complained about stomach pain. He also lost five kilograms in less than three weeks, even though he wasn't overweight to begin with. Two days ago, I took him to the hospital, where they made this diagnosis. Everybody hopes that surgery might help if there is no metastasis, but we're preparing for the worst. We know that very few people survive pancreatic cancer beyond a few years if it isn't caught early enough. And because he already had serious symptoms at the time of the diagnosis, it looks pretty bad." Indeed, the prognosis wasn't promising at all. Helga received the news with great consternation. She could imagine what a mournful place her parents' house had suddenly become. All she felt at first was intense nausea; a little later all she wanted to do was just to scream, and she did.

Helga had a very interesting relationship with her mother. While Monika, her sister, was still at home, things between her and her mother were great. But as Helga entered her teenage years, her sister left the house and she was the only child remaining at home. This was the time when her relationship with her mother began to founder. Her father worked long shifts at the railway, and he had very little time to deal with her. In contrast, her mother had too much free time on her hands. This meant that she gave Helga extra scrutiny, exactly at the time when the teenage Helga should have been left to grow and exercise her independence. There was always something that Helga did wrong. When she tried to study or just simply read a book, her mother made all kinds of accusations, like she was backing out of housework or withdrawing from the family. But when Helga tried to help with cooking or washing the dishes, her mother snubbed her, saying that she wasn't doing it right. At the time, Helga's personality was characterized by quite a bit of diffidence, so she generally didn't oppose her mother's wishes or opinions. But most probably this was the time, characterized by all these indignities, when Helga developed a subconscious but incredibly intense thirst for independence that both Ulrich and Peter failed to correctly diagnose. This longing for independence that she carried to the US then became associated with her longing for some aspects of life in Germany, which recently began to include her mother as well.

As the years went by, she began to see her mother in a different light. Separation from her brought forth the positive aspects of her characters, rather than the negatives. Interestingly, she went through the same thing after she separated from Ulrich and Peter. As soon as she realized that she was safe from the hurtful aspects of these relationships, she became more understanding and even sympathetic toward them. In the process she also came to better understand herself and the people she had separated herself from. She started to see her mother as a stalk of bamboo swaying in the harsh wind but never broken. And now, more than ever, she felt she had to forget her bad memories of her mother and support her through this terrible time. It wasn't an easy task by any measure.

The news about her father shook Helga up tremendously, even though the final verdict hadn't yet been rendered; she was afraid, and with good reason, that it would be rendered soon. She promised to call her mother the next day. By then she would decide if she still wanted to go on her vacation to the Appalachian Trail or instead fly immediately to Stuttgart. It took her only minutes to parse the situation from all sides and decide that she would cancel the hiking trip. After she canceled all her motel reservations, she called Ulrich, who was just about to leave for work. He still had to complete a patent application prior to the Appalachian vacation and his trip to Switzerland, although the company's coffers were nearly empty, and he still had no idea how he would pay for the patent lawyer's fee. "Are you getting excited about the trip?" asked Ulrich. Helga heard him laughing over the phone; the nearing trip had put him in good spirits, despite all the difficulties he faced. "I scraped the barrel for my part of the expenses, and I already started to put away my hiking gear," he said.

"Ulrich, we aren't going on the trip," Helga told him abruptly. "I'm sorry. I just learned this morning that my father has been diagnosed with pancreatic cancer. I really can't go on holiday when my father is in pain and his days are almost certainly numbered."

This announcement hit Ulrich like a cold shower, and for a few moments he didn't even know how to respond. Then he said, with deep compassion in his voice, "Helga, I'm so sorry. I fully understand you, and I'm with you all the way. This trip is the least important thing right now. Would you like me to accompany you to Stuttgart?"

"No, no," replied Helga, "but it's nice of you to offer. Now I need to buy an airplane ticket, the sooner the better." This was her way of saying goodbye to Ulrich.

Ulrich's heart felt like lead as he hung up the phone. "Is this going to be the second time that I lose her?" he wondered. "Or maybe this time, Helga will need me more than ever." He felt terrible for Helga's father, but he coveted this latter scenario more than anything. He wished he could once and for all get through to Helga and convince her to come back to him. At the same time, he didn't want to seem needy for Helga's love either, which could ruin any chance they had at a new beginning. He wanted to be subtle just like Helga was, although it was difficult sometimes.

After such conversations, Ulrich often asked himself what it was he truly wanted from Helga. He had no clear answer. His brain cautioned him: "It didn't work the first time. Where is the proof that it will if you try a second time?" His determination almost oozed away at this thought, but then his heart countered, "If you really love her, go for her. If this time around you do things the right way, you two may indeed live happily ever after yet." In the end his heart won, but the circumstances wouldn't allow his love to bear any fruits just yet. But he wasn't about to let this second chance slip from his grasp.

Because of the relationship between diabetes and pancreatic cancer, Ulrich was quite familiar with the situation that Helga's father was doomed to face; he was in big trouble, whatever the German doctors might say. This was the kind of disease that rapidly converts even the strongest, bravest man into a weak trembling skeleton. For surviving cancer patients, their diagnosis is always a watershed moment in their lives. They tend to divide their lives into two periods, before and after their diagnosis.

Ulrich didn't sever all his ties to the Mayo Clinic. He managed to retain collaboration with a research group there with the purpose of using his compound to improve the efficacy of the best available antipancreatic cancer drug at the time, gemcitabine, to stop the growth of experimental tumors developed in mice. His compound worked, but even the two drugs together couldn't completely stop tumor growth or prevent its regrowth. The best that Helga's father could hope for was that the tumor had not metastasized yet, and in that case, a surgical intervention could be beneficial, lasting at least for a few years. But if he already had serious symptoms, as he seemed to, it could be assumed with great probability that it was too late. The absence of symptoms during its early stages, when the patient could still be helped, is what makes pancreatic cancer so exceptionally insidious and dangerous. Ulrich didn't want to share this terrible truth with Helga, although, being a nurse, she likely knew this already.

Helga still had twelve days of vacation, but she decided to use only five days for her trip to Stuttgart, in part because she wanted to save at least three days for a Washington, DC, trip to visit the Smithsonian sometime after her return. She wanted to keep the rest of her vacation days in case she soon had to return to Stuttgart, which seemed very likely. Her travel agent found a relatively cheap ticket for a Sunday flight through Amsterdam, which left some time to investigate how good the Mayo Clinic was at treating pancreatic cancer. Helga was thinking that maybe it would be better to bring her father to the Mayo Clinic, unless it was already too late. She had no idea how frail he might have become. There wasn't much time remaining to ruminate about this possibility; most likely her father's tumor grew relentlessly.

The next morning Helga woke up early, and the first thing she did, after drinking a coffee, was to call her mother. She told Helga that the doctors were still assessing the situation to see if the tumor was operable or if they should begin chemotherapy right away, to attempt at least to prolong his life. Helga asked her mother to fax the medical certificate with the diagnosis, which she would then discuss with a specialist at the Mayo Clinic. Her mother promised to send a copy of it right away.

Still feeling very depressed, by the early afternoon she finally gathered enough strength and visited unbidden Ulrich in his office. He was still in the middle of his perennial struggle to bring his company to success; just then he appeared to be organizing a slew of papers needed as references for his new patent application. It was a tedious job, but Helga's visit electrified him. The birthday of Ulrich's mother was coming up, which was one of the few personal things in his life that he kept track of. He had bought several small gifts for her that he asked Helga to take with her to Germany.

They also began to talk about pancreatic cancer and what could be done at the Mayo. "I suggest that you talk to Dr. Dean Andersen," Ulrich advised her. "You must know him very well because when you worked in Doug's laboratory, you used to tell me how he overwhelmed you with requests for blood and other analyses. Now, it's time that he returns the favor. To my best knowledge, pancreatic cancer isn't his specialty, but he knows virtually all the doctors at the Mayo, and he can give you access to them. Since Mayo doctors are extremely busy, it is very unlikely that through any other means you would be able talk to any of them, unless of course you have a medical appointment."

"Thank you so much for your advice. I'll talk to Dean about this tomorrow," Helga told Ulrich. "Now give me the gifts you bought before we forget the reason I came for. I'm sorry, I still have so much to do. Besides, I see you're working, and I don't want to pester you for too long."

As Ulrich handed over the gifts to Helga, he remarked, "Before you talk to Dean, I think I should warn you that doctors generally don't like to poke their noses into the matters of other doctors, unless they received a request to do so."

"Meaning that Mayo specialists may be unwilling to take over the case?"

"Yes, exactly that's what I meant. But don't get dispirited. Go ahead and try."

Before Helga left, she absentmindedly scanned Ulrich's desk and noticed two glass containers full of pills. "What are those for?" she asked with a cursory glance.

"Oh, you mean the pills in the glass containers? Well, the tablets in both containers are made from a compound very similar to PDC and some minor ingredients. I just call it PDCM, short for 'modified PDC.' If you recall, at the mall we talked about science and I explained that this drug is good against diabetes. In the container to the right, the tablets also contain zinc. If you take only one tablet, then it doesn't matter which one you take. But if you want to take more than one, then you must take the ones on the left."

"Why is that?" asked Helga.

"Well, the amount of PDCM and zinc in one tablet is just right for the treatment of diabetes. But if you triple both ingredients by taking three pills, it causes serious stomach aches. I tried it, so I can testify to that."

"Why these pills cause stomach aches?"

"PDCM and zinc together start killing cells in the intestines, and this causes pain as a first warning that something went wrong, and no more medication should be taken for a while. Eventually, the material in the three pills would be diluted and degraded before seriously damaging other tissues. However, if someone took, let's say, six pills, such an amount would lead to an overdose, and almost certainly death. But in that regard, this pill is not much different from many FDA-approved pills, because by taking six of almost any of them in a short time could also kill you or at least make you sick for a while."

Helga was mildly interested, but she really had to go. They stood hesitating for a moment, and then they briefly kissed each other on the cheek before Helga set out to leave. Although Ulrich moved toward Helga, she didn't resist him. None of them thought that something unusual had just happened, although Helga clearly wouldn't have allowed him to go any further. This new calmness and understanding between them was so different from the period of quarrels just a few years ago. Helga again had Ulrich under her spell, and she knew it. In fact, when she wasn't thinking about her father; she enjoyed it. It was like she had put an all-powerful salve on an old wound.

Their kiss also filled Ulrich with hope; they could start making up for the inanity of the past few years.

Two days before her trip to Stuttgart, Dean helped Helga arrange a meeting with a highly respected expert on pancreatic cancer, Dr. Fogelman. After reviewing the medical report signed by a doctor at the Stuttgart hospital, he saw no reason to transfer Helga's father to the Mayo Clinic. "I know the doctor who signed this report personally," he said. "I met him last February at the clinical oncology meeting in Paris. He is a very conscientious doctor, and I know for a fact that he can do whatever we could do here for your father. In his condition, such a long trip would be too taxing on him."

"Is it that bad?" asked Helga, alarmed.

The doctor tried to backtrack a little, realizing that he had probably said too much. "I don't know, and at this point I think that any specialist would be reluctant to state how bad his condition is or what to expect. Everything depends on the ongoing tests that may or may not show metastasis to the bone or elsewhere. In the latter case, there is a chance that the tumor hasn't yet spread beyond the pancreas, and a surgery might be curative for several or perhaps many years to come. The reason I worry," he added, "is that in virtually all cases where there is pain and such weight loss, the tumor has already advanced and spread. But it is very hard to say for sure without running more tests. What I can assure you is that the Stuttgart hospital can do everything we can do here, and they don't need our help. Now, if you will excuse me, I have a consultation in two minutes to prepare for my next surgery."

"Thank you for your time and advice, Dr. Fogelman," said Helga. Then she thought, *Ulrich was right when he warned me about doctors' nonintervention policy.*

She had wanted to do something good for her father, but she only wasted her precious time. It was only 9:00 a.m., but she already felt so dizzy that all she wanted to do was go home and take a long nap. Instead, she had to go back to the Stewartville hospital to complete her work schedule for the day and then start preparing for the trip. The next two days would be very tiring, but if everything went well, she would be in her parents' house the day after next.

CHAPTER 15

Events Leading to Peter's Death

In Stuttgart, Helga found her father very ill. Unfortunately, as was expected, his cancer had already metastasized to the liver and bone marrow, and he was still losing weight at an alarming rate. The doctors had decided that surgery wasn't an option anymore, and they apparently prepared for the worst. Upon Helga's inquiry over the phone of why surgery would not be attempted, the surgeon first tried to skirt the issue, not really knowing how to respond. Finally, he gave in and explained that there was some communication between the original major tumor and the smaller metastases derived from it. By using some unknown mechanism, the large tumor smothered competition by the smaller secondary metastatic tumors and prevented their rapid growth. Thus, in many cases, when the large tumor is surgically removed, this accelerates the growth of the smaller tumors, which could lead to the patient's earlier death. In addition, the tumor's anatomic location within the pancreas was very unfavorable; the doctors involved judged this location as inoperable. This meant there was a high probability that even the operation itself might kill Helga's father. After all these considerations, the doctors determined that the best course of action was to stabilize or at least slow the growth of the primary tumor via the best chemotherapy available, which would also slow the growth of the secondary tumors.

However, even this strategy might not save her father, as the doctor reluctantly acknowledged.

Some doctors in Europe aren't too willing to detail the diagnosis to close relatives or even to the patients. But the doctor Helga talked to knew that she was a nurse, and this made him more communicative. He admitted that her father might have only a few weeks or months, not years, to live. "A big problem is that because of his serious weight loss, your father's body has been weakened to the point that he is less able to tolerate chemotherapy. We don't really know what dosage he can tolerate. Thus, we need to be cautious with the dosage. We would rather give him less chemotherapy than he needs, than kill him with too much. Because of all this, in his condition chemotherapy is far less effective and more dangerous."

Helga went to the hospital and witnessed in person the heroic effort his father was making to cling to life. He still hadn't given up the fight, which clearly availed him. His determination to live and the strong painkillers helped him stay guardedly optimistic. In her practice, Helga dealt mostly with patients of heart disease, so she had never seen a pancreatic cancer patient before, even though the Mayo Clinic saw more than two hundred of them every year. But she had heard stories that after a while even the strongest painkillers become ineffective, and the pain can become so excruciatingly strong that at some point almost all patients just give up and let themselves die. When she asked the doctor assigned to his father how long he might live, he confirmed his earlier estimate. "It seems he can live anywhere from one to five months, depending on how effective the chemotherapy turns out to be. We started it just one week ago, with no tangible results so far. But it also depends on whether other complications arise or not," the doctor added.

Helga was devastated. This also meant that she would soon have to return to Stuttgart for her father's funeral, and from that point on, this was in the center of her thought.

During her short stay in Stuttgart under the pressure of the evolving tragedy, she matured quite a lot, and she began to see many things in a different light. There was no escape from reevaluating her attitude about personal relationships, and this really transformed her personality. She was tired of it all, and for the foreseeable future, she wanted to keep only a friendship, and nothing more, with Ulrich. After her return, she would let him know this change in her priorities. "I owe it to Ulrich to be honest about this new reality," she thought.

While she was scaling back the strength of her feelings for Ulrich, she simultaneously felt herself grow a little closer to Peter. Not yet as a lover but as a friend. In a few days, she went from conceit and self-pity to self-reproach when she thought about the way she had treated Peter over the last few months. During those months she had shielded herself with the excuse that she couldn't help it, but now she knew that in a way she could have. Even though she couldn't have forced herself to love Peter again or countermand her desire to divorce him, she still could have shown him the respect, care, and support that he deserved. But while being freed from the "rigid" ties of the marriage brought her closer to Peter, he in the meantime had outgrown his relationship with Helga; he didn't really need Helga anymore for support. But Helga was unaware of this turnabout, that she became superfluous for Peter; she still only saw the suffering side of Peter.

During her flight back from Stuttgart, she thought about how to help him, and she decided to talk to Ulrich to see if he would still be willing to meet with Peter and explain the benefits of his drug. Last time she had turned Ulrich's offer down, but now she thought that since both he and Peter were struggling with diabetes, perhaps Peter could relate to Ulrich, which could make easier to convince him to start taking Ulrich's pill for a limited time, just to see if it made a difference. She decided that after a good night's sleep she would call Ulrich right away and accept his previous offer. Whether Peter would accept was a different story, however. After all, there are so many demonstrably effective and relatively safe diabetes medications available, all tested on millions of people; why should Peter take Ulrich's?

Helga was prepared to do some subtle persuasion of her own. It was quite strange and inexplicable that Helga now trusted Ulrich so much that she didn't think for a second that what she was about to offer to Peter, an unproven and unregulated drug, might be dangerous. On the contrary, Ulrich's drug seemed to her to be safer and more efficient than anything else available for diabetes. Perhaps because of the trauma of her father's illness, her judgment was somewhat diminished, but in her view, if the drug worked for Ulrich, why wouldn't it work for Peter as well? And if the drug helped Peter with his diabetes, this would also reduce his risk of pancreatic cancer. Maybe this was the greatest gift she could give to Peter.

Helga hoped to sleep at least eight hours, but on Thursday morning, she was already up at 4:00 a.m. In Stuttgart the time was already 11:00 a.m., and it seemed that Helga had spent enough time in Stuttgart to readjust to the German time zone she had been born in. She had previously noticed that after arriving to Stuttgart, it took her only two days to adjust to the German time, while after arriving back to the US, it took her at least a full week to gradually readjust. Years ago, she had mentioned this paradox to Ulrich, to whom it wasn't a paradox at all. "Our body remembers forever which time zone we grew up in," he explained to Helga. "But fortunately, physiologists haven't found any unfavorable long-term effects of relocation to a place even many time zones away from the place of one's childhood, as far as the relocated person gets plenty of sleep. So, Helga, you shouldn't worry about relocating to the US."

After drinking some coffee, she checked her mail. One letter immediately stood out to her; the divorce decree, which finally made her marriage to Peter a thing of the past. The ruling stipulated that in full agreement of both parties, Helga would be given $200,000 in cash compensation within ninety days and that she would also receive all personal items back, as they agreed. In a way, this was good news for Helga, because she got what she considered a fair settlement. But another side of her felt strange; now that she had achieved this ruling, she realized that she wasn't entirely ready for it. For almost a year she had wanted to become fully independent of Peter, and now that she finally was, she was uncertain of what exactly to do next.

Despite her confused feelings, she decided to stick to her original plan of bringing Ulrich and Peter together for a face-to-face meeting. Helga knew that Ulrich usually woke up at around 5:00 a.m., so she called him soon after that. He informed her that he had finalized his travel plans and would fly to Switzerland the following Wednesday. Because they had canceled their vacation, he had brought his trip forward and now he had less time than he had expected to prepare and clear his desk. But he promised that he could spare one hour to meet Peter that evening and explain what exactly insulin resistance is and why his medication could help Peter more than other medications. He also added that in all honesty, he was somewhat uncomfortable with the idea of meeting Peter, but he would not back out and would do it for her. Still, he added, he didn't want to engage in "contortionist exercises" either. Helga had no idea what he meant by that, but she let it go. It was more important to her that Ulrich agreed to meet her and Peter in the Starbucks just across from the Mayo's main building at eight.

They all arrived separately. Peter got there first, and then came Helga about two minutes later. Helga had just enough time to say hello to Peter when Ulrich arrived too. The meeting among the three started with understandable coolness but with no open hostility between the two men. Helga wasted no time trying to warm up the icy atmosphere by almost immediately getting to the point that Ulrich had come up with a kind of drug for diabetes that Peter might want to learn about and perhaps even try.

"Well, if I understand correctly, you want to help me, and that I really appreciate very much," said Peter. "Helga explained to me that you have a pill that is not FDA approved, but you've been taking it for quite some time and it works well in keeping your blood glucose in check. But you see, I don't know much about medications, although I have heard here and there that often they do more harm than good. Because of that I'm not overly inclined to trust any of them. I want to be frank with you from the outset, and I'm sorry to say it, but you have to understand where my reluctance comes from." Peter searched in his memory for some evidence against drugs. "I'll give you an example. My doctor told me that I should start the antidiabetic drug Avandia right away. I refused to comply, and you know what? Right afterward, I read in a newspaper article that some scientists concluded from clinical trial data that Avandia may cause heart problems, even deaths. After that news became public knowledge, which I still don't know if true or false, how could I, a layman, ever trust Avandia? Even experts apparently can't tell what's safe, so how can I? My question for you then, Ulrich, is why should I trust your medication? What specialty does it have that makes it different from the other drugs? Can you explain to me in simple terms how it works? You see, I have been trained in many things but not in the medical sciences."

Peter said all this in a sullen voice, making no secret of how much he loathed drugs. Many people are obsessed with medicines, literally hoarding them, but Peter was just the opposite. He only liked vitamin tablets, which he didn't think of as drugs.

For a moment Ulrich was taken aback by Peter's pungent critique of drugs in general, even though he had similar sentiment toward them. Helga also began to sense that her exquisitely wrought plan might be falling apart. But then Ulrich seemed to regain his composure.

"I fully understand your concerns," Ulrich replied with full confidence in his voice. "I try to explain the context for my drug's action, and hopefully I can do it in layman's language. Helga mentioned to me that you have an insulin-resistance problem, which is probably due to stress and the extra weight that you carry. When you eat too much fatty food, certain types of fats, like saturated fatty acids, they begin to work very insidiously in almost all your organs. By the way, when you ingest more sugar than you need, this also produces extra fat. Now, in your fat tissues, these fatty acids start to kill fat cells, which triggers inflammation, which in turn causes all kinds of problems leading to obesity and eventually insulin resistance and diabetes. Stress further contributes to this process. This is a vicious cycle because obesity, itself a stress to the body, results in more inflammation. All this leads to the pancreas producing insufficient amount of insulin. As if this wouldn't be bad enough, insulin works less and less efficiently because of the fat accumulating in the muscle, and liver makes these tissues resistant to the insulin actions. Also, my own research and several other scientists' research right here at Mayo suggest that because of the inflammation, there is a greater risk for pancreatic cancer. Since inflammation is triggered by fatty acids and stress is the key to obesity and diabetes, many scientists are working to discover drugs that can control it without causing stomach ulcers or other serious side effects."

"So how does your drug relate to all of this?" Peter asked.

"My pill blocks a key regulator of inflammation, which is the harbinger of diabetes and other medical problems. Thus, I believe that if fatty food, stress, and a lack of exercise in combination caused your diabetes or prediabetes, my pill should help you more than anything else. Helga told me that you like New York steak and other fatty foods, and because of your frequent trips, you don't have the chance to exercise enough either."

"I think I'm beginning to understand what you're saying," said Peter, "although my doctor once mentioned that the latest research indicates that eating fatty food is not all that bad. But putting the fatty food issue aside, I still don't get something. Why is this drug better than all the best prescription drugs out there?"

"Because it can also do something that no other drugs presently on the market can. It protects your insulin-producing cells in the pancreas from dying. You see, these cells are bombarded with killer substances like sugar and fat, and therefore every day fewer of them remain alive. Eventually most of these cells will die, and even before that point, you will have to start taking insulin to compensate for their loss. You can avoid that scenario by taking my drug. At the same time, you'll reduce your risk for pancreatic cancer, which is what Helga's father is going through. In fact, I have the exact same problem that you have. That's why I'm also taking my drug."

After a short silence that allowed all this to sink into Peter's brain, Ulrich continued. "Helga also told me that you like to eat natural food and wine to help with your insulin sensitivity problem. This drug is chemically close to a compound present in garlic that reduces blood glucose in diabetics. My drug has the same effect, except it is stronger, and you don't have to be afraid of talking to people from a close distance."

"But I can brush my teeth after eating garlic," Peter retorted. He was still ransacking his brain to find reasons to oppose Ulrich's drug.

"No," Ulrich replied, "for the same effect, you would have to eat so much garlic that no amount of toothpaste would help you."

"I have to acknowledge that with the stress and lack of exercise, you are right on target," admitted Peter. "And indeed, who would want to eat garlic by the bushel if you can replace it with a harmless, odorless pill? You know, I'm very thankful to you and Helga, but despite everything you said, I'll need some time to make up my mind about this. How about if I call you tomorrow? This decision is a big deal for me, because if I choose to accept it, then it would really go against my principles that I have stuck to so far. But if I say yes, when could you give me the pills?"

"Well, tomorrow is Friday. If you say yes, I can bring thirty tablets here tomorrow around 5:00 p.m., and you don't have to pay anything. All you will have to do is sign a short note saying that you consent to take these pills without outside pressure and under your own responsibility and that you understand that you cannot take more than one pill a day, preferably in the morning, during or right after a meal. I'm compelled that you sign such consent, I hope you understand."

"Then it's settled," said Peter. "Actually, you know what? You mentioned that many diabetics may have to start using insulin at some point, which I am afraid of more than anything else. I just realized I actually made up my mind while you were talking. Please bring those tablets tomorrow, and I'll also sign the consent."

"Good choice. I'll certainly do that, but if you'll forgive me, I need to leave now. See you tomorrow, and, Helga, I will call you tonight. Good night, Peter."

After Ulrich left, Peter and Helga had only a brief conversation about how things were in Stuttgart and what her father was going through with his pancreatic cancer. Then Peter apologized, saying he still had to prepare for a meeting the next day, but Helga stopped him.

"When Ulrich gives you the tablets, will you start treating yourself?"

"I certainly will for a week, and then I'll see how I feel. In any case, I'm really grateful to you for trying to help me." As he left, he thought to himself, *How often does it happen that someone first wants to destroy you and then tries to save you?* There was a clear difference in Helga's behavior before and after their divorce. The situation Peter was in was a little awkward because a while ago he had stopped caring about Helga's feelings toward him, but he was too polite to say so. Besides, this new drug appears to be natural. If Ulrich was correct, then this drug might prevent him from becoming dependent on insulin and, God forbid, even pancreatic cancer. If he would have asked Doug or Steven for a second opinion, they would have told him to be extremely careful, because even small modifications of a molecule can result in a completely different array of effects. A small modification of Ulrich's molecule compared to the one present in garlic might mean the difference between life and death.

As convincing as Ulrich sounded, and as persuasive as Helga was, all this wouldn't have been enough to convince Peter to take Ulrich's pill if it hadn't been for symptoms he had finally started to notice. They were getting quite alarming, and Peter probably would have eventually accepted drug treatment anyway, if they continued to escalate. Very recently, one morning when he got up from his bed, he briefly fainted, waking up minutes later sprawled on the floor. Then, as he was driving to work, he briefly turned around to see the traffic behind him, but when he turned back forward, he again fainted for perhaps a second or two, although he was able to regain full control over the wheel and avoid an accident. But he was sweating and shaking with chills, and at the earliest opportunity, he parked his car to give himself time to recover from the shock. These events began to convince him that something was out of whack in his body, and his refusal to accept the doctor's advice and start taking a medication might be tantamount to a death wish.

So Peter asked his doctor for a checkup, albeit hesitantly, but the earliest appointment he could give him was three weeks away. He decided to call a small private clinic, where they checked his blood pressure and blood sugar on the very same day. He had no serious problems with the blood pressure except for a slight elevation, but his blood sugar was way up. The doctor asked him not to eat any food or drink anything sugary after dinner and early the next day come back to the clinic for another blood test. This second test also showed that his blood glucose was almost 80 percent over the normal level, which clearly called for treatment.

"What you have is more than insulin resistance. It's diabetes. Your blood sugar is apparently permanently high, and you need treatment right away," the doctor told him. After seeing Peter's reluctance to start treatment, the doctor put him on a strict diet coupled with an exercise regime and agreed to wait for two weeks, but not any longer, before starting the treatment. If at that time a new test would once again show high blood glucose, then he would recommend for Peter to go back to his original doctor in the Mayo system who would probably advise him to start treatment with the antidiabetic drug metformin.

This grim diagnosis alarmed Peter. The doctor made a good case for metformin, explaining at length how safe of a drug it was, but Peter suspected that diabetes treatment might be knottier than the doctor would have him believe. So the first thing he did at home was to search the internet for metformin's side effects. Well, he found a fair amount of them, including nausea, bloating, diarrhea, and loss of appetite. Even though none seemed to be too serious, he abhorred the thought of any of these side effects, which seemed to occur often unexpectedly and with a surprisingly large frequency—one out of three patients. He thought of scenarios in his head, like having to somehow overcome the urge to visit the restroom while in a traffic jam or in the middle of a dinner with a pleasant companion. Despite these inconveniences, he might have settled for metformin, had he not also read that it can cause lactic acidosis, which, although it only occurs in one of every thirty thousand patients, is fatal in 50 percent of cases. What really caught Peter's attention was that dehydration is a serious risk factor for lactic acidosis. He immediately made the connection that during his frequent long-distance flights, he could easily become dehydrated. So he reasoned, "Metformin could kill me if I take it during transatlantic flights." This prospect made him quite jittery for days. His other problem with metformin was that once he took it, he couldn't drink alcohol. He made up his mind that at his next doctor's visit, he would ask for another medication. "If I need to take medication, I would still would like to drink a little wine, or at the very least I shouldn't die from it. There has to be something else available that is less dangerous," he reasoned. If he would have continued his internet search, he would have found that metformin was one of the safest antidiabetic drugs available. In fact, it was found that it might also enhance the effects of certain cancer treatments, and it might even be good for people with serious heart conditions.

As soon as Peter stepped into his house at around 9:30 p.m., his secretary and lover, Agnes, called and innocuously asked if she should come over to finalize the program for tomorrow. Agnes never expected and never received the answer no, but she always called first when she wanted to visit Peter. They had a lot of things to talk about in preparation for his meeting with his boss the next day, which was a Friday, and for a business trip to Los Angeles next Wednesday. Tomorrow he also had to go to the airport at 11:00 a.m. to pick up two visitors from Chicago, have lunch and a meeting with them, and then take them back to the airport around 4:00 p.m. Then he would just have enough time to pick up the pills from Ulrich at 5:00 and do a little shopping at SAM's Club, his favorite place for groceries and household items. Then he hoped to spend some time with Agnes because for Saturday and Sunday he planned to be in Madison to see his children, the prospect of which made him quite happy. So far, he saw nothing on the horizon that could contravene his plans.

Peter had hired Agnes several months ago and slowly became attracted to her, although not so strongly that he would have lost sleep if the relationship didn't work out. It was certainly not something that people call "love at first sight." Agnes felt the same way; she wasn't overly entranced by Peter, but she had always held him in high esteem. From the very moment when she first entered Peter's office, she had an unswerving commitment to him. She knew that Peter was married, and it took a while until she figured out that he was in fact free to do whatever he pleased. When she learned this, it changed the context and the way she looked at him, and he noticed this change; from that point on, there seemed to be a simmering chemistry between them. Soon an opportunity presented itself for them to test if they really wanted to be in a relationship or not. They went to a small party at a downtown restaurant with their IBM colleagues and decided to walk home together because they both lived in the same direction, not too far from each other. That evening Agnes exuded pure sensuality, aided by her evening dress, which left Peter almost powerless.

At that time, Agnes had been renting an apartment, and Peter casually noticed that his next-door neighbor to the right was selling his small but charming house. "His wife already left him, and my bet is that the house will become dilapidated in a couple of years if somebody doesn't take care of it. But now, it's still in good shape." Throwing caution to the winds, he tried to convince Agnes that in few years she would be better off financially if she bought the house, particularly now that housing prices were still dropping. But Agnes was well versed in economics, and she wasn't at all sure that it was a good idea for a single person to own a house. "Who knows what the housing market will be like by the time I might want to resell it?" she told herself. Besides, she suspected that Peter probably mentioned this house idea on some pretext. Nevertheless, she agreed to see the house right away. As they drew closer, the house and its surrounding left a good impression on Agnes. She felt that she was in a forest, yet at the same time in a civilized environment; there was such harmony between the large trees and the manmade structures. *This might not be a bad idea after all,* she thought. She was happy that Peter had given her honest advice and felt a little ashamed that she had ever considered that he might have motives other than trying to help her.

Peter then invited Agnes over to his house, and this time he found himself even more attracted to her, although nothing physical happened between them that evening. Perhaps they both wanted something to happen, but both decided it would be better to think it over carefully. The fact that Peter was close to divorce certainly helped to push them toward each other, but there was still some distance to cover. The next day, Agnes visited the house and decided to buy it. She felt happy that she would be close to Peter at home and work alike. Within two weeks, Peter and Agnes were neighbors.

After moving in and buying the most important pieces of furniture, Agnes held a little neighborhood party that Peter helped organize. It was during this party that Agnes and Peter truly got to know each other and began to recognize, perhaps for the first time, each other's values. Agnes liked the way Peter entertained her and the guests with stories and sometimes gut-wrenching jokes; his personality was endearing to both her and the people from IBM. Peter hadn't done this kind of thing for a long time, and he enjoyed Agnes's party tremendously, always trying to remain close to her. That evening they subconsciously treated each other almost like newlyweds. Peter was the last guest to stay and offered to help with the dishes and cleaning up. Agnes accepted his offer, and at one point when they were cleaning next to each other, Peter slowly drew close to Agnes and kissed her. With that, all pretense went out of the window. That night Peter stayed with Agnes, and from then on, whenever Peter's travel schedule allowed it, each week they spent a couple of nights together.

During the following weeks, the relationship between Agnes and Peter had evolved in an interesting way. With time, Agnes got closer and closer to Peter, although deep inside she was always wondering if that was the true, all-encompassing love that she had been waiting for; on occasions she had doubts about her feelings. Yet it felt good to be with Peter, and she was ready to build with him a strong relationship, just short of marriage for the time being. As for Peter, at the house party he was simply crazy about Agnes, and this lasted for weeks. Then came a period when the fire of his passion cooled off quite a bit, although he never even considered breaking up with her.

Among the many things Peter liked about Agnes, one stood out; she very rarely used profane words. Peter was known among his friends to avoid using profanities whenever possible, because he simply found them repulsive. Agnes also noticed this similarity between them and considered it a rare treasure.

But that Thursday night, their meeting didn't play out as usual; it was one of those rare occasions when both were too tired to stay together. So after they discussed the programs ahead of them, Agnes left. Later, she often revisited that evening in her mind from different angles, but she always drew the same conclusion: "That evening was arranged to happen that way, in preparation for better things to come."

Friday progressed for Peter almost exactly as planned. Although the guests from Chicago landed thirty-five minutes later than expected because the airplane had to dodge a small storm, they caught up with the program. Peter met Ulrich at 5:00 p.m., who gave him thirty tablets in a small container labeled with black ink as '30-PDCM.' Ulrich also gave instructions: "Take one pill during or soon after breakfast or dinner every day, and keep the container in a safe place, far from sunlight. It is important that you take the pills twenty to twenty-four hours apart. I also heard from Helga that you like Bordeaux wine. I brought two bottles, each containing PDCM. If you take the tablets with dinner, you can replace it with one glass of 30-PDCM wine instead of the tablet. Just keep the bottles in your refrigerator and take them out about thirty to forty minutes before you want to drink from them. Shake the bottle well to make sure the PDCM is mixed well with the wine. And I still have something else for you. Take this plastic bag with you too. It has a white powder in it, but don't be afraid, it's not some kind of recreational drug. It contains a special chemical that will immediately cancel the pill's effects if for some reason you go overboard with it. As soon as you feel anything uncomfortable in your stomach, mix two spoons of this white powder into water or any soft drink, and drink it immediately. It should relieve you of discomfort within twenty minutes. The powder isn't completely soluble in water, but don't worry. After thorough mixing, it absorbs very well. I hope you'll find nothing confusing about this treatment plan."

Around the middle of Ulrich's instructions, Peter's attention began to veer away, but by the end he had refocused his mind; he assured Ulrich that he was okay with the plan.

"Excellent. Now before I leave, please sign this paper for me. As I said before, this is a sort of agreement stating that you are taking the pills and wine voluntarily and that you'll adhere to the regimen of one pill or one glass of wine with a meal, once each day, in the morning or, in case of wine, evening." Peter signed the agreement without even reading it; he was in a hurry too. But Ulrich still had one more thing to say: "On Monday, I will call you to see if you have had any signs of a side effect, like diarrhea, headache, or cramps in the stomach or muscles. In fact, if you notice any of those symptoms, please stop taking the pill, and like I said, take two spoons of the white stuff mixed in water. Oh, and one more thing. I ran out of containers, so I just emptied a Centrum A-Z vitamin tablet container and put my pills in it. You see, on the other side, you still can read all the information about the vitamin tablets. I didn't have time to remove that label, so be careful not to mix it up with your vitamins if you happen to have this brand." He paused. "You know, Peter, the news that I had diabetes was devastating, and you must have felt the same way. But I'm getting through it, and unless you mess with the dosage and the schedule, you will be too."

It was interesting that the two could talk to each other almost as old friends, without exhibiting any signs of animosity. In fact, Peter knew nothing about Helga and Ulrich growing closer again, but he probably wouldn't have cared if he did; Helga no longer interested him.

"Okay, Peter, now I really have to leave, but again, good luck with the pills. And please, don't forget what I said about the dosage and what the white powder is for. I keep coming back to these points for a reason. You and I may react to this drug differently." With this final remark, Ulrich left, without noticing that Peter's thoughts were already elsewhere.

A heavy rain accompanied by frequent nearby lightning began to lash the city. This was one of those times when Ulrich really appreciated that the downtown had skyways, giving him shelter from the wrath of the storm while walking to the parking lot. As he got into his car, it occurred to him he should have warned Peter not to take a vitamin tablet and a PDCM pill at the same time. "Well, I'll just call him on Monday and warn him then," he decided. Certain types of vitamin tablets, incidentally including the Centrum brand, contain a lot of zinc. If in the morning Peter took the vitamin and PDCM pills simultaneously, something practical people like him would likely do, these two drugs together might result in the death of cells in the intestine, causing discomfort and pain. According to Murphy's law, it is immutable that if this can happen, then chances are that sooner or later it will happen.

After their meeting, Peter went straight home, and he casually put the two wine bottles into the refrigerator on the main floor. Then he went to the basement and placed the container and the plastic bag into the medicine cabinet in the bathroom. He was absentminded and didn't pay attention to what he was doing. He put the container on the shelf, with the side that showed the vitamin composition and not the real content facing front. Peter was crazy about vitamins, particularly since he had been diagnosed with diabetes, and he had three other containers full of vitamin tablets, each one from different manufacturers; one of these containers read Centrum A-Z just like one side of the container Ulrich gave to him. His mind was really on Agnes, who was to come over at around 9:00 to make up for the previous night. He couldn't wait to nuzzle up to her. During the day they had talked to each other only briefly, just enough time to whisper that both needed a little romance that night. But he still also wanted to shop at SAM's Club before that. It was only 5:40 p.m., and he had a lot of time left to go shopping. The next day he wanted to visit his sons in Madison, and for that occasion, he had already bought some Legos and a few other toys for them. In the SAM's Club he went straight to the grocery section and put three large oven-roasted chickens into his cart, which were one of his favorite foods at the SAM's Club. Another item he loved and always bought was oyster. A couple of weeks ago, Helga had recommended it to him, saying it is by far the richest source of zinc, and it has been shown to have effects against diabetes in animal experiments. The allure of oysters' health benefits was so great that he replaced cereal with oysters for breakfast, which was a major upheaval of his lifestyle.

By around 7:00 p.m. he was done with shopping and was bringing his cart to the register to pay when he ran into Sarah. She was also about to finish shopping, and they agreed to chat a little bit outside afterward. They hadn't met for almost two years, and they both were happy to see each other again. Sarah told Peter that sooner or later she would have called him. She had a friend back from her University of Minnesota years who taught English at a St. Paul high school and was looking for computer programs suitable for data management. Peter said that he had information about this subject at home, both in his computer and in printed form, and suggested that perhaps it would be best if Sarah followed him to his house and took the information with her. "You're just as nice as you used to be." Sarah smiled, nodding in agreement. This simple sentence and the way she said it bemused Peter, which lasted until they got to his house.

Despite the information deluge that was evident all over the place in Peter's house, it took him only about ten minutes to find a detailed description of the relevant programs. In the meantime, Sarah had the opportunity to look around, and to say the very least, she wasn't impressed by what she saw. It was hard not to notice that Peter had neglected the house after Helga left. She even suspected that in the kitchen, some foods might have become stale in their containers. After Peter handed over the thick package, she delicately suggested that in return for his favor, she could ask a cleaning company to come by and put the house in order. She said that in her house this is what they had been doing twice a month, and between the major cleanings, it took little effort to keep order, even with small children around.

Peter wasn't offended at all; instead he was moved by Sarah's concern and jumped on her offer. "Yeah, it would be nice to get some cleaning done before the dirt suffuses the house completely," he said brightly.

On the spot, Sarah called the same cleaning company that she used to hire, and to her surprise, the lady said that they could come next Monday. "Would next Monday be okay?" Sarah asked Peter.

"Perfect! I can't wait to have them here," answered Peter. He was happy that he would have one less thing to deal with, and the house will be cleaned without reprieve.

Sarah took the documents about the available software and promised Peter that she would come back on Monday. "I'll probably come around 4:00 or 5:00 to see if they did a good job," she said. "But are you sure that you can be here around 8:00 a.m.? The cleaning ladies will come around then."

"Sure, I actually never leave for work before 8:30."

"Great, then see you on Monday." With that Sarah was about to leave, but before she opened the door, Peter noted that perhaps it would also be useful if he emailed her some of the information he had on his laptop. He brought his laptop to the couch and in minutes sent the information to the address Sarah had given him. Then Peter proposed that they exchange their cell phone numbers as well, which they did. Sarah had no idea that however short her visit was, it didn't go unnoticed by someone worriedly watching from the neighbor's window.

This meeting with Sarah evoked a lot of poignant memories in Peter, memories from very deep layers that he had successfully suppressed over the years and that he wasn't even aware that they still existed. But the layers that covered his memories about Sarah were thin and suddenly cracked. In Sarah's presence, he once again felt the same titillation he had felt when they had been together. He was surprised that even after so many years, Sarah still held sway over him, and he was unsure of how to deal with his newfound affection toward her. One moment he was determined to brush aside the past, but the next moment he suddenly wished to embrace her and hold her tight with no regard to Agnes or Doug.

As in the following few minutes he thought more about their meeting, he realized that something had bothered him, something that was difficult to put in words. Sarah seemed to radiate a kind of weariness of life, so well-concealed that perhaps only he could sense it because only he knew her so well. Although on the surface Sarah seemed to be satisfied with her life, Peter could tell that in fact she wasn't truly happy. That evening he kept mulling over his intuition about her, but he couldn't decide whether it was real or illusory. He had no idea that Sarah's short visit had just ruined his love life with his only remaining partner, Agnes, who in just a few short minutes had also undergone a whole range of emotions.

Agnes was an avid bird watcher, and the location of her house was perfect for this joyful activity. Her large window provided a very good vantage point with which to watch the birds flying by or resting on the branches of the several large trees around the house. She always kept a military-issued set of binoculars close to the window so she could study the birds' behavior in close detail. She often took pictures of them on her cell phone. Sometimes she opened the window and tried to mimic the birds' songs, and to her surprise, she occasionally got a reply. This always brought her great happiness; she imbibed the beautiful birds' songs. Then almost every day a group of deer came by, often encamping for hours in the shelter of pine trees and raspberry bushes. But she occasionally also saw some horrific events, like a crow killing a small rabbit and then a whole flock feasting on the carcass. Once she shared her conclusion with Peter: "Life isn't idyllic all the time in the animal kingdom either, even in an urban environment," she said. "And we humans share our genes with animals, so then why are we surprised that there is so much aggressive behavior in human societies? Perhaps we just should accept it and learn to live with it," she added.

That Friday evening, Agnes too did some shopping after work and came home late. The first thing she did was take her binoculars and look out the window, hoping that she could spot some cardinals, her favorite birds because of their beautiful red color and playful behavior. Suddenly a larger group of deer came from behind the house and headed toward Peter's garage. She took a picture of them, along with a white car parked in Peter's driveway that she had somehow missed while driving home. Seeing a strange car there was unusual because since she had moved into her house, she has never noticed anyone visiting Peter except Helga, and Agnes knew that this was not her car. Agnes's house was only about thirty yards away from Peter's, and she had a good view of Peter's living room through the large windows that faced her kitchen. She saw Peter seating a woman in the living room. Then Peter was out of sight for about ten minutes, but after his return, the two conversed for quite a while. Agnes was sure that she had never seen the woman before, but even from the distance, she could see she was very attractive. The movements and gesticulations of the two also made it clear to Agnes that they had known each other for a long time. Agnes grew a little worried at the scene. Her feminine instincts suggested that this woman meant to Peter more than a casual visitor would, and this thought vexed her.

In any relationship, no matter how deep it might be, Agnes's basic principles were fair play, commitment, and honesty. When she became Peter's girlfriend, she made it clear in no uncertain terms that she expected him to strictly adhere to these principles. Peter jokingly called Agnes's terms 'the rules of engagement.' Peter was not to fool around, and he really didn't mean to. This was fair, because Peter expected the same from Agnes.

An advantage that Agnes had over Peter's previous partners was that in many ways, she was more intellectual. She could easily converse with Peter for hours about such wildly different topics as the nature of mind, the big bang, and the structure of the universe, often citing most recent astronomical findings and theories. She had a disposition to new ideas in almost any area of life, whether she agreed with them or not. As a tourist, she also traveled a lot, both in America and Europe, which was fascinating to Peter because he also visited some of the same places, although in truth he rarely had time for anything besides business. Agnes couldn't believe that although Peter had been in Paris several times, he had never found time to visit the Louvre or even the Eiffel Tower. "I bet that in Paris you hadn't engaged in any kind of indulgence in sensual pleasure either. You really are fully devoted to your work," she told Peter once. He laughingly admitted to these deficiencies.

She loved to talk about medieval European towns and their history, and one evening she made Peter commit to a three-week trip to Europe, just the two of them, the following summer. To force such a commitment out of Peter wasn't easy. He bristled at Agnes's idea first and then murmured for a while that taking a three-week vacation is something only Europeans can afford. But since he wasn't as recalcitrant as sometimes he pretended to be, and this planned trip was still almost a year away, he agreed to the plan.

Toward the beginning of their relationship, Peter asked Agnes about her religious beliefs. Agnes knew that Peter had been going to church since his early childhood, so she was a little bit afraid of his reaction when she admitted that she was a kind of semiagnostic, tilting toward the acceptance of an omniscient, omnipotent, and omnipresent being but not willing to accept at face value the otherwise beautiful stories and miracles in the Bible. But she liked and accepted all of Jesus's teaching about how humans should behave. Agnes's approach to religion was much more acceptable to Peter than Helga's fierce, almost blind, atheism. In fact, Peter also admitted to Agnes that he was very close to becoming a semiagnostic himself, at least with regard to the literal meaning of the Bible. His conversation with Doug about cosmology had left a big impact on him, even though he was well aware of the science fictional nature of Doug's theory. When he told Agnes that, according to his friend Doug, God might have a permanent material self and thus He could communicate with people through some matter-based signals, she was thrilled.

"That's it!" she exclaimed. "I love that idea. It is really pushing the boundaries into a new direction that has not been traveled yet. Finally, here is someone who steps out of this mythical spiritual cloud and materializes God and His realm. I hope your friend is correct. Honestly, the communication between God and humans has always been a problem for me, which so far has prevented me from embracing any one religion. But if I could be convinced that there is some substance behind this claim, I tell you, even I would become a devoted believer. I would like to meet your friend once," Agnes begged Peter. "I'd love to talk to him about this."

"That's no problem," he said. "As soon as we make our relationship official, I'll introduce you to all my friends, including Doug. I can guarantee that you will enjoy his ideas, however far-fetched they may be. And you need to know that he is only raising the possibility of these ideas. He doesn't necessarily believe in them. He is aware that there may be an infinite number of other possible explanations, all different from what is in the Bible, to describe God, His realm, and His relationship with us."

On some other occasions, Peter revealed about Doug that he thought God has both spiritual and materialistic sides, so that He is a dualistic being. Although the mention of God's spiritual side somewhat cooled Agnes's enthusiasm, it didn't deter her from wanting to meet Doug.

"After all," she thought, "Doug clearly thinks about God in an unconventional way, and perhaps what he calls 'spiritual' is also different from the conventional meaning of it."

After the woman left Peter's house, and Agnes was through the first stressful reaction evoked by the sight of her, she took a deep breath and slowly managed to calm herself down. She needed to put her worries to rest so she could think logically and develop a plan for how to address this situation. She concluded that the first thing she needed was Peter's voluntary explanation of what had happened. "He must be forthcoming, or I'm done with him," she decided. "He knows that under our agreement, he should feel duty-bound to clear up this situation without any prompt from me. I'll call Peter, and we'll see if he mentions the woman's visit." It took her a good hour until she gathered enough courage to call Peter. They made small talk for a while, but he said nothing about the woman's visit. After hearing so many extraneous details about his weekend program, but not the information she was expecting, Agnes said that she had a headache and would rather just go to sleep. Being an honest woman, she usually didn't like to make up excuses when she wasn't in the mood to do something, but this time she didn't see any other way to avoid the meeting with Peter.

The fact that Peter deftly avoided mentioning his visitor hurt Agnes. She wasn't a control freak, but she expected Peter to at least mention the woman. She didn't want to know any details; just one sentence would have been sufficient. It was equally hurtful that Peter didn't seem to worry a bit about her headache even though she faked it. The more she thought about this whole incident, how Peter had besmirched their burgeoning love, she grew furious. "Now, Peter, we're done," she cried out. Later in her life when she remembered that evening, she again concluded that this was how this event had to happen. She, Peter, and Sarah were just following a script that evening; they were rowing in a boat that somebody else was steering.

On Saturday morning, Peter woke up before 6:00 feeling feisty, and in less than an hour, he was already on his way to Madison where his ex-wife and his children moved to from New York soon after Peter moved to Rochester. He arrived just around noon, and his two sons, Peter and James, by now seven- and nine-year old, were ecstatic when they saw him. For a few moments Anna and Peter stood in front of a large window facing the backyard. The sun filtered through the window and lighted the side of Anna's face. Peter noticed that she still had the same chiseled features she had had before, except for few shallow wrinkles around her eyes. In fact, Anna looked quite ravishing and was exceptionally friendly, much to Peter's surprise. She wasn't the quarrelsome woman who was incensed by his smallest failings and who caused him so much stress.

When Peter asked if her boyfriend was home, Anna acknowledged that they had broken up about two months ago, and at that moment she was alone. She was about to finish preparing a meat soup and steaks for lunch and wouldn't let Peter and the boys go anywhere else to eat. As they ate, Peter suggested that they all go to the theme park, which greatly excited the children and even Anna. Anna seemed to have been transformed by the recent events she had gone through. Breaking up with her abusive boyfriend was liberating to her. But at the same time, she really didn't know what to do with her newfound liberty. She also began to see Peter in a much more positive light, although she kept this entirely to herself. Peter was never even close to being abusive; in fact, if either of the two of them had ever been abusive, it was her. And she retroactively recognized that she had often been hysterical too, turning their home to a chaotic place to live in. While she began to understand Peter's reasons for leaving her, she still was trying to mend the hole the divorce had blasted into her heart.

A week ago, Anna and the boys had moved into a new house in a different suburb of Madison. Luckily enough, Peter had called her few days ago and Anna told him the new address. This was a smaller home than the previous one, but the surroundings were much nicer and quieter. Both a school and a public swimming pool were within walking distance, and the major traffic was directed away from the area. Anna had already made plans to take the boys for regular swimming lessons. It seemed to Peter that Anna was all set, except that she perhaps missed a male presence in the house. But at least she apparently had the wherewithal to carry on with her life, in part thanks to the child support she still regularly received from Peter.

In the park, Anna was voluble the whole time. She gazed deep into Peter's eyes and asked him many questions about himself while the boys enjoyed the rides. Just as at the time of his arrival, she was so smooth and agreeable the whole time that Peter felt like he was dreaming. He again had to conclude that this wasn't at all the same woman he had left. Later in their marriage, Peter came to see Anna as a perpetually furious woman, liable to blow up at the slightest perceived provocation. For a moment the thought flashed across Peter's mind that this new calm demeanor was only a facade. But in any case, Peter greatly enjoyed Anna's company. There was enough food left over to take care of dinner, particularly because the boys got large ice creams that took their appetites away. Anna and Peter had a long talk after the boys went to bed, and eventually they decided that it would be better if Peter stayed for the night, albeit in a separate room.

They had had a lovely day together, but that wasn't nearly enough for them to sleep in the same bed again, and in fact neither of them even considered this an option. But as Peter prepared for bed, he did toy with the idea of getting back together with Anna. That would certainly make his children very happy, and he had the feeling that Anna probably wouldn't oppose the idea either. But, Agnes aside, the big questions for Peter were if Anna had really changed for the better and if they could overcome their past bitter feelings. Would they be able to free themselves from the unpleasant events of the past? Could they forget that during the last year of their marriage they each viewed the other as an enemy? Could he forget that Anna was often so petulant over essentially nothing? It was also true that no other woman could purr into his ears and placate him so gently after a dispute as she did, on the rare occasions when she was in that mood.

He didn't dwell on the past and their possible reunion for too long; by the time he fell asleep, his thoughts inexplicably centered around Sarah once again. During the day he could tell himself the noble lie that he was no longer interested in Sarah, but by the evening when he was tired, his brain couldn't fend off the truth that he was still in love with her and not with Anna and perhaps not even with Agnes. The brain at night, being freed of the forced controls and expectations that it accumulates during daytime, is less sanctimonious and more truthful, and Peter was no exception to this subtlety of human physiology.

It was interesting that whether it was Anna, Helga, or Agnes, Peter could deal with these relationships quite rationally. However, his recent meeting with Sarah, which he became convinced wasn't a fluke, brought to the surface a long-buried, intense longing for her that seemed completely irrational, no matter how he looked at it. He made a great effort but could not rationally repress it, even when from time to time he forcibly reminded himself that Doug was his best friend who had two kids with Sarah. Peter became mired in remorse just by thinking about Sarah and Doug. How could he possibly mess things up with his friend and even break up his family? Besides, even if he was willing to pursue Sarah, there was no way she would even entertain the idea. His dueling rational and irrational selves clashed with no clear victor. If his rational self would have been fully functional, he would have been terrified of meeting Sarah on Monday because he couldn't be sure if he would be able to refrain from saying or doing something that he would regret forever; however, at that point, rationality hardly controlled his brain. He was fully aware of how pernicious the situation might become, and it was a good thing that he was. It helped him to retain a shred of logical thinking, and to at least plan to approach Sarah only if she gave him a strong signal. But then he changed his mind again, thinking that approaching her wouldn't lead anywhere anyway. "There is no end game here, except perhaps a short and shallow affair with my best friend's wife," he concluded. His last thought before falling asleep was that he shouldn't even try.

Early the next morning, the boys were very restless and woke Peter up at around 6:00. He would rather have stayed in bed a little longer, but Peter Jr., who was older and had more authority than little Jim, wanted to play with him until breakfast, which he couldn't possibly refuse. The boys tested the electric cars and the big Lego set that Peter had bought them, and then kept bringing out more and more toys to impress their father. The boys acted like little con men, always asking for just a little more of their father's time. Which father could say no to that? Peter first told the kids and Anna that he couldn't stay for lunch because he had a lot of things to arrange at home for some meetings on Monday, but leaving was hard with the boys pleading with him not to go, and he eventually caved in and stayed for lunch. Staying longer with his kids also helped to ease his angst, which occasionally surfaced.

After lunch Anna walked Peter outside, but as he started toward the car, she asked him to first talk with her a little bit. "Peter, I have something to tell you. I didn't want to say it in front of the children. During my last checkup, my doctor informed me that I have a rare disease. Apparently, the Mayo Clinic is the best place to go to take care of it. Next Tuesday I have an appointment there at 2:00 p.m. and my flight back is at 7:30, so I might be able to pay you a short visit if you can make it home a little earlier. If it's okay with you, could you give me your address?"

"I'm truly sorry to hear that. I hope it's not life-threatening," said Peter with deep compassion in his voice.

"No, the doctor said at this stage it is 100 percent curable, but if I wait any longer, it could become cancerous."

Peter handed over his business card. "It has all of my information," he said. "I would be very glad if you could make it. I promise I'll be home all day on Tuesday."

It was almost six in the evening when Peter arrived home, and there was no way to see Agnes that evening because it took him until midnight to put his work and presentations in order. But around eight, he had a short phone chat with her, and they agreed not to see each other that evening. Although Agnes had already decided to terminate her relationship with Peter, she still would have given him a second chance to explain the unknown female visitor she spotted in his house. She knew most of Peter's faults, but his honesty had never been in question in her eyes. That is what made this situation especially hurtful for her; he had broken her trust. It was more painful for her than a knife in the back.

Peter's continued silence about his visitor was further proof for Agnes that he had really double-crossed her. This made her decision to break up with Peter final. Peter on his part didn't make any attempt to propitiate her because he wasn't even aware that he was guilty in Agnes's eyes.

On Monday morning, Peter ate two cans of oyster for breakfast and was about to leave when he remembered to take one pill from the container Ulrich had given him. On Saturday morning he had taken two pills at once because he knew that on Sunday morning he wouldn't take any. But he forgot all that, and to make up for his perceived lapse of memory, on Sunday evening he again took two pills. He was already at the door, ready to leave for work when the cleaning crew arrived. He would have forgotten about them if they hadn't shown up just then. They looked around the first floor and quickly concluded that this was going to be a full day's work, and in the worst case, they would have to come back the next day as well. Peter said that either way it was okay with him. Two lengthy meetings took up most of Peter's day, one in the morning and another in the afternoon. Around noon he began to feel a slight pain in his stomach, and this made him a little flustered. He complained about it to Agnes, but he decided to go out with a guest and two of his colleagues for lunch anyway.

In the restaurant he ordered his usual New York steak and red wine. He was in the middle of his meal when Ulrich called and asked if he was feeling anything unusual. He answered that he had a slight stomachache and in general felt a little languor, but as he started eating, these symptoms went away. "I was probably just hungry, or maybe I'm a little nervous because today I have two important business meetings, and unfortunately there are already signs that neither may work out." Ulrich was not fully satisfied with Peter's explanation of his stomachache and warned him to be very careful with the dosing.

"Don't take more than one pill a day for any reason," he said firmly. "With this kind of drug, unexpected things can always happen. And please don't forget the white powder in the bag! If you feel any more discomfort in your stomach, stop taking the pills, then immediately make a nice suspension of the white powder and drink it as we agreed."

Peter didn't admit to Ulrich that on Sunday evening he had taken a double dose and that he might not have kept track of the dosage and timing on Saturday. Normally, he would have paid much closer attention to the details of Ulrich's instructions, because a careful approach toward drugs was in his nature. But suddenly there were way too many other important things going on in his life, and on the drug front, he left his guard down. Of course, he should have been much more attentive to the danger signs; his stomachache foreshadowed what was to come. In a way, pain is our friend. Pain tells us that something out of the ordinary is going on in our body, and it is better to find the source as soon as possible and treat it, than be sorry later. Heeding Ulrich's warning and taking the white powder right away would have been the correct response to Peter's stomachache.

Peter concluded his meetings by midafternoon and asked a colleague to take the guests to the airport. This allowed him to get home just before five, when the cleaning crew and Sarah were just about to leave. It turned out that the cleaning ladies had finished their work half an hour earlier and were just waiting for Peter to arrive. After the ladies were paid and left, Sarah stayed to explain what they had done. She had stopped by the house at noon and then again around 4:00 p.m., and she found that they had done a superb job.

Peter thanked Sarah for her involvement and jokingly said, "Now I availed myself of a good connection." Then he suggested that they sit down a little bit to talk. He felt better now, although his stomachache hadn't gone away completely. For a few moments, both thought of what to say, then Sarah suddenly asked him with her usual suave voice about Anna, the children, and Helga, all in one sentence. While Peter searched for an answer, she added that she hoped these questions weren't discomfiting to him. Even before they met, she had found it quite surprising that Peter hadn't been able hold on to these women, but she didn't know any of the details. Now that they were together, she was dying to know the reasons, but she was polite enough not to try to force out any information from Peter that he didn't want to share. Peter sensed that, and he tried to be as concise as possible.

"Simply put, I'm almost embarrassed to admit this, but I don't really know why my wives weren't satisfied with me. I can tell you the same thing in a more eloquent way too. My acumen was clearly insufficient to understand the subtleties of a woman's soul. Clearly, I didn't pass some sort of aptitude test required to be a fitting husband."

"I like it better when you talk simply," said Sarah, smiling. "When you speak eloquently, you sound almost like Doug, although lately he has been cutting back on using words reserved for high society." This small snide remark again signaled to Peter that the relationship between Sarah and Doug might not be as picture-perfect as it might seem to outsiders.

After another short silence, Peter said something that he later wished that he had never said, because of the unintended consequences it led to. "You see, my problems with women might have started with you, Sarah," he said half joking. "I was madly in love with you, and I probably never could have gotten over that you dropped me for Doug. Maybe this event remained in my subconscious and has prevented me from giving all myself to my wives, and they probably sensed it with their sixth senses. You see, Sarah, the truth of the matter is that even after all those years, however much I would like to suppress it, I'm still in love with you. This feeling slumbered in me for years, but last weekend it conquered me once again. I can't wipe out the thought from my brain that if you wouldn't have left me for Doug, maybe today we would be the happiest married couple in the world, but then again, maybe not. We'll never know. I know I'm one of Doug's best friend. I don't want to put you in an uncomfortable position, and I'm truly sorry if I'm out of line, but this is how I really feel, and I can't help it. I just had to tell you this once. Now that I did, even breathing seems to be easier."

"Don't apologize," said Sarah. "When we met on Friday, you also made me feel something I haven't felt in years. Seeing you again sent mild quivers through my spine that were both pleasant and disturbing at the same time. Last weekend I tried to confront and analyze the meaning of my feelings in the context of the reality of my life, and this is what I found out. Doug is a wonderful and very intelligent man, but the naked truth is that lately, I have begun to miss a certain spark from him, a certain level of romanticism, which he used to give early on in our marriage. I miss it so much, sometimes when I'm alone, I would like to bawl why this is happening. Unfortunately, as time goes on, it is only getting worse. I think he's too serious. Probably his daily drudgery at the Mayo wormed its way into his personality. In any case, it seems we never have enough quality time for each other either. He always tells the next day we'll have more time to spend with each other, but that day never seems to come. We haven't been on a good vacation for three years, and I don't see anything shaping up for next year either. Once we had agreed not to vacation because he had to prepare an important and complicated grant application. Then he couldn't take a vacation because he had to finish some work required for his next grant application. Then, last summer he spent all his time on a patent application that is so far only costing money but gaining no income. I'm sure that something will come up next summer as well. I know that he loves me and cherishes me in his own way, and I don't really harbor anything against him either. I only feel worn out, which has a ripple effect on all my other activities as well. I don't know how much longer I will have the will or the desire to handle this situation within the constraints of our marriage."

Once again, a short silence fell on them—Peter wasn't sure what to say—and then Sarah continued her confession. "I've gotten to the point where I'm no longer even sure that having more family time together or occasionally getting flowers from him would make our marriage work. And as I have been going through these very confused feeling, you suddenly came back into my life, which only adds to my confusion. I would be lying if told you that I don't miss the long hours you and I spent together talking about silly things. Yes, I would love to be more open to silly things and not always deal with serious matters as Doug does. I admit I also miss your gentle touches." She gazed into Peter's eyes but then looked down at her feet. "But despite all of this, my family is still the most important thing for me. Also, if I look at it with more realistic eyes, there are many good things that I'm still getting from Doug, and I suppose no person can give another person everything she or he wants. Besides, I'm probably also doing something wrong too. To give you an example, I know that recently he has begun to deal heavily with God's nature and role in the creation of the universe, and perhaps I could be more involved in that. But over the years, I have lost my zeal to deal with philosophy and discuss these kinds of matters with him. I'm sure he has noticed that, and this doesn't help us to remain close either."

"And tell me, honestly, Sarah, how do you feel about me?"

"Recently I used to moon about the time we were together. As for now, all I can say is that you caught my attention again, big time. But my present mental state conjures up so many different and competing feelings, that at this moment I really don't know for sure how I feel about you. Are you starting to see what a predicament I'm in?"

"Yes, I can see it, and I understand," said Peter with a deep sigh, which sounded like a sigh of relief. "I believe that on the whole, Doug is the best man for you. But I also believe that it was good for both of us, and certainly for me, to talk about this. Last weekend my mind had been whirling around you quite a bit. After our last meeting, I sensed that you weren't entirely happy. That's why I felt that we needed to talk, to get certain things off from our chests. And I think that we just did that." Peter wasn't satisfied with the way he had expressed his deep feelings, and he added with his emollient voice, "And if you ever feel that your marriage is no longer tenable, you will always find me here."

None of them knew how to continue the conversation. After all their confessions, nothing substantive remained to be said, and Sarah decided she should leave. Peter stood up, confused, and searched for one of the wisecracks that he was so famous for in high school, but he couldn't come up with anything. He stepped closer to Sarah with the intention of shaking her hands, but then he just couldn't control himself and his hand touched Sarah's hair. He slowly drew Sarah's face toward his and kissed her. Sarah didn't resist him, and they kissed for a long time. After they unfurled from the hug, Sarah sighed and then said very slowly, "I will always remember this kiss, and there may be a time when I indeed will knock on your door wanting to stay with you. But now it's better if I return to my family. This is perhaps also best for you too, Peter. We can't keep toggling between two realities, being at the same time here and there. We must live our lives in one. Otherwise, it gets so confusing. After all, we're not kids anymore, although there are times when I wish I could be one."

After she had said all this, she turned her face toward the neighbor's house and noticed a shadow moving away from the window. For a moment Sarah was shocked but then forced herself to calm down; finally, in her most casual voice, she asked Peter who his neighbor was.

"She is a young lady, Agnes. She moved in not long ago. She has a business degree from the University of Michigan and worked for two years here at IBM, but then she was laid off just after she moved into the house. A few months before she lost her job, she had broken up with her software engineer boyfriend, who incidentally also worked at IBM. Agnes told me that the primary reason for the breakup was that at the time she still had kept contact with her previous boyfriend in Boston. In any case, her parents couldn't financially support her, even temporarily. She was desperately looking for a job when I offered her one about three months ago. I used to work alone, but as I began to travel more and more, I sorely needed a secretary to keep up with things, particularly when I was away. Thus, my boss grudgingly allowed me to look for a secretary, and that is how I was able to hire Agnes. She is a comely girl and does an excellent job," he added. "I am very happy with her work. I have never heard her complain or seen her slacking." Peter failed to notice that he went a little too far in praising Agnes.

"This perfect secretary of yours was watching us," said Sarah abruptly, in a matter-of-fact manner. "Tomorrow, quite a few tongues will be wagging."

Peter was shocked to hear this sudden change in Sarah's tone and the sarcasm by which she reacted to seeing a shadow. But Sarah also realized she sounded harsher than she had intended and asked her next question more calmly, "Do you go to work with her in the mornings? That would make sense." What she really wanted to know with her barely veiled question was whether Peter had a romantic relationship with Agnes. Peter got the message. Sarah had clearly moved from a romantic to an irritated mood, and now to a more investigative one.

"No, no, we go to work separately, and we come home separately. We have a different work schedule. She goes and comes earlier than I do." With this explanation, he hoped to allay Sarah's concern, or perhaps jealousy, which he found strange in the first place. *Why should I owe Sarah any explanation?* he thought to himself.

He tried to get back to their initial conversation. "But I would rather talk a little more about you and me. After our kiss, I realized that perhaps in both Anna and Helga, I had been actually searching for you."

Before he could go any further, Sarah interrupted him. "Peter, I propose that for the time being we keep what just happened in our memories, and trust God to show us which way we should go from here." She turned to leave, and although Peter made a weak attempt to kiss her again, she just said, "No, Peter." She once again let him know that she could be tempted only up to a certain point.

After Sarah left, Peter dropped into the nearest chair. His mental state, full of confusion about his relationships with women, was upended. Helga had left him without giving a good reason for it. He thought he had loved Agnes, but with Sarah coming back from the past, he suddenly became unsure of that. Then the thought crossed his mind that maybe he should just straighten things out with Anna and have a real family again. He was certain that Anna would happily take him back. On the other hand, he had basically no chance with Sarah, he was done with Helga, and he was far from ready to tie the knot with Agnes; in truth, Agnes probably wasn't ready either. He had heard a lot about clinical depression, and he knew that was exactly what he was going through. The tremendous pressure he was under, both at IBM and in his personal life, had gradually led to a stress level that he couldn't bear with a straight mind anymore; that night something had to give. He felt completely and irreversibly defeated. He felt he walked on the edge of a deep abyss.

He knew that at 7:00 that evening he had to go play poker with his friends, and he only had half an hour left. He felt neither the strength nor the desire to go. Instead, with trembling hands, he opened a wine bottle that he took from the refrigerator and drank from it without using a glass. It felt good as the alcohol began to make its way into his brain. He wanted to parse a little more about his exasperating relationships with women and where to go from there. He felt like a horse tethered to a guardrail; he wanted to roam free, but to where? His hands were flailing, as if he were about fighting an invisible enemy. Then a sentence, which he read or heard somewhere and found so true that he often had thought of, suddenly came to his mind: "Our path through life is strewn with a mixture of good and bad memories, and it is in our character to decide which one we are going to draw from as we move along." "But which of my memories are good or bad?" he lamented.

As he drank more, his thinking became more incoherent, but his mind still could remain focused on women for a while. "Maybe I should just liberate myself from women in general and live alone for a while, until I sort things out? If Agnes saw me kiss Sarah, as she probably did, then my relationship with her is over." What depressed him was the naked truth that even with all these women around him, he still was lovelorn, a miserable situation for anyone particularly at his age. All these women oozed sex appeal, but Sarah was the one he was still really attracted to. But his infatuation with Sarah was clearly a dead end; it would never result in anything. Not long ago he had found Sarah's decision to keep her distance liberating, but now it again left him feeling dejected. This evening he could not free himself from Sarah's spell, and he had a strong feeling that he never would.

After drinking a little more wine from the bottle again, he rapidly lost his willpower to even move, despite still being fully conscious. Some people going through deep depression often have suicidal thoughts, or they just sit helplessly with glazed eyes, losing contact with their environment. Peter had none of these mental states. All he wanted was to forget about everything that connected him to the world. He kept drinking, finishing the bottle in only a few minutes. Then he opened the other bottle sitting in the refrigerator. After a long gulp from it, he suddenly went through a seismic change. Pulling his remaining conscience together, he went down to the basement to take some pills. At that point he only remembered that he had to take one pill, but he had no idea if it was morning or evening. After he got back to the first floor, he again drank quite a bit until he entirely lost track of time and everything else. He had long given up on the poker meeting, and now he couldn't concentrate on any of his thoughts for longer than a few seconds. After finishing the second bottle, the only thing he felt was a strange sensation of hurtling toward a black hole, and then he literally passed out on the couch in the living room, with the lights still on and the front door unlocked.

He woke up around 3:00 a.m. First, he drank several cups of water to quench his terrible thirst. Then he turned off the light, but again forgot to lock the front door. He had a headache, and his only thought was of getting back to the bed. He had no memory of what happened during the next several hours; he was in a deep swoon. Eventually he thought he heard someone moving around the house, or maybe it was only the vibration of his cell phone. He couldn't open his eyes to check which one was the case, although he struggled to do so. Finally, after a lot of effort, he partially woke up sometime around 10:00 a.m. Still very tired and started to suffer from stomachache, he ate two cans of oysters and then went down to the basement to take two vitamin pills to make up for Monday. He didn't make any effort to verify whether he really took vitamin tablets or Ulrich's pills and completely forgot that he had already taken one pill the previous evening. He went back upstairs to his bed and was in a half-conscious state when he heard again some noise, but he had no strength to verify the source. He woke up early in the afternoon sometime between 1:00 and 2:00 p.m. because he felt the need to vomit, which he indeed did three times in quick succession. By then he had terrible stomachache and his arms and legs were trembling, but he had now awakened enough to remember that he still had things to do before his trip tomorrow.

After some self-goading, he went to the bathroom to take a long soothing bath, hoping it would relieve him from the stomachache, which every minute became more and more painful. He had spent less than five minutes in the bathtub when he could no longer tolerate the pain and decided that it would be better to come out of the water and make the solution with the white powder. As he took the towel in his hand, he again had the feeling that someone somewhere in the house was making noise, but by then everything was becoming foggy around him. Maybe the cleaning crew had come back to finish some work or pick up something they left in the house? Or could it be Sarah? Perhaps she had come to return the key to the house?

Despite his rapidly deteriorating condition, Peter still clearly remembered that Sarah had kept the key. Perhaps he remembered it because he had always hoped Sarah would come back? The key was now in the forefront of his brain. Agnes also had a key to his house, he remembered, although she would never come before calling first. With the towel still in his hand, he flopped on the floor beside the bathtub because he was losing consciousness and felt extreme pain in both his chest and stomach. He had completely lost both the physical strength and the will to stand up, but his mind was still functioning and told him that he was in big trouble. While the world faded around him, and he was struggling for air, he faintly became aware of some steps close to him, but he could no longer react to anything. His last thought was that he was flying down a very dark deep tunnel, a yawning abyss, with frenetic speed from where he never returned. His soul became a buoy and beacon, afloat in the vast cosmic ocean, searching for heaven. He soon would know the answers for the mystery of God, a knowledge the living may never gain.

Anna's examination at the Mayo Clinic was over by 3:00 p.m., and she still had time to find a taxi and visit Peter. Inside the taxi she dialed his number but got no answer. The taxi parked close to Peter's driveway, and Anna paid the driver for an extra twenty minutes to wait for her. She crossed over to the house through the lawn and pushed the doorbell, but again, there was no answer. On a whim she tried the door and, finding it open, cautiously went in and called Peter's name loudly, but to no avail. As she turned to leave, she heard a scarcely audible voice coming from the bathroom. She went in and found Peter on the floor, clearly in pain and gasping for air. Peter's body was wet, and Anna tried to clean his face and nose with the dry towel that he still clutched in his hand, but it was all too clear to her that he was in his final moments of his life. And indeed, after few more minutes of struggling, he was dead. Anna quickly cleaned his face once more and then ran to the taxi through the lawn again. She didn't inform anyone. The last thing she wanted was to be implicated in Peter's death. As she learned at the Mayo, she would have to undergo a major surgery in three weeks, and her immediate concern was what to do with her sons during that time. After the taxi pulled away and was about one hundred feet away from its parking spot, she instinctively looked back and saw that a white car had just stopped in front of Peter's house. She didn't see what happened next because the taxi made a sharp left turn, which blocked her view.

On Monday evening, Sarah put the children to bed at 9:00. Doug was at the poker game, and Sarah couldn't help herself to slink out of the door without making noise and drive to Peter's house to see if he was home. The full meaning of what had happened between her and Peter began to slowly percolate in her brain, and she felt a sort of guilt. She regretted that she had not stayed a little longer with Peter, and when she did leave, she didn't do so on a more positive note. She saw in Peter's eyes that he was deeply hurt, but she hadn't done anything about it; she had only showed the stony side of her heart that she didn't even know existed. Now something suddenly told her that she should be apprehensive for Peter's safety. Since Peter had nobody else, she felt that the onus was on her. The sadness in Peter's eyes that evening were telltale signs of trouble, but it took some time, perhaps too much, for her to correctly read them. Sensing that Peter might be in trouble was another indication that even after so many years they still were on a very similar wavelength.

Sarah stopped her car in front of Agnes's house, because from that position she had a better view of the lights and movements in both houses. She was surprised to see that Peter was at home, or at least his lights were on, although the time was only about 9:40 p.m. Doug used to come home from their poker parties only after 11:00 or even midnight, which meant that Peter hadn't gone to that night's meeting at all. Perhaps he was preparing for the meeting in Los Angeles he had mentioned to her? In Agnes's home the light was also on, and Sarah could see someone's silhouette through the light curtain. So, apparently, Agnes was also at home, and thus not with Peter. For the time being she had all the information she needed. She was tempted to go in and see Peter but then gave up the idea; somehow her worries about Peter had subsided. Besides, she didn't want to leave the children alone for too long either.

The next morning, when Sarah heard from Doug that Peter remained absent from their meeting unannounced, she pretended to be surprised but immediately sensed danger again and wanted to do something about it. Although she felt a little seedy, soon after Doug left for work, she once again jumped into her car, and around 9:00 a.m. she entered Peter's house. She hoped to see him up and working; instead she clearly heard Peter snoring loud in his bed. She saw the two wine bottles on the table in the living room. "Clearly last night Peter got drunk and that's why he wasn't at the poker meeting," she concluded. Instinctively, she knew the reason for his heavy drinking; this was how Peter tried to allay his inner pain after she left. But he seemed to be all right. His breathing and snoring appeared to be normal, and at that point that was the most important thing for her. She again left, reassured. Once again, she let her guard down.

The garage, built recently, was attached to the house, and the door to it was just two feet away from the front door. Last Friday Sarah had noticed that Peter kept several empty wine bottles in the garage, and she decided to maintain a little order by adding the two emptied bottles to the pile. She wasn't a particularly good observer, but after looking at the wine bottles, something seemed to be a little strange. On closer examination, she found what disturbed her; all the wine bottles in the pile had the same label, except the two new bottles. All of them contained Bordeaux wine, but the visibly distinct labels on these two new bottles indicated a different appellation.

Sarah also knew that Peter was planning to work at home the whole day and that Agnes wasn't expected to come home until a little after five. She decided that after she took the children home, she would pay another visit to Peter before his secretary returned. Perhaps she could lift his spirits before his trip to Los Angeles. She again and again returned to the idea that there was a connection between their meeting and Peter's unusual behavior. Just like Peter, she also had deep insight into human behavior. The way they had separated the previous evening affected her tremendously, and she could imagine that the same had happened to Peter too. She was ready to assuage Peter's and her own pain by going back to his house in the afternoon. She was always contemptuous of other married women who cheated on their husbands, but this time she felt different; she wouldn't let Peter down a second time. She felt justified to see Peter again and do whatever was needed to bring him back to normal. At that moment, her fidelity to Doug became of secondary importance. Thinking all this through and reaching that conclusion, which she now was sure was the right one, filled her heart with peace. She was so overwrought and distracted by all these thoughts in her head that she forgot to lock the front door.

    She returned to Peter's house around 3:50 p.m. She was not too far away when she saw a taxi parked close to Peter's driveway, which then pulled away before she got there. But she didn't pay any attention to the taxi. She was fully consumed by the expectation of meeting Peter, who was certainly working and preparing for his meeting tomorrow. She was ebullient and prepared to embrace Peter and go to any distance he wanted to go. She pushed the doorbell twice, but Peter didn't come to the door. Still confident, she entered the house and called Peter's name out a couple of times without getting an answer. She intuitively peered into several rooms and then the bathroom, and there he was, lying on the floor, dead. Her brain stopped working for a moment while she processed what she saw. After that first shock, her next emotion was seething anger. "He was murdered, and I will find out who did it," she murmured deliriously. Thinking quickly, she gathered Peter's laptop, his two cell phones, and all the other items from his study that might give her a clue about the person who might have done this, and then she staggered back to her car without even considering calling the police. She spent only ten minutes in Peter's house. While she was driving home, sweating and shaking, she thought about the situation over and over again and tried to come up with a suspect. It didn't take her long to find one: Agnes. "She obviously saw me and Peter kissing, and she killed him out of jealousy." This was her verdict, which she never changed, but she had never told anyone.

After Peter's death, Sarah could never forgive herself for not visiting him on Monday night. "If only I had visited him," she often lamented to herself, "I certainly would have stayed with him the next day, and nobody would have had the opportunity to kill him." Living with this thought, which was an absolute certainty for her, took a toll on her mental state. That tragic event changed her psyche very much, placing all her future, ever-worsening human interactions in a very different context.

For Agnes, Monday night turned out very differently than she originally expected. Her previous boyfriend from Boston, Robert, whom she jilted long ago but still maintained occasional and very innocuous email contact, was visiting her. Robert had emailed her on Saturday that the next day he and his friends would leave Boston for a motorcycle tour with his friends to Yellowstone and that their plan was to spend Monday night in a Rochester motel. Once in Rochester, he would have time to visit Agnes in her house if she would have him. The next day they would ride on to Sturgis, a small town in South Dakota that was the annual meeting place for hundreds of thousands of motorcyclists. Although the grand meeting of the motorcyclists was held in early August, this would be a follow-up meeting combined with a festival in Deadwood, close to the beginning of a scenic canyon drive in the Black Hills.

Agnes blissfully ignored the fact that their relationship had recently been very low-key, and she agreed to the visit. The abrupt change in her feelings toward Peter certainly contributed to her otherwise inscrutable decision. Robert was to arrive around 9:00 p.m., and she kept looking outside. She wanted to shepherd Robert into the house as soon as possible to keep him unseen by Peter. This was when she spotted the mysterious woman's car in front of her house and noticed that the light in Peter's house still was on. She found this strange because Peter was supposed to be at the friends' poker meeting. Clearly, the woman was visiting Peter again. But she also noticed that after about fifteen minutes, well before Robert finally arrived at around 10:00, the car pulled away. "Well, that was a short visit if that's what it was," Agnes thought. She couldn't guess the meaning of this, except that perhaps this woman was really interested in Peter. Interestingly, this thought didn't elicit any emotion from her; she had already decided she was through with Peter.

Earlier in the day, she had had a little time to analyze her relationship to Peter. For some people, jealousy has the destructive power to turn their rational brains insane in no time. Once this process has begun, it is very difficult to stop. But she always knew that she was in a very different category. She wasn't jealous by nature; she was only a very honest person who expected the same level of honesty from Peter. Now she found she couldn't trust him, and because of that, she decided to break things off. Agnes, above all, preferred clarity to sanctimony, particularly because she was confident that eventually she would find her true love. "I'll tell him there are no hard feelings," she decided, "and that I've completely forgiven him. I will add that I'm thankful for giving me the chance to get to know him and to learn from this relationship."

Agnes and Robert had started their studies at the Harvard Business School in the same class. In their second year they became good friends. But much to Agnes's chagrin, Robert spent more time in bars with friends than in school. Such behavior was far from Agnes's expectations of a man. Eventually, after the first semester of their third year, he dropped out of school and worked at an insurance agency for two years. Then he went back to the same school and just recently finished it after passing a final exam in the middle of August. He had received his diploma only a week ago, and this trip was in celebration of his achievement.

Even after their breakup, Robert wanted to keep contact with Agnes, and he was very tenacious in that. In fact, he still was in love with Agnes, and he hoped that in a face-to-face meeting, she would recognize that he really had changed for the better. In his email letters to Agnes, he never failed to mention, among some other triteness about her beauty and sentimental remarks, how much he wanted to repair their relationship. The desire to reconnect to Agnes rankled in his heart whenever he thought of her. Agnes usually didn't pay too much attention to these letters, and she let Robert know that she wasn't too crazy about the idea of reuniting with him. But in her most recent letter, she exonerated Robert of his previous failings, although she had still tried not to give him too much hope. Until a few days ago, Peter was in the center of her life, but now she had a much more open mind to consider new possibilities.

In a sense, Robert's visit came at just the right moment for Agnes. Having Robert around helped her deflect her attention away from Peter and regain her balance. She offered Robert something to eat and drink, but he asked for only a glass of wine, which was followed by several more, with Agnes eventually joining in. Perhaps because of the wine, Agnes told Robert everything about Peter and how he had betrayed her.

"Seemingly betrayed," Robert corrected her. "I don't think that during the one hour or so that you saw them together they were involved in sex," said Robert, attempting to convince Agnes that she was probably exaggerating the situation and was too hasty in drawing conclusions. "Not that I want to talk you into going back to him, because I still love you, but I do think that nothing serious happened between them. If I were in your place, and if that is the only thing you have against Peter, I would just ignore what you saw. She was probably just an old friend visiting him, perhaps seeking some advice. Whatever is going on between them, I don't think it's necessarily romantic. You even said that you couldn't see it clearly if they kissed or not. Maybe they only kissed each other on the cheek. That isn't unusual between friends. And perhaps the rest is only a figment of your imagination." Then Robert smiled widely. "But you see, if you would only come back to me, you wouldn't even have to deal with any of this. You and I are so congenial. We'll always get along. Come on, Agnes, let's not trifle away our precious time. Life is so short."

"Not now, Robert," Agnes answered, with the assured voice of one who holds sway over another. "Rob, you are incorrigible," she added in an almost approving tone that was just enough to indicate that his plea wasn't entirely hopeless.

Around midnight, one of Robert's cyclist friends called him up wanting to know if he still intended to come to the hotel.

"What should I do?" he asked Agnes.

"You can stay here if you want," she replied. Even she was surprised when she said that, but she didn't want to take it back even though she was aware that she was sending mixed messages to Robert. Was it the wine, or did she really find Robert attractive again, against the backdrop of her relationship with Peter?

"I will stay here and see you guys tomorrow at the motel," Robert told his friend.

When they had gotten ready to go to their separate beds, Agnes noticed that the light in Peter's house was still on, although she saw neither movement in the living room nor a car parked anywhere in the nearby. That seemed to rule out that Peter had a guest. In any case, she once again decided that after Peter's return from Los Angeles, she would call or visit him and tell him about her irreversible decision to break up with him. She no longer had any interest to find out whether Peter was innocent or guilty of cheating on her. Although the allure of the not-so-distant past that she had spent with Peter was still undeniably lingering around her, her indifference toward him was solidly growing with every passing hour.

The next morning, Agnes woke up at 7:00, and she was about to leave for work when Robert also opened his eyes. Agnes instructed him not to forget to lock the door before he left. Robert went back to sleep and only woke up again when one of his friends called around 11:00 to let him know that another guy in the group had gotten sick and they now planned to leave Rochester only early in the morning the next day. Then he added, in a taunting voice, "You don't mind staying another night with the girl, do you? Or will you be back to the motel tonight?"

Robert replied that things hadn't worked out as well as he had hoped and that he would indeed stay in the motel; he promised to meet them there around 3:00 p.m. Then he again fell back to sleep for a couple of more hours.

After a quick shower, Robert made some scrambled eggs and left the house for the motel a little after 3:00. He was driving toward Highway 52, when for a moment he glanced at the car to his left; his impression was that it was Agnes behind the wheel, but the light rain that started to come down reduced visibility. He continued driving but was so deep in his thoughts that he didn't see the stop sign at the last crossing avenue before the highway, and a policeman stopped him. He was lucky to only receive a written warning perhaps out of sympathy because the officer also rode a motorcycle. At the hotel, Robert learned that his companion was feeling better and they could indeed continue their ride tomorrow. Robert had three companions, and they all were busy cleaning and inspecting their motorcycles while devouring pizza. Before joining them, he decided to stop back at Agnes's house. But when he got there, he realized she had left already. He made a turn in the neighbor's driveway; that neighbor happened to be Peter.

It was already 5:35 when Robert called Agnes on her cell phone, and he was surprised to hear that she still was at work. "I thought you were already home. After I left your house, at one intersection I saw someone driving in the direction of your house, and I could bet it was you. Later I drove back, but your car wasn't there. In any case, I want to ask you out for dinner."

"Yes, that could have been me," Agnes acknowledged. "I went home for a minute to pick up a file that I couldn't postpone until tomorrow. However, I ask you to please listen to me very carefully. A few minutes ago, someone called me with the news that Peter is dead. Yes, you heard me right, he is dead. Apparently, he died this afternoon. Even though I obviously had nothing to do with it, please never mention to anyone that you were in Rochester or that you saw me at any time here. But wait a minute, how is it that you're still here?"

Robert expressed his shock and offered his condolences. He then explained that one of the guys had gotten sick but now he was all right, and they could continue their trip the next day.

"Good," Agnes replied, "and for us, the sooner you can leave Rochester, the better. Please, from now on, don't try to call me. Once I'm in Boston again, I'll call you."

"Fair enough, see you soon in Boston then," said Robert, still digesting the news. Over the phone, Agnes sounded a little nervous but at the same time amiable. For him this was a good sign for the future reunion with her he still hoped for. He would have liked to meet up with Agnes to provide security for her, but from then he kept his promise not to see or call her.

That afternoon, Agnes hadn't gone home to pick up a file, as she told Robert; instead she wanted to visit Peter. When she left for work in the morning, she noticed that in Peter's house, the lights were on, although he already should have been at the airport. Her first thought was that Peter had probably postponed his trip to Los Angeles and worked from home instead. This would have given her an opportunity to bring forward her planned meeting with him to tell him about her decision. Around 2:30 he called Peter's house but received no answer. Despite her new attitude toward Peter, she began to worry that he might be sick, which made her even more determined to pay a visit to him, even if it was uninvited. Without anybody noticing, around 3:10 p.m. she slipped out of the office and drove to Peter's house. On her way she saw Robert on his motorcycle but hoped that he didn't recognize her. Unexpectedly, a taxicab was parking in front of Peter's house. This ruined her plan to see Peter, but at the same time, the sight of the taxi reassured her that Peter had one or more guests, so he should have been okay. Thus, she drove back to IBM without stopping and decided to postpone her visit until the evening. Then, already back at IBM, she received the fateful call from Mike, which shook her up rather badly, although she managed to stay relatively calm during their conversation. This wasn't easy; she was surprised how she could keep her countenance.

When she went back home around 6:30 p.m., she saw several cars parked at the curb on the road in front of Peter's house, and she didn't need a lot of imagination to visualize what had been going on inside. The investigators were probably scouring the house for evidence to reconstruct how Peter's death might have happened. At this hour, she would normally be thinking of preparing dinner, but not today. Instead, she started to delete all her email correspondences with Peter from her laptop, and when she was done, she deleted all the messages from her cell phone as well. Then she turned off her phone because she wasn't in the mood to get a call from somebody in Peter's house, or from anybody else for that matter. After having a tea and nothing else for dinner, she turned off all the lights and went to bed a little after 8:00. "Aloofness never felt so good," she whispered to herself with a wry smile. She made up her mind to resign from her job tomorrow, before she would be fired in any case, and to return to Boston as soon as she could. Now she really regretted that despite her business acumen, she hadn't been able to resist the temptation to buy that house. The housing market had only partially recovered, so she would have to find a very talented real estate agent to sell her house on a good term. And, barring a sudden upswing in housing prices, she would probably end up losing money. She suddenly had a glut of problems to solve.

CHAPTER 16

MORE FINDINGS, MORE ANALYSIS

On Thursday morning, Doug had barely sat down in his office when Carl called him and asked if he could see him immediately. Doug begged him to come by after 10:30 because he had to catch up with his work. Although he was determined to help solving Peter's death, he also was aware that it was high time to refocus on his real job. In recent weeks his research practically had come to a halt, and he was even behind with his official duties. He didn't get rattled easily, but this time he was nervous. He couldn't make a living from just helping Carl's investigation; it was time to start working in full pelt again. That morning, he had to produce three documents for a panel that helped Mayo doctors make decisions about treatments. He simply couldn't wait to act on this matter because the treatment of three seriously ill new patients depended on it. With every passing day he became slightly more irritated and stressed, because if he didn't perform well at the Mayo, he could lose his job; there was nothing equivocal about it. He was very much afraid that Carl's visit meant that the work on Peter's death wouldn't be over for a while. Carl soon confirmed his prediction.

The other reason Doug asked Carl to come later was that by then Joseph was expected to finish the wine analysis, and thus he could give this information to Carl directly. Carl showed up a little later, and while looking at Doug with telling eyes, he placed a barrage of new items, one by one, on Doug's desk: two wine bottles with some wine left in each, one large plastic container labeled "PDCM," a plastic bag that contained about twenty spoonfuls of white powder, and a smaller container labeled "30-PDCM," most probably indicating that initially there were thirty tablets in it. These new items were exactly what Doug was afraid of.

Carl chose to ignore the worrisome look on Doug's face and meticulously described each item. When he wanted to secure someone's collaboration, he could usually persuade that person without antagonizing them. His favorite method of persuasion, delivering his wish in a calm and objective tone, focusing solely on the importance of the task ahead of them, had always worked on Doug. Carl began to describe the items. "The wine bottles are from the refrigerator in Ulrich's house. The labels on them match the labels on the two distinct bottles found in Peter's garage. The larger PDCM-labeled container is from Ulrich's laboratory. It contains exactly 136 pills. Both Ulrich's apartment and his laboratory are closed, but I received permission from the judge to search it. The container was sitting on Ulrich's office desk along a Dell computer which contained a lot of information on PDC and PDCM and their intended uses. As it turns out, the *M* in PDCM stands for 'modified,' meaning that PDCM is a modified version of the PDC compound.

"I also conducted another search of Peter's house, including the medical cabinet in the basement's bathroom. It was there that I found the container labeled '30-PDCM.' During my previous searches, I missed this container because it was a Centrum A-Z vitamin bottle, with the front label still intact. Obviously, the container came from Ulrich, and to save money, he must have used an empty vitamin container to put his tablets in. Peter then probably put it on the shelf with the side showing only the vitamin label turned to the front. At first, I also mistook this white powder for the stuff some people use to whiten their teeth. But as I examined the bag, I recognized that it was different. Now, I have no doubt that this PDCM stuff came from Ulrich. The question is now if the stuff in the tablets that came from Ulrich is the same as what we found in Peter's house and if any of the wine bottles contained PDCM, perhaps along with zinc. Doug, I ask you to please analyze each item to see if the PDCM you find is the same as what you found in Peter's body."

"I see that once again I can't duck out of this," said Doug with resignation. He promised Carl that for one last time, he would ask Joseph and Mary to perform these analyses. Doug really hoped that with the latest experiments, they had been able to bring this project to a conclusion, but the case for examining these new items seemed too strong to refuse. He would have to broker a deal with Joseph and Mary, and if they would have to come in during the weekend, he would have to find a way to add these days to their vacation time. But he knew that he couldn't continue to demand extra work from them for too long without consequence.

Just then, as if Joseph knew his name had been mentioned, he entered Doug's office. "It's nice to see you again, Carl," he said. "I have some new information for the two of you. We checked all six wine bottles, and in two of them we found quite a bit of the PDC-like material that was found in Peter's body. I calculated that each of these two bottles had about two milligrams of this compound per one milliliter. Now, if someone were to drink a whole bottle of this wine, that is, seven deciliters, this would mean ingesting about 1.4 grams of this supposedly dangerous material. If Peter emptied both bottles during the same evening, he ingested 2.8 grams. And these two bottles were the ones that were different from the bottles sold by Apollo Liquor."

"Is that correct?" asked Carl, clutching Joseph's arm. "Are you absolutely sure?"

"Yes," answered Joseph.

"I know that you will hate to hear this," said Doug, gesturing to the items on the table, "but we have few more samples to analyze. Carl brought them here, and he even knows that the compound you found in the wine is called PDCM, meaning modified PDC. Now the question is if the same compound shows up in the tablets and the wine and what the white powder is composed of. We need to have this information."

"Nice," said Joseph, "and I suppose the deadline is within one hour?" He said it in a way that one could sense a kind of derision in his voice. In fact, inside, Joseph was livid at the idea of working again overtime; he was wondering, just like Doug did earlier, when this would stop.

Doug expected this kind of reaction from Joseph, and he remained calm. "No, I'm more reasonable than that. I ask you to do this by tomorrow, late afternoon. But because we have other analyses going on as well, I ask you and Mary to work some extra hours this weekend. I will arrange for either extra pay or extra vacation time, depending on your preferences. Carl, can you come over tomorrow around 4:00 p.m. to hopefully settle this PDCM matter once and for all? Joseph, I suggest that you take two samples from each bottle, then randomly take twelve tablets from both containers and analyze half of the samples separately for PDCM and zinc. One sample from the plastic bag will be sufficient for the analysis because it's likely to be composed of only one substance."

"I'm sorry, guys," Carl intervened. "This final analysis really needs to be done to prove whether the drug you found in Peter's body came from Ulrich or not. If there ever will be a trial, both the prosecutor and the defense lawyer will have to know the truth. Unfortunately, our crime lab isn't equipped to handle this kind of sophisticated analyses. It is mostly specialized to analyze DNA. In the meantime, I will search for evidence that might clarify whether Peter took the drug knowingly and voluntarily or if he was poisoned. In the former case, Ulrich probably wouldn't even be charged with manslaughter, although I can't know for sure. There is a lot of queerness going on in the courts."

Joseph, seemingly appeased, was about to take away the new items when Carl stopped him. "I was waiting to see," he said, "if either of you would notice a difference between the two bottles."

"Oh yes, I do," Doug exclaimed; he now had a good direct view of the bottles and recognized that the seal on one of them was clearly cut. He also had a theory of what happened: "It looks like someone, perhaps Ulrich, removed the cork, and after adding something into the bottle, he forced the cork back. To ensure tight closing of the bottle, he sealed the opening around the cork with scotch tape. I guess that the person didn't want to lose the wine's quality. In any case,

Joseph, please make it sure that you mark the two bottles differently, so we know which sample belongs to which bottle."

"I believe that there is no need to spend any more time with this," said Carl. "I'll be back tomorrow at 4:00 p.m., and let's hope that you won't need to do any more analyses. I know this came out of the blue for you, and I can imagine that it ruined some of your plans. All I can say is that I'm very thankful to you for doing all this. But think about it this way: after you finish this last analysis, we might finally be reasonably sure of what killed Peter and whether Ulrich manufactured the killer pills or not. And from that point on, it will be on my shoulder to find out if there was foul play involved or if he died accidentally because someone was careless. Something tells me that a successful ending to this tantalizing story is within our grasp now."

Finally, Doug had almost two uninterrupted days to deal with the tasks he was paid for. It was high time; he was behind on all fronts. He opened his emails, and one immediately stood out. A research council in California asked him to review a grant proposal on stem cells. The state of California provided a lot of money for stem cell research to find out, among others, whether embryonic stem cells can be replaced by stem cells from nonembryonic tissues, or if they can be altered to make them safer for humans. Most stem cell researchers agreed that using embryonic stem cells to heal diseased human tissues without altering them wasn't feasible, because after transplantation, many of these cells can eventually become cancerous.

The applicant showed that certain types of stem cells derived from the bone marrow do not become cancerous, and thus they can be safely used for transplantation to repair damaged tissues. Since there was a previous publication that such stem cells can become cancerous, this issue needed to be dealt with. Thus, after reading the proposal, Doug immediately decided that this was an egregiously important project that needed to be funded. The grant seeker was in a very good position to move ahead with the practical applications; he had a productive and technically cutting-edge laboratory with a very good track record. Doug decided that early next week he would write a very supportive report to the granting agency. He enjoyed very much that he was back in the normal life of a scientist. By Friday afternoon, he had reduced his backlog of urgent matters to zero, although at the expense that he had no time to supervise his assistants.

That same Friday Carl showed up at Doug's office at 4:00 p.m. sharp. Doug was a little uneasy because he had no idea what results his assistants would come up with. But his worries were uncalled for; the always-reliable Joseph entered his office with a sheaf of printed records in his hand. "We mopped up all the samples, and we couldn't do any more analysis even if we wanted to," he said. "Do you want me to summarize the data for you, or do you also want to see the original records?"

"First, just go ahead with the summary, and then we will see," Doug advised him. The thickness of the package was too intimidating for Carl and even for Doug.

"Well, the story is the following," Joseph started his presentation. "The wine in the bottle that someone opened and then resealed contained a lot of PDCM, about the same amount that we found earlier in two bottles taken from Peter's house, and that weren't from the Apollo store. Obviously, someone put PDCM into all three bottles. Two out of six tablets that came from Ulrich's laboratory and three out of six tablets that came from Peter's house also contained large amounts of zinc, close to fifty milligrams. In addition, each tablet contained a lot of PDCM. It seems to me that all the PDCM tablets came from the same source, about half of them supplemented with zinc and the other half not. However, it seems to me that zinc wasn't added to the wine. The wine in these new bottles contained about the same amount of zinc as did the wine in the bottles from the Apollo store. We also analyzed other brands of red wines. In general, the red wine that apparently came from Ulrich contained four to seven times more zinc compared to the other brands."

"That's an awful lot of zinc," Doug commented. "But the amount of zinc in the tablets is staggering too, considering that we need only about fifteen milligrams of zinc a day. Now if we consider that he also took a vitamin tablet every day, which alone supplies close to twelve milligrams, that he also ate oyster, which is very rich in zinc, and on top of all that he apparently drank two bottles of zinc-rich Bordeaux wine, during the day and night before his death, he could have ingested up to one hundred milligrams of zinc, or perhaps even more. And we don't even know how many tablets he took. But even without the tablets, this amount of zinc is way too much, particularly since he also ingested PDCM, which helped the zinc to enter his cells. Now I'm almost sure that zinc engendered his death. What an unfortunate situation poor Peter got himself into."

"Yes, you nicked it," Carl agreed.

"And how much PDCM was in the tablets?" Doug asked.

Joseph was ready with the answer. "I have an approximate calculation, and it shows that there is about two hundred and fifty milligrams of PDCM in each tablet."

"Okay, let me do a little calculation," said Doug while he pulled out a paper and a pencil. "We can assume that the maximum amount of PDCM in the blood is about 10 percent of the ingested amount, which is twenty-five milligrams. We have about five liters of blood, so one liter could contain as much as five milligrams of PDCM. I know from Steve's search data that one to five milligrams of PDC in one liter of blood can effectively kill cells if there is at least the same amount of zinc in the medium. What I am getting at is that Peter could have gotten enough PDCM from just one tablet to kill him if there was sufficient zinc in his blood. We don't know the amount of zinc in his blood after ingesting one hundred milligrams of it, but it is reasonable to assume that it reached at least the level that PDCM did."

Of course, Doug was aware that he could be making several mistakes in his assumptions. For example, he could only approximate, very crudely, how much PDCM would be inactivated through its binding to blood proteins. Also, both PDCM and zinc are eventually removed from the blood, but Doug had no idea how fast this process was. It could be that it took more tablets or a bottle of wine spiked with PDCM to cause death. He felt it was pointless to share these uncertain elements of his calculation.

"Just one last question on this PDCM," Carl began. "Can we be absolutely sure that the PDCM you found in the tablets and the wine is the same material that showed up in Peter's body?"

Carl's question exasperated Joseph a little bit. "Absolutely, no question about it," replied Joseph tersely. "Our data very consistently shows that. Believe me, our instrument provides such accurate data that it doesn't leave any room for error. This instrument would also immediately pick up any indolence in the sample preparation."

"And what have you learned about the white powder? Could you figure out its composition?" Doug asked Joseph.

"Initially, we had some trouble identifying it with anything. But then we refined the measurements, and it turned out to be identical to a compound called phytic acid, or phytate. This is a relatively simple molecule that has six carbon atoms in it with a phosphate group attached to each of them."

"What on earth can someone use phytate for?" Doug searched his memory, and in a little while he remembered that he had often encountered this compound in his nutritional studies. "Now I get it," he said with a beaming face. "This compound is present in large amounts in cereals and other wheat products. It is known to bind certain metals, such as zinc, very strongly. If someone eats a lot of cereal in the morning, this could lead to zinc deficiency because the body can't use the phytate-zinc complex as a source of zinc. Thus, cereal lovers need to be a little cautious about this because they may end up with zinc deficiency. But you can avoid it by consuming zinc-containing foods or pills a few hours after the cereal was consumed. During the intervening hours, the phytate is metabolized and will not interfere with zinc. I can tell you more about phytate if you are interested. I just have to find the corresponding file."

Doug was very conservative with his filing of scientific papers. Unlike most of his colleagues, he wasn't in thrall to computer filing systems because he found them to be less of time savers than they were advertised. Besides, just as he didn't like electronic books, he also liked to read scientific papers on a hard copy. Doug had close to 150 boxes on the shelves in alphabetical order, each containing up to 100 printed articles. He collected articles in the areas of cell communication, cell metabolism, enzymology, cancer, diabetes, obesity, wound healing, cardiovascular diseases, and even bacteriology. Each box was devoted to a subcategory within these larger areas. For example, within the larger area of Metabolism, he had subcategories like obesity, insulin action, and so on. Under Cancer Treatment, he had a file on phytate that he found in less than a minute. He opened the file and showed it to Carl and Joseph; it contained around thirty papers, each devoted to the anticancer effects of phytate.

"You see," Doug commented, "several research groups claim that phytate reduces the risk of colon cancer, and that it is beneficial against other cancers as well. Some other papers describe phytic acid as an antioxidant. But Peter had no cancer. Besides, it is nearly impossible that he would have read papers on phytate. Could it be that a knowledgeable person advised him about the benefits of phytate? Maybe the same person was also aware that Peter liked beef and that red meat increases the risk of colon cancer?"

As Doug went through the whole sheaf of articles in the file, he also found one describing how phytate binds zinc much more avidly than PDC does. This lead Doug to an important conclusion, which he announced loudly: "Another possibility is that the person who gave the tablets and wine bottles to Peter was worried that something might go wrong with the PDC-zinc combination. Thus, as a precaution, he also gave him phytate to bind and offset the toxic effects of the zinc. According to this paper, minutes after oral intake of phytate, all zinc in the body's circulation would end up binding to it. Once bound to phytate, zinc cannot enter the cells and thus remains harmless. Possibly, Peter might not have had time to dissolve and drink the phytic acid solution, or maybe he simply forgot about this advice. But if this was the case, then Ulrich, the most probable provider of all these drugs and wine with PDCM, had no malicious intent. He just wanted to impart Peter maximum benefit with the least risk. This might be a critical new twist in this sad story, and I really pray for it to be true. Such an outcome would also fit Ulrich's character, as far as I know him." He paused, and then added, "It is hard to imagine that Peter was somehow forced to consume PDCM and zinc. He must have consented to it. Listen, Carl, you said that you want to look for evidence of whether Peter took the pills voluntarily. I believe that a good start would be to search Ulrich's computer and his house, along with Peter's house, for a document that would indicate Peter's consent to take the drug."

"I agree, and as I said earlier, I meant to extend my search in that direction as well," responded Carl. "And I don't think that Ulrich or Helga, even if they were involved, had any malicious intent. After all, Helga was already in the process of separating from Peter, and they had lawfully settled everything. She couldn't get any more from the settlement, so she had no incentive to kill or help someone to kill him. And I also have trouble seeing any motive on Ulrich's side. Yet one thing bothers me very much. If we assume that Ulrich was the one who provided the drug, how could he possibly be so sure that the drug was not dangerous? This is a real vexing problem for me."

This was a difficult question, and it took quite a bit of time until Doug came up with a plausible theory. "You know, it often happens that young drug developers are overly optimistic about the safety and efficacy of their product. They follow a hefty goal and forget that safety comes first. My impression is that Ulrich was probably a little bit careless, and instead of giving Peter detailed instructions on the drug's use, he treated the matter too casually. However, even if this were true, there still must be more to it. I know that Peter was very careful with the drugs. Thus, I believe Ulrich had to present some very convincing and astute arguments to him about the value of his product. Just think about it. There were no clinical trials. Again, knowing Peter, the most convincing argument Ulrich could have come up with was that the drug is very similar to a natural substance present in some food item. Peter liked the idea of using the health benefits of certain foods over drugs. Ulrich's other argument could be that he also had diabetes and the drug had helped him better than any of the approved drugs. He could have even cited some evidence that taking PDCM was the only way to avoid eventually using injected insulin, which he was afraid of. If all this is true, then obviously Ulrich also has diabetes, which should be easy to verify."

"That's a nice theory," Carl agreed. "But going back to Helga, she may be holding the cards in this matter. I'll talk to her as soon as she comes back from Washington and ask her about the eating habits of Ulrich and Peter and if Ulrich also had diabetes. You know, just thinking about phytate, I think it would be interesting to know if these guys generally ate cereal for breakfast or not. Perhaps Ulrich had cereal for breakfast but Peter didn't. Who knows? This alone might explain the difference in their reactions to PDCM and zinc. Peter could have had less phytate in his blood, and this could leave more zinc available to bind to PDCM, thus becoming a cell killer."

"Carl, bravo, you are amazing," Doug exclaimed. "It seems to me that in the process of this investigation, you have become a scientist."

"I'm just trying to use some logic here," said Carl.

"Yes, we need that." Doug smiled. "But there is still one point we differ on. The questions we raised and the conclusions we reached all seem sound to me, but I still firmly believe that Helga had a hand in all this. I know the girl. She is nice, but her behavior can be very erratic, and I believe that she still holds sway over Ulrich. How otherwise would you expect Ulrich to know Peter? We might never be able to get Helga to admit to this, but I just can't see any other possibility. But I agree with you that first you should ask her more neutral questions, like Pater's eating habits, and then you can see how far you can go. When will Helga be back from Washington?"

"Last time we spoke, she said that she would be available to talk to me next Tuesday," Carl replied. "But before I leave, I have another news item for you. I tried to talk to Agnes, but I only found the For Sale sign at the front of her house. I inquired at IBM, and they told me that she had resigned from her job, which she would have lost anyway, they told me, and relocated to Boston. At this moment I don't read anything into this." Carl added, "She simply might not have had another option. Although at some point I must interview her as part of our standard protocol. By the end of next week, I will let you know how much more I will have accomplished. Until then I'll try not pester you unless absolutely necessary."

CHAPTER 17

AGNES AND HELGA MAY GET OFF THE LIST OF SUSPECTS

In Washington, Helga received a call from Ulrich that he apparently made from a public phone. Helga quickly informed him that Peter had died and that Doug was working very hard to figure out what had caused his death. "I hope this has nothing to do with your tablets, although I know from my forever-faithful Joseph that they found something structurally close to PDC in Peter's body and that Doug presumes this was related to his death, but I have no idea if they will abide by this theory. I'm in Washington now and I really don't know the very latest news, but I have a bad feeling about this case and not looking forward to going back to Rochester. I'm supposed to be having a good time here, but how can I after this?"

Ulrich usually tempered his emotions well, but this uncanny news was a crushing blow to his psyche. Initially he was paralyzed with the shock, but then hot anger started to build within him, directed at Peter. His instinct told him that Peter had done something very stupid; he might have misused either the tablets or the wine or perhaps both; such oversight could have easily killed him. Peter's death and the likelihood that his role in providing the drug would be obvious seriously threatened his future in the US and perhaps even in Europe.

"Are you still there?" asked Helga.

"Yes, I am," replied Ulrich, with a voice suggesting that he was regaining control. "Listen, Helga, this news changes everything. I think that for the time being, the best we can do is not tell anyone about my—in fact, I should say *our* dealings with Peter or even that I ever met him. Fortunately, I'm in Switzerland, which is a relatively safe place. But if someone asks you about my whereabouts, please say that you know neither my forward address nor my phone number. However, you aren't safe there in Rochester because the detective on the case will certainly interrogate you. Detectives can be very tenacious, and they will pursue you. If this happens, please simply tell the detective that we last met long ago, and thus you know very little about me, and nothing about my company. After all, we're not in a relationship, are we?"

Ulrich took a moment to organize his thoughts about their strategy moving forward. "Occasionally I'll call you from a public phone, but please remove even that number afterward. I admit that this situation doesn't look good, but now that we're in it, at least you should be free from interrogations. You see, if they have found this compound in Peter's body, as you say, and they figure out that I produced it, there is only a small chance that I can avoid legal consequences, because I wasn't authorized to give those tablets to anyone. On a positive side, I made Peter sign a declaration in which he took full responsibility for taking the tablets and the wine, acknowledging that I required from him full adherence to my detailed instructions. This document is with me, and I also left a copy in my office in Stewartville. I still don't get it how Peter could be so oblivious to my instructions, but at least this declaration might get me off the hook."

"That would be fantastic," said Helga. The existence of such signed document put her in a much more positive mood.

But Ulrich wasn't through with his warnings just yet. "Now, Doug and the detective will almost certainly assume that without your mediation, I couldn't have known Peter. Thus, they will treat you as a co-conspirator. That's why you have to be very careful to not reveal anything to anyone about our recent interactions."

"But how would they be able to figure out that we met recently?" asked Helga.

"Perhaps the detective will not be able to find a witness to confirm our meeting, but the telephone company could confirm our contacts. The more I think about it, Helga, the more I am convinced that you should consider coming after me to Switzerland or Germany. I'm just about to close a business deal, and we would have enough money to start a new life, you and me together. For quite some time we won't have to live our lives on a shoestring budget. Or is this a bad time to think about these kinds of things right after Peter's death?"

"To be frank with you, I never felt the deep sorrow that I perhaps should have felt after hearing the news. I simply acquiesced to what happened. But his death did make me reevaluate myself and what I did wrong in that marriage. Instead of helping him through his struggles, I only beset his life with more difficulties. I would make a huge deal out of the smallest issue, for no good reason. I also often think of Peter's children, and I fault myself for not allowing him to interact with them as a father should. I was too self-indulgent at that time. I can see now that he was torn between his children and me, and he primarily chose me while getting little from me in return. I prevented him from being a good father, and in the retrospect, this was a terrible thing to do. No matter how I think about it, it was my attitude that led us down this path, leading to our separation and eventually his death. And it's not just Peter. My father is dying too. So honestly, I'm still not in the mood to think about reuniting with you. All I want is a peaceful life, at least for a while, not a new relationship with yet another unpredictable outcome. You might think that I'm hesitating to join you because I'm a craven person, but this where I stand now."

"No, no, I fully understand you," Ulrich replied, "and I'm very sorry to hear that the doctors can't help your father. I hope you know that the last thing I want is to cause you more problems. But under the circumstances, here is what I think we should do. I will give my phone number to your mother, and then when you come to Stuttgart, call me. I won't ask you to reunite with me, but we'll need to discuss calmly what to do about everything." Ulrich paused. "I didn't leave a lot of things in Stewartville, and I wouldn't have any problems starting again a new life in Germany. I suggest that at some point you try to do the same, although you are in less danger than I am. Have a good day, Helga, and no matter how you feel about me, I am still in love with you."

"Thanks." That was all Helga could say at that moment; she wanted to enjoy Washington and its galleries the best she could, and she wasn't in the mood to continue with the conversation.

Yes, Helga knew that eventually she would have to open a new chapter in her personal life, perhaps alongside Ulrich, but she wasn't ready yet. She needed some calm first so she could assimilate the richness the galleries had to offer and continue this new chapter of her life as an artist, granting her independence both psychologically and financially. Thus, for a while, she would distance herself from Ulrich and Peter and all other men in the world. Carl wouldn't be able to prove anything to implicate her in Peter's death. Besides, she wasn't the one who had manufactured the drug and given it to Peter. So she advised herself, "When you meet Carl, don't be jittery, because that makes you a suspect of wrongdoing."

Even though Helga had the physical stamina to stand in galleries for many hours, by Sunday evening she was totally exhausted. She was ready to go home to Rochester the next day even though she wasn't looking forward to her meeting with Carl. She was scheduled to leave Washington at 11:35 a.m., but because of bad weather, the flight was delayed by four hours. As a result, she missed the connecting flight from Minneapolis to Rochester, but fortunately the airline paid for her hotel. Around noon the next day, she had barely entered her house when she received a phone call from Carl, who was hoping to meet her that very afternoon. "Carl, please, I can't. I just arrived from Washington, and after taking a shower and eating something, I still have to report to a night shift at the hospital."

"I understand," said Carl. "Then how about meeting tomorrow after lunch?"

"That's better. Where would you like to meet me?"

"I don't mean for this to be a very official meeting even though I'll take notes that you will sign. Thus, I would prefer to meet you in the house where you and Peter lived together, if that is okay with you," said Carl.

"All right then," agreed Helga with clear urgency in her voice; she couldn't wait to shower and eat. She would have had time for the meeting because she had to report for work only later in the evening, but she wanted to call her mother in Stuttgart, and since she expected bad news, she wanted to start looking for an airplane ticket.

The news Helga received from her mother was indeed very bad; according to the doctors, her father was unlikely to live beyond one week. She immediately contacted Delta Airlines and bought a one-way ticket to Stuttgart through Paris for the next Monday. Then she called her hospital to request three weeks of unpaid leave, with the understanding that she may come back earlier if circumstances allowed it.

She also called Joseph to inquire about any new findings in the lab that she might not have yet heard about. Joseph could only confirm that the data showed without ambiguity that Ulrich was the one who provided the PDCM-containing tablets and wine to Peter. They had also found that Ulrich gave Peter a white powder too, probably to counteract the side effects that might be caused by the PDCM and zinc.

"Do you think that Carl and Doug are certain that Peter was poisoned with this compound?" she asked.

"Yes, I think they are pretty much convinced that this is the case. But I also overheard that they don't think the drug was given to Peter for any malicious purpose. They think that Ulrich may have had good intentions, although at the same time he had obviously been reckless, and then things somehow ran amok. I assume this is where they are now, although I admit that Carl and Doug usually don't share their conclusions with me. I only gather them from hints they drop when they fling a lot of work at me and Mary. But how about you, Helga? Are you doing fine?"

"Oh, I just came back from Washington, where I was visiting galleries," Helga responded with melancholy in her voice. "It is such a different world there, so refreshing. I had a chance to see a lot of new art, many abstract paintings, sometimes mixed with more traditional ones in the same gallery. I think I got a good cross-section of today's art, some of it ridiculous, some of it rather fascinating. In short, it's clear to me that in modern art, anything goes. But in any case, I can certainly make good use of these exhibitions in my own work. I wish I could have stayed there longer, but frankly I don't have the money, and my time is also very limited. Unfortunately, my father is dying, and next week I'll fly to Stuttgart, purportedly for his funeral. Because of all these events, I haven't even touched a brush for weeks now, and that makes me so irritated. In Washington I became obsessed with an idea for a painting that I can't wait to stand behind my easel again."

"I am terribly sorry to hear that your father is dying. Is there anything I can do to help?" asked Joseph with heartfelt sympathy in his voice. Helga didn't want to say no; she suggested that she might need a ride to the airport. Joseph was glad to oblige.

Helga wanted to finish the conversation, but with Joseph it was never easy. "If it's not a secret, what is your idea for the painting you mentioned?" he pressed her.

"Oh, please not now, Joseph," she protested. "I have to go to work shortly, and besides, the image of the picture that I have in my brain is still quite amorphous. I need to work a lot more to give structure to it, before I'll be able to explain it. It's like your clouds. You need to look at them before you can make full sense of them."

"I understand perfectly what you mean," Joseph assured Helga. "Then I'll see you soon, if you need a ride."

"Thanks again," replied Helga with a sigh of relief; she had finally managed to not only end the conversation but end it amicably. Now she could relax a little bit before her night shift.

Agnes, after returning to Boston, had a lot of time to think about Peter's death. She was at peace with him. Although she didn't overly mourn him, she felt no anger either. From her vantage point, she could now see clearly that Peter wasn't the right partner for her; they were not destined for each other, spiritually or otherwise. This would be so even if Peter were still alive and even if that furtive woman would have never showed up at his house.

Agnes's major concern now was how to avoid becoming a suspect even though she had absolutely nothing to do with Peter's death. But some people in the investigation might find out that something had been going on between her and Peter. She might be even put on the list of suspects if she wasn't there already. There were times when she was devoured by fear; she had to do something about this. After a sober assessment of her current situation, she concluded that an interrogation with questions coming like a barrage wouldn't feel pretty, even if she would be cleared in the end, which might take quite some time. She had to avoid such investigation—but how? She took stock of her options based on the likelihood that even if Carl heard some rumors about the relationship between her and Peter, there was nothing incontrovertible about them. The best course of action, she decided after considering all the possible pros and cons, was to be proactive by directing the detective's attention elsewhere in a way that seemed genuinely helpful but subtle at the same time.

Agnes still had Carl's phone number, and the very Tuesday afternoon when Carl wanted to talk to Helga, he instead received a call from Agnes. She explained to Carl that the reason she returned to Boston was that she had had no other option other than to return there. She also suggested that she had an observation in Rochester that might help the investigation.

"What kind of observation?" Carl asked with great curiosity.

"As you know, my house was close to Peter's, and my kitchen window faced his living room with a good view of a large open space, including his driveway. I'm an avid bird watcher, and many different birds used to flock on the nearby large trees. I used to spend a lot of time looking out through that window, and sometimes I saw certain things unintentionally. Four days before Peter's death, more exactly late that Friday afternoon, I saw a woman in Peter's house who stayed there for at least an hour. Her car had a Minnesota license plate number BGE 178." She said all this in the most affable manner she could come up with.

"You must have a very good memory remembering this number," Carl noted. Agnes felt that Carl was somewhat jeering in his remark, but she remained calm.

"Well, being an accountant, I do have good memory at least for numbers. But in this case, it wasn't my memory. I took a photograph of a blue jay in front of Peter's garage, and I happened to capture the front of the white car with the license number. But while we're on the subject, I also noticed that the same car was parked in front of my house late in the evening on the following Monday, just a day prior to Peter's death. I didn't see her, and I wasn't really interested to know whether she went into Peter's house or not. She could have, because the light in Peter's house was on, and he was apparently home. I don't know if this will help your investigation or not, but I thought it was best to share this information with you before I forget."

"Thank you very much for sharing this information with me. It may turn out to be extremely useful," Carl said with a raspy voice that he had been struggling with for days now. "But once I have you on the phone, may I ask you a personal question? Would you say you had a relationship with Peter that maybe went further than what is usual in a workplace?"

Agnes was in a tough spot; her honesty was taken to task. She had expected this question, but now hearing it she felt very uncomfortable. She had to respond quickly to avoid raising Carl's suspicion. "Perhaps we had been somewhat closer than most bosses and secretaries, but nearly not so close that we ever discussed living together or marriage—far from it." In a way, her statement was factually true, particularly in the light of her recent epiphany that Peter wasn't the right partner for her and that they would have broken up in any case.

"Thank you again," said Carl. "I assume you wouldn't object to it if I were to call you again just in case we needed further clarifications?"

"Of course not," replied Agnes. Carl was convinced that Agnes and Peter had been in a relationship, but he couldn't prove it, and even if he could have, perhaps it wouldn't provide any new leads. So for the time being, he put Agnes on the sidelines of his investigation and instead turned his attention to the license plate number.

Agnes put the phone down with a great sigh of relief. She was happy that she appeared to have avoided getting entangled in further investigation. She was pretty sure that she had avoided prolonged interrogations by being forthcoming and providing only minimal information about herself, deflecting the attention from her to the unknown woman. She blanched at the idea of Carl prying into her romantic relationship with Peter in the middle of her new effort to get her life back on track. She hoped that this interview with Carl, an old stiff in her mind, would be the last.

It took Carl only five minutes to access the name of the car owner with the license plate number Agnes provided. He was shocked when he saw the name on the computer screen: Sarah Lowry.

"Oh, for God's sake," he muttered to himself in exasperation. "Doug's wife is involved? How she is embroiled in this? This is turning into a complete farce. I wish I could retire right now. Perhaps someone else would have more luck with this investigation. Or could it be that the meeting was innocent? Maybe Sarah and Peter just met as old friends?" In any case, he decided he would ask Helga if she knew whether Sarah was an old acquaintance of Peter or if they had any professional relationship.

When Agnes called Carl, he had been in the middle of calling cleaning companies to ask them if they had recently worked in Peter's house. He figured that only a professional maid service could have accomplished such a nice order in the house as what he had found there. So far, none of the companies gave him an affirmative answer, but after his conversation with Agnes, he still decided to call the last three Rochester companies on his list. To his joy, the secretary of the first company he called, Molly Maid, remembered doing a job there. She even volunteered to tell Carl that Sarah Lowry, an old customer of theirs, was the mediator. She then looked up her record and told Carl that they had done the job two weeks ago on a Monday. Without any prompt from Carl, she added that the crew had found the first floor almost derelict, but they had managed to clean the whole house in one day because the basement was in a much better condition.

"Thank you so much, you helped me a lot!" said Carl; at that moment he wanted to kiss this talkative secretary for the information. He put down the phone with a great sigh of relief. The secretary's account seemed to explain why Sarah had been in Doug's house that Monday, although it remained unclear to him why she had parked her car in front of Agnes's house that evening. This is something he still needed to find an explanation for. Besides, what if she had also visited Peter on Tuesday, the day of his death? This was yet another prescient guess for which he might never be able to provide evidence. Carl was reluctant to talk to Sarah for the simple reason that she couldn't possibly know about the PDCM tablets. Contacting Sarah would also bring Doug into the investigation in a very inconvenient manner, which he wanted to avoid at all cost. Clearly, he first had to clarify what roles Ulrich and Helga might have played, and then move on to Sarah and Agnes if he still deemed it necessary.

Helga arrived at the house at the agreed-upon time, where Carl was already waiting for her. Entering that house, that not long ago was also hers, sent chills down her spine. According to the settlement, she wouldn't own any part of the house; she assumed that it would eventually go to Peter's children. In the throes of the divorce process, she forgot to take some of her belongings. After her return from Stuttgart, she would still have to pick up all these, including books and a laptop that was sitting undisturbed in the same corner she had left it in.

Carl didn't allow Helga too much time to dwell on her thoughts; he meant business. After some mandatory small talk about the weather, her father's health, galleries in Washington, and her upcoming trip to Stuttgart, he soon shifted the conversation to Ulrich and Peter.

"Now, Helga, we are both short of time, so let's get to the point. This case clamors certain clarifications, which I hope you can help me with. My first questions are, do you know where Ulrich is, and did he contact you? Secondly, I checked Ulrich's and Peter's earlier medical records, and at that time, both had diabetes of various degrees and origins. Ulrich's doctor told me that he self-treated himself, which he didn't agree with, although whatever he did seemed to work for him. Helga, do you think that Ulrich gave Peter tablets and wine to treat his diabetes?"

These pointed questions sent shudders through Helga's body; Carl was clearly on to something, and he was very close to the truth. She had to be very careful not to fall into his trap.

"To answer your first questions, I know indirectly that Ulrich is now in Switzerland, but I have no idea exactly where or for how long, because we haven't been in contact for a long time. I've had no intention of maintaining a relationship with him. That's the reason why we split up in the first place. As for your second question, at the time when we still were together, I knew that he was working on something that could be used to treat diabetes, but he told me nothing specific about this drug, and I wasn't really interested enough to ask. After our separation, we still had small conversations when we met occasionally by chance, but otherwise we weren't nearly close enough to share, to any detail, our work or personal lives."

After few seconds of silence, Carl asked her again, "Do you know if Ulrich gave drugs or wine to Peter?"

"I don't know anything about that," answered Helga resolutely. "But I hadn't talked to Peter for quite a while before his death."

"All right," said Carl. "Then I establish for the record that after your separation, you and Ulrich haven't been in regular contact, and you don't know if Ulrich gave any tablets or wine to Peter." He said this while sharply looking into Helga's eyes.

"I agree with your conclusion," Helga said while looking straight into Carl's eyes.

"I see. Then I have yet another question. Do you have any idea whether Ulrich knew Peter, and if he did, then how?"

"While I was with Peter, I told him some stories about Ulrich, but I don't know if they have ever met before or after our divorce. But they conceivably could have met recently without my knowledge. I know that German business contractors used to come to IBM, and maybe Ulrich and Peter hung out with them at a party thrown for the guests."

Carl was in a difficult position, because short of Helga's admission, he had no other way of finding out how Ulrich and Peter met and if Helga was the mediator between them. Since this track of questioning was unfruitful, he again changed the topic.

"Helga, it might help us if we were to know what Ulrich and Peter used to eat for breakfast."

She was very much surprised to hear that question but immediately realized that there was no danger in giving an honest response this time. "When we're together," Helga began her answer propitiously, "Ulrich used to eat a lot of cereal, but I don't know if he changed his diet since then or not. Peter also used to eat cereal for breakfast, but last time we met, he boasted that he had now switched to oysters. He was fascinated by the enormous amount of zinc oysters contain. Someone once explained to him, perhaps Doug, during one of their poker sessions, that zinc was good for enhancing sexual desire and performance and fending off Alzheimer's and infections, among other things that I don't remember. In fact, zinc was the reason that he also liked to drink Bordeaux wine, recently in larger amounts. He admitted this to me when we're still working on the divorce documents." She couldn't bring herself to say that she had advised Peter to drink Bordeaux.

"Thanks, Helga. I have just one more question, and after this I'll let you go. Do you know if Sarah and Peter have ever been friends?"

So far this was the best opportunity for Helga to demonstrate her willingness to collaborate with Carl without any danger; she couldn't possibly flub that question. "Well, what I know is that in high school, Sarah was Peter's first girlfriend, before she dumped him for Doug. He was deeply hurt by that, and in my opinion, he has never fully recovered. But while we were together, to my best knowledge, Peter never met Sarah, which is in fact strange, because he was such a close friend of Doug's despite what happened. I met Sarah once several years ago, at the time I worked in Doug's lab and he invited me and Ulrich to their house. We had a long discussion about faith and atheism, but Sarah rarely opined. Maybe she revered Doug too much to oppose his opinion, or maybe she just agreed with everything Doug said. All I can say is that she is a beautiful woman, and I can understand why Peter was very sad when she left him."

"Well, for the time being, that's all what I need," said Carl. "However, if you happen to meet Ulrich in Germany, or if he calls you, please tell him that if he wants to come back to Rochester, he should call me first. I also authorize you to tell him that we know that he's the one who provided the pills to Peter, and I'll need his cooperation to convince myself and others that no malicious intent was involved. If he can prove that, this would significantly dial back the seriousness of his involvement, so he has no reason to be afraid of talking to me."

"Should I try to contact him?" asked Helga.

"I think you should. It would be good for him to know what we're just talking about. You see, if he remains in Europe for long enough, we'll be forced to consider him a fugitive, with all the undesirable consequences, including taking away his green card. He's under active investigation, and he may still be found guilty of providing an unauthorized drug to another person, leading to accidental death, but that's really nothing at all compared to a murder charge. And there's yet another point which you should discuss with him. Would you please ask him if he ever made Peter sign a declaration that the drug was used voluntarily, according to detailed instructions, and with the full knowledge of the potential dangers of overdose. Such a statement signed by Peter would even further lessen the seriousness of Ulrich's role in this tragic event. I can imagine that right now Ulrich is under onerous psychological pressure and your call along the lines I suggested could help him to ease his worries."

"I'll try my best," Helga promised. She was ready to leave, but Carl still had more to say.

"You see, I do not want to be looked upon as negligent or pandering to Ulrich. This would be against both my personality and professionalism. But at the same time, I would like to conclude this investigation as amicably as possible, provided no one meant to cause any harm. Have a safe trip, Helga, even if it's not going to be a pleasant one. I'll see you after your return. I'll be waiting to hear if you could meet Ulrich, and if so, what his response was. Be a good mediator, Helga. This is in everybody's interest, and particularly in Ulrich's." While Carl said all this, he carefully watched Helga's face, but he saw no wincing or any other emotional reaction. He still wasn't sure what to make of Helga's performance, but his instinct told him that Helga was the mediator between Ulrich and Peter, and she had only wanted to help, but somehow everything went terribly wrong. It is very difficult to admit to an accidental wrongdoing when the initial intention was to help someone; Carl could identify with Helga's dilemma and understood why she chose to cover up her and Ulrich's act instead of telling the truth.

Once in the car, Helga felt awful; against her true nature, she had lied to protect Ulrich. Had she been the only person involved, she would have told the truth. "But now we're in this deep," she thought. "I have to do some damage control. But how? Carl was dead accurate in his assessment of Ulrich's involvement, but he had no idea of how Ulrich and Peter had met. If it came to a trial, this information might come to light." A trial was obviously in neither Helga's nor Ulrich's interest. Helga reasoned that the only way to avoid this was to fully terminate her contact with Ulrich, at least for the foreseeable future. The next day she acted on this plan. She went to her cell phone provider to change her number, without leaving a reference to the previous one behind. Now it was Carl's duty to find Ulrich. She was confident that with a bit of luck, she would emerge from the situation unscathed, and so might Ulrich.

CHAPTER 18

DOUG AND ATHEIST-TURNED GEORGE DISCUSS GOD AND MULTIPLE UNIVERSES

The next Monday, Doug went to work for the first time in several weeks knowing that he didn't have to deal with the investigation of Peter's death. What a relief; he could again deal with his own professional duties without worrying that some new demand from Carl would once again disturb the otherwise highly organized fabric of his working hours. As far as he was concerned, the technical aspect of the case was closed. He was about to leave for a seminar held in Minneapolis but then decided to check his mail beforehand. He immediately spotted a letter from the NIH. He hesitated for a while to open the letter and ruin his day with the bad news sure to be found within, as he had received so many of it recently, but he eventually overcame his worry and tore it open. He couldn't believe his eyes; his grant application was ranked so high that it would be funded without question. This was essentially an award letter, except that the word *award* hadn't been spelled out; the details of the award would be given in a following letter a few weeks later.

Doug hadn't felt such joy in a very, very long time; that moment was the single most triumphant one of his entire career thus far. He had been inching toward this goal for years, and he had finally gotten there. This was the consummation of his wish to elevate his science to a higher level. Now it was also certain that his job would be secured for at least another four-year term, or maybe even five, depending on the fine print of the award. Suddenly all his bad opinions about the funding process at the NIH went out of the window. In his eyes, the NIH was no longer a penny-pincher organization impervious to scientists outside the inner circles, but one which funded research projects on merit. Now, even the "ostentatious" NIH bureaucrats, as he had thought about them earlier, suddenly became nice, generous fellows. It is amazing how differently the same scientist can view NIH after getting a rejection or an award letter. Now it was important to start working diligently on the awarded project from day one after receiving the funds and not to become giddy with success.

He had a quick lunch and then drove to the University of Minnesota campus for the seminar. After he had accomplished the most difficult challenge of the day, finding an empty parking space, he got to the seminar room just in time to greet George and still catch the opening remarks of the seminar speaker. The topic was about how the daily light-dark rhythms are regulated by the retina and hypothalamic nerve cells and how all this is related to various bodily functions like body temperature, eating behavior, hormone secretion, and others. Doug had absolutely no prior knowledge of this subject and was happy that he came. He was surprised to hear that if people disturb their internal clocks by regularly sleeping at daylight instead of during the night, they are at risk of developing serious metabolic problems that may end in obesity and diabetes. He immediately related this information to Peter's case; since he had frequent long-distance trips, could that relate to his insulin-resistance problem?

Then the seminar speaker went on to cite studies that unequivocally proved that people working night shifts are also at increased risk of breast cancer, prostate cancer, colorectal cancer, and cardiovascular disease. In fact, when animals had been exposed to dim light during the night, they showed clear signs of depression. Although the physiology of mice and humans is somewhat different, he urged the audience to shut the blinds tight before going to bed. The speaker led a research group that was beginning to identify the links between the diurnal rhythm and metabolism as well as how all that could relate to cancer. Doug's attention ebbed and flowed during the presentation, depending on how deep the speaker ventured into the biochemical details, but overall, he was very satisfied because he learned a lot. It felt tremendous to be back in the position of a full-time scientist. Today's news about his practically certain NIH funding was the icing on the cake.

It was already almost 5:00 p.m. when the unusually long seminar ended. Several faculty members were badgering the speaker to provide more details, which most of the time, after some demurring, he couldn't. Before they left the seminar room, they chatted for a little while with the seminar speaker; then George suggested going over to the restaurant in the Radisson hotel for dinner. During the short walk, George admitted, out of nowhere, that his wife had filed for divorce and their hearing would be next Tuesday.

"It's been pretty hard," he said, "but I really hope I'm going to get through this ordeal okay."

But he was clearly not the same blissful George who had visited Doug not long before. Doug was very sad to hear that George now belonged to the almost 50 percent of Americans whose marriage eventually goes belly-up. He didn't want to pry, but George volunteered to elaborate a little more. A few months ago, he had noticed his wife began acting very friendly with her boss at the insurance company she worked at, which pretty much foretold what was to come. When he questioned her, she admitted to having an affair with the man and that she had wanted divorce.

"In deference to her wishes I agreed, and we filed for divorce together. Since then I have been trying to reassure myself that my divorce may still turn out to be a blessing in disguise. Well, in a nutshell, that's what has happened to my marriage," said George, ending his woeful account with a deep sigh.

"Well, it looks like nobody's marriage is a sure thing these days. Divorce is rife perhaps because it's too easy to do, and I'm sorry that now you are part of the 50 percent of divorced men." Doug tried to say something, but that was the best he could come up with at that moment. While he tried to show compassion to George, he found himself thinking about Sarah and was happy to know with absolute certainty that this kind of thing couldn't possibly happen to them, ever. Their deep love and honesty with each other completely obviated any possibility of such thing.

They arrived at the restaurant, and after sitting down in a quiet corner, they immediately started to discuss Peter's case. Doug thanked him for his insights, which had helped greatly to identify PDCM.

"And what will Carl do with the case? Is he building it up with the goal to prosecute Ulrich, or the evidence is still too inconclusive to charge anyone?" asked George.

"I don't really know," replied Doug. "I think right now he is stuck between the two possibilities of either leaving Ulrich alone or pressing for lesser charges. My preference would be to leave the case as it is, as far as Carl can convince Ulrich to not return to the US for a while. I feel for the guy. What he did was clearly wrong, but I truly believe that he only wanted to help. By the way, yesterday Carl informed me that among Peter's letters he found a signed declaration, clearly stating that he was going to take the PDCM solely on his own responsibility, strictly adhering to the treatment protocol and safeguards as described by Ulrich in the same document. Ulrich was smart to do that. This declaration and the specific instruction he provided might take him off the hook completely. According to the prescription Ulrich had given, Peter should have taken only one tablet a day, around breakfast or dinner time, always at about the same time. Or he could take the PDCM dissolved in wine that Ulrich provided. The signed declaration even contained instructions for Peter on what to do if he noticed any irritation in his stomach. He should have immediately dissolved one spoonful of a white powder, phytic acid, in water and drink it. I don't believe that Peter obeyed any of these instructions, and he probably didn't pay attention to the warning signs either, or if he noticed them, it was already too late. But my hunch is that Carl is giving up on the possibility that somebody else killed Peter, and he's now focusing all his attention entirely on Ulrich and Helga."

An older waiter slouched over to their table and took orders for beverage. Since both were driving, they settled on tomato and orange juice. George admitted that in recent weeks he had become a teetotaler, to make up for his hard-drinking past. "You know, my father was an alcoholic," he explained to Doug, "and since this trait can be inherited, I felt like I was also going down on that path. So I decided to go cold turkey."

"Then you weren't always the altar boy I thought you were," joked Doug.

The waiter brought their drinks, and they ordered the same fish plate to satisfy their weekly requirement of omega-3 fatty acid. "In this restaurant the fishplate is excellent," George assured Doug.

While they were waiting for their food, Doug casually asked George if he still attended the same church as he did before. As it turned out, this was a thorny issue for George, and for a while he was hesitant to respond.

"You know, Doug, I was a well-behaved boy and extolled God along with the others in the church with all of my heart. My wife had also been a regular churchgoer, although I didn't see a lot of conviction in her. In fact, she ended our marriage in a rather unchristian manner. But while we're together, she felt it was important to observe Christian traditions. As for me, as I had begun to think about religion more, I gradually transformed into an atheist. I know that you're a practicing Catholic, so please don't let this affect how you think of me. Unlike some pugnacious atheists, I don't hate believers, and I don't want to change them. I fully accept anyone's belief in God."

"This is fantastic," Doug exclaimed; it wasn't immediately clear if he meant to refer to what George said, or because the waitress just then laid down two gorgeous plates of fish on their table. "This is the second time tonight I must say that you are not immutable. Man, oh man, you've changed a lot."

After the waitress had left the table, Doug explained the reason for his exclamation. "You abandoned the church because, as you began to think more deeply about nature, you gradually came to believe that there was no need for a Creator. In contrast, when I began to think about the cosmos and life as an atheist at the time, I gradually and somewhat grudgingly concluded that neither could exist without a god. But my God is different than the one described in the Bible. My God has a permanent dualistic nature, both a spiritual and a material one, and it is the latter that can communicate with us. I must add that for me the word *spiritual* has a different meaning than to traditional believers, and I only use it because I have not yet found a better word for it."

"This is an interesting concept," George acknowledged. "A material God who communicates with us, I assume through some physical sort of signals, am I correct? But in a way, wasn't Jesus supposed to be the material God, and will be again at His next coming? What's really new in your theory?"

"This will take a little longer to explain. Let us start eating first. When I'm hungry, I can't think and explain things clearly."

They ate their fish in a contemplative mood, not talking for quite a while. But then George, with his eyebrows furrowed, asked Doug what considerations led him to conclude that God existed.

"Very simple. When I considered how many different physical constants had to have the exact values they do for the universe to exist at all, I came to believe that the way our universe was born couldn't have occurred by chance out of a chaos. Creation required a Creator."

George thought for a while. "Well, as an old friend I don't want to disparage your theory, but I'm afraid I don't agree with you. There are hypotheses that postulate that there are many different universes out there, each endowed with a different set of physical constants. Most of them, probably the overwhelming majority of them, aren't suitable to host any forms of life because their physical constants are so different from ours that they wouldn't even allow the formation of galaxies, stars, and planets, let alone the elements needed for life. However, an astrophysicist who is also an expert in probability theory calculated that it would probably take about formation of 10^{123} universes to find one with the proper physical constants, allowing galaxy formation and eventually life. You see, Doug, we don't need to invoke God for the creation of our universe. It could have occurred by a random event repeated many billions of times until all the physical constants required for our universe were in the correct range. As I see it, a lot of trials and errors led to the creation of our universe."

Doug had of course read about the various multiverse hypotheses. The common thread of these ideas is that universes continuously arise, either from nothing or from the collapse of previously existing universes. But for him this multiverse possibility was even more hypothetical than the idea of God's existence.

"Listen, George, well-known physicists acknowledge that we'll never be able to obtain any evidence for the existence of any of these other universes. Thus, the multiverse theory becomes an issue of faith as well, and scientists proposing it are its prophets. Technically, these theories should be listed under the name of 'science fiction.' At present, my theory about God and Creation is also purely science fiction, and I'm honest about it. But scientists who promote their multiverse theories have positions at various universities and research institutions, so they can't admit that what they are proposing is science fiction as well. In my opinion, taxpayers aren't supposed to pay for science fiction theories."

"I agree with that. Then tell me about your 'science fiction' theory. What makes it different from other theories?" George asked.

"To start, I believe that another universe exists that is actually interwoven with ours. The way I think about it is that this other universe was the space, composed of forms of energy and particles that we still don't understand, into which our universe could have expanded after the big bang. The initial structure of that preexisting space also provided the template for the formation of galaxies."

"Okay, but then explain to me how the whole creation process started."

"Well, no matter how many universes are out there, we can't escape the conclusion that at some point, energy and matter were formed, and perhaps are still forming, from nothing," replied Doug. "In my view, the spiritual side of God is part of nothing, or is nothingness itself, although I again must add that I think about 'nothing' differently than most people do. In any case, just as energy and matter can arise from nothing, or perhaps a whole universe can arise from nothing as Hawking and other physicists think, the material side of God also arose from the spiritual God—in other words, from nothing. In fact, I believe that both spiritual God and material God coexisted forever. But mankind will never have the capacity to understand such duality of God, just as we'll never have the capacity to know if there is a single universe or billions of them. The point that I'm making is that the super-intelligent physical God was the one able to prepare for and execute the big bang to create our universe, for which He had provided the template in a preceding big bang."

"Okay, slow down, Doug. Let me ask you this. If your God has both spiritual and physical entities, how do these two entities communicate with each other? Who does what? Do they always agree?" George frankly thought that Doug's idea was rather childish, and this simple question would instantly unravel it.

"I wish I knew," replied Doug. "I know it is a somewhat silly comparison, but how do the two sides of a coin communicate? They obviously do not, because they don't have to. The two sides are part of the same coin. When you look at the coin, you can look at either side and know that both sides are parts of the same coin. Or better yet, an electron can behave both as a wave and as a particle. How do the two representative forms of the electron communicate? Again, they don't have to. They are two different forms of the same. If you shoot one electron toward a disk that has two slits next to each other, it will simultaneously pass through both, revealing its wave property. But if you use an instrument to observe it, the electron will pass through an open slit as a particle. There are many other dualities in nature that are virtually incomprehensible. Have you ever noticed that matter and energy are different versions of the same thing? Or that space and time aren't distinct, they only exist together? So, like an electron and all these other dualities, God might also have a dual nature. The spiritual God, or you can call Him the 'invisible' or 'matterless God' if you like, may be composed of something that could yield a physical form, the two forms being in symbiosis, representing the same entity. In fact, we humans also have a spiritual and a physical being, inseparably intertwined. But I admit that I don't have enough imagination to describe God's physical form, except that it certainly doesn't resemble anything that we're familiar with. Because the spiritual and physical sides of God represent the same being, there is no disjunction between them, and therefore the question of how they communicate with each other doesn't even arise in my mind."

Now George, silently conceding that Doug managed to avoid his trap, posed another question that he meant to be a fatal blow to Doug's theory. "Tell me, Doug," he began with an archly smile, "since time is infinite, wouldn't it be logical to assume that more than one God could be born from the nothing? This would lead us back to the Greeks and many other cultures worshiping several gods."

"I see that you won't connive my theory any longer and you want to kill it by any means possible," Doug replied smilingly. "Listen, I don't expect that there had been a profusion of God creations for the simple reason that spiritual God existed forever. Then, perhaps, from the totality of spiritual God was material God formed. Let me illustrate this idea. Just as for every defined matter particle there had been, or maybe still is only one antimatter particle, I believe that the spiritual God has only one counterpart. But, eventually, I may have to find better arguments."

"Why are you so sure that God has existed forever?"

Doug paused for a while. He had struggled with this question many times before, and it had taken him quite a while until he figured out his beliefs. "Indeed, my position is, just like religious teachings generally hold, that God had existed forever. However, for me forever again has a different meaning. To better understand my line of logic, let me start by saying that time started with God. Furthermore, time and space can't exist without each other, as Einstein proved so brilliantly in his theory of general relativity. In this context, it doesn't make any sense to talk about whether anything was before God, because there is no such thing as 'before' Him, because that would have required that time had existed before God. Do you know what that would entail?"

"Yes, it would mean that God didn't exist forever, so He Himself had to be created," answered George.

"Exactly, but how? You see, such a scenario is a total nonsense to me, so I remain with my theory that time and space started with God, and neither could ever exist without the two others. A logical extension of my theory is that the universe started from this trinity, explaining that God can be everywhere at the same time, including within us. This also explains why instant communication between us and God isn't a problem." At that point Doug glanced at his watch and said, "I think continuing our discussion in this vein would get us too involved, and I don't think that we have enough time for that tonight. Maybe on some other occasion."

But George was far from ready to leave; he still wanted to make sure that he understood Doug's concept about a spiritual God. "Doug, you said earlier that the spiritual God represents the *nothing*, probably in space-time, but you think about *nothing* differently than other people do. Tell me what exactly *nothing* is in your view?"

"I told you that this was a very tough one," acknowledged Doug. "But what I think is that *nothing* is a kind of vacuum, containing only information. In this sense, information is not an abstract idea. It is an entity that is entirely different from any other entities that we know, like energy or matter. Information has always existed in space-time, and the totality of information equals the spiritual God. This kind of *nothing*, which is in fact something in space-time, then can give rise to another kind of entity, which I think of as a special kind of energy field. This energy field can then give rise to the forms of energy and matter particles that formed right after the big bang, and that we're acquainted with in our universe. So, to summarize, there are two kinds of *nothing*. There is *absolute nothing*, or the spiritual God, which exists together with space and time and which consists of all the information needed for the creation of anything, including our universe and us. However, the totality of information, if it remains random, isn't sufficient to create anything. Thus, I believe that there is a very highly organized structure of information, with a practically limitless ability to plan and execute anything. This is what I call the material God. If you like, this is another kind of nothing, and I'm using the word *nothing* only from our perspective, merely indicating the fact that in the absence of any relevant experience, we have no word for it."

Unfortunately, there was no more time for their discussion. The waiters began to strongly hint that it was time to leave. As they left the restaurant, George loudly admitted that he was rather shocked, and he was unsure if Doug was simply joking, if this was just an intellectual exercise, or if he seriously believed in what he had said.

"Well," said Doug, trying to assure George that he was still in full control of his mind, "I suppose that you could say that the ideas I just explained were mostly an intellectual exercise, and I'm still early in the first round of many others to follow until I fully develop and present them."

"I wonder how the Catholic Church would relate to your ideas about the two sides of God," asked George.

"The church probably wouldn't like my ideas, but as I just said, I'm far from ready to come forward with them either. But really, let us talk about information and how it might relate to the absolute nothing, God, and the big bang some other time."

"I'm starting to see that you really are serious about this," said George with an incredulous expression on his face. "I can't wait to hear more about what you mean by information. Well, good luck with your efforts and your new theory."

"You meant my science fiction theory, correct? I really want to emphasize that aspect. But in any case, I don't believe I managed to change your views about God even a little, am I correct?" asked Doug, smiling. "Maybe when next time I'll talk about how evolution is in the center of God's plan, you will change your mind."

"Well, to be honest with you, until I hear about your evolution theory, I probably will stick to my own belief," George admitted. With that, they said good night and mutually promised to keep in touch and continue this conversation sometime.

CHAPTER 19

CARL'S REPORT

For the next few days, Carl worked exclusively on a drug-trafficking case. The Minneapolis police had successfully implanted a mole into a gang with apparent ties to a Mexican drug cartel and a drug-distributing ring in the Midwest, including Rochester. Carl organized plans for drug raids at multiple locations on Saturday. The well-planned and coordinated operation resulted in numerous arrests in three states, and this gave Carl some time on Sunday to once again turn his attention to Peter's case.

There wasn't much recent evidence that could propel the investigation forward. Although he wanted to convince himself that Ulrich's drug had caused or at least contributed to Peter's death, he had to admit that the only thing he had evidence for was unauthorized use of PDCM. But even that appeared to be on shaky ground when the document signed by Peter was found. It clearly stated that Peter agreed to use the PDCM, and phytate as a safeguard, voluntarily and according to Ulrich's instructions. In that respect, Ulrich didn't show indolence, as the document made clear references to both the tablets and wine bottles, all containing PDCM. Helga could have organized the meeting between Ulrich and Peter, but this would be difficult to prove and might even be irrelevant to the case. Sarah, the one-time high school sweetheart of Peter, probably just wanted to help him by supervising the work of the maid service, although he would have liked to know if Doug knew about that; and he still had to figure out why Sarah was around Peter's house that Monday evening, one day before Peter's death. Also, how did Sarah know that Peter's house was a mess without seeing it? What brought her to Peter's house in the first place? These uncertainties kept Sarah on his list of suspects, particularly since traces of a woman's shoes had been found in the bathroom.

As for Agnes, he checked with several people at IBM, and nobody had noticed her leaving the building during the afternoon on the day of Peter's death. One person testified that she had met Agnes in the building at around 2:00 p.m. and just before 5:00 p.m. as well. All this made Agnes's role in Peter's death highly unlikely. And then there was this mysterious motorcyclist. Why was he in Peter's driveway? And once he was there, why didn't he go in? He clearly wasn't delivering some advertisement catalogue because there were no similar traces in the driveways of any of the six neighboring houses. So, unable to get strong evidence for anyone's involvement, the case seemed to be at an impasse.

On Monday morning, Carl received the report from Bill that eventually summarized the findings in and around Peter's house. Only two things stood out that he was not aware of before. First, one of the women's shoes that left traces in the bathroom also left traces outside the door in a different position compared to the other woman's traces. Thus, it appeared that at two different times, two different women were independently wandering in and around the house, trying to find Peter. One of them could have harmed Peter, specifically the one who left traces in the bathroom. Second, comparison of the traces made it clear that the other woman had only been in the kitchen and maybe in some of the carpet-covered rooms, but not in the bathroom. However, the identity of the two women remained unknown, so identification of the shoe traces didn't help much.

Almost two weeks had passed since Carl and Doug last spoke. Doug was back to his scientific activities full-time. Before yesterday he had gotten detailed information from the NIH declaring that his grant was for four years, starting on March 1, 2012, and that the amount was $360,000 for each year. This allowed him to hire two postdoctoral fellows and two assistants. One morning he was in the middle of composing an advertisement in the journal *Science* for the postdoctoral positions, when he received a call from Carl, who offered to come to his office for a short visit because he needed to discuss something important relating to Peter's case. They agreed to meet at three in Doug's office. "Just please don't bring me anything to analyze," Doug begged. Carl promised him that much.

A few minutes before 3:00, Carl arrived at Doug's office, looking as weary as Doug had ever seen him; this and other investigations had taken a lot of energy out of him. He didn't waste much time with formalities and immediately directed the conversation to Peter's case. "A few days ago, I mentioned that I had in my possession a document, apparently a copy signed by Peter, in which he acknowledged that he had received PDCM in tablets and dissolved in wine as well as a white powder from Ulrich Siebert, and that he volunteered to use it for his diabetes treatment entirely under his own responsibility. Here is another copy of the document," Carl handed it over to Doug. "You see, there is even a full sentence stipulating that Ulrich had no responsibility whatsoever, whatever the ultimate outcome of the use of this drug."

"Well, this is crystal clear to me," Doug opined. "Ulrich cannot be held responsible for Peter's death. And he can also get a statement from his doctor that he had used the same drug for months with good results and without noticeable side effects. In fact, as it stands now, there is no evidence to prove that PDCM killed Peter in the first place, even if I'm almost certain that it did, because we don't know when he took the drug or how much. Also, based on the examiners' and our own findings, no one at the Mayo will sign anything stating that this drug caused Peter's death or even that it contributed to it. The evidence is too circumstantial, and there is a huge difference between suspecting something and proving it. And even if we somehow could prove that PDCM killed him, this letter certainly exonerates Ulrich from the murder charge. He still could be charged with unlawful use of a drug, although he would have a good defense against that as well. For example, he could say that he didn't accept any money in exchange for the drug. In my view, although I'm not a lawyer, there is a very weak case for any charge. Overall, as I see it, this is an auspicious development for Ulrich," Doug concluded.

Carl nodded in agreement. "I came to the same conclusion on all counts, and I have no other choice than to write my report accordingly. As for you, Doug, all along I appreciated your work very much. I know that toward the end of the investigation, it became quite annoying for you when I kept bringing more samples. But even if we can't prove it in court, deep inside, you and I at least know with reasonable certainty what really happened. With or without extra help, PDCM killed Peter, but I'm sorry to add that this was entirely his fault. He apparently messed up his life so much that he lost control of himself. As for the potential of someone else's involvement in his death, it is still there. After all, we still can't account for the fact that there were shoe prints in the bathroom that didn't belong there and that Peter's laptop and one of his cell phones are still missing. Taking all this into account, in the last paragraph in my report, I will have to indicate that 'should new evidence surface, we will reopen the investigation.' Does that sound fair enough to you?"

"Yes, it sounds fair enough, as long as reopening the case won't mean new samples for analysis."

"Then goodbye, Doug. I hope some other time we can meet under better circumstances."

That very night, Carl wrote his concise report, stating that although Mayo Clinic's toxicologists identified a drug in Peter's body, this couldn't be proven to cause, or even contribute, to his death. Also, there was no material evidence of foul play, and there were no identifiable suspects, including Agnes, who would have had motive to harm Peter. Further, he emphasized that he had found no evidence against Helga or anybody else in Peter's closer circle or beyond that would justify further investigation. He consciously chose not to mention Sarah's name. He decided that at that time this wouldn't add anything to the case. Finally, he alluded to the missing items; if, and when, they will be found, the case might be reopened. His report was accepted, and a few days later, the case was closed. On the surface, the incredible amount of work that had been put into solving the case proved to be for almost nothing. Yet, at the end, neither Doug nor anybody else involved regretted the time and effort they had poured into the case. The friends at least were reassured that Peter hadn't been murdered, and from the many resulting discussions, everybody grew intellectually.

CHAPTER 20

A Sudden Turn in Doug's Life

The grant that Doug had received from the NIH allowed him a little more flexibility with his time. It almost was an invitation to make changes in areas of his life that had recently gone out of control. He had developed bad eating habits and had almost completely given up exercising. Consequently, he began to put on some weight; in less than a year he went from his usual 180 pounds to 210. He had noticeably slowed down, and as a self-taught nutritionist, he was sure that his weight gain would sooner or later have consequences. He, like Peter, had always disparaged other people who weren't able to control their weight, and now as he looked in the mirror, he was livid with anger at himself. But for a while he didn't have the strength or the motivation to act.

Then Sarah started to make remarks about his waistline, first ostensibly jokingly and then more seriously. He knew that she was right, and he should stop denying his weight problem, but several months had passed before he could face up to it. Finally, he gathered enough motivation to start doing something about both his eating habits and lack of exercise. As a first step, he decided that instead of eating out in restaurants and fast-food places, he would go home for lunch. This way he could have far better control over his diet. Then he decided to pick up playing tennis again, which he had had played at Harvard, but in Rochester he had no partner to play with. Now an opportunity presented itself; one of the postdoctoral fellows he hired turned out to be a good tennis player and suggested they play together. It was the end of March, and snow still covered the outdoor tennis courts, so they bought a season ticket at the local YMCA and started playing there twice a week, usually after dinner. Initially, Doug was unsurprisingly rusty, but he didn't give up, and by their third game he was almost at an equal footing with his partner. His partner also had tennis player friends, so occasionally they could play doubles. By around mid-May he had already lost some fifteen pounds, but he still had a long way to go. And from his nutritional studies, he was aware of the constant struggle he would have to wage with his body for the remainder of his life.

These studies revealed to Doug that the human body has a very interesting regulatory mechanism for body weight. In his case, once he had reached 210 pounds, this remained the set point that his body 'remembered' and wanted at all cost to return to. Nutritionists still don't have a clear answer as to why the body doesn't remember the lower body weight as the set point, which in Doug's case would have been around 180 pounds. Perhaps there is an evolutionary reason to maximize body weight. Many thousands of years ago, when during most seasons food was scarce, excess body weight could have prepared humans for periods of starvation. This insistence of the human body to only remember the higher weight as desirable is responsible for the fact that even if a nutrition program is initially successful, most participants eventually gain all or most of their weight back unless they keep a strict regimen. Doug could expect a real struggle against his recalcitrant body, which he didn't look forward to, if he wanted to slim down to 180 pounds again and stay there.

One day, in late May of 2012, Doug went home for lunch a little earlier than usual because at the Mayo they had a 1:00 p.m. scientific meeting that he didn't want to miss. He gathered the letters from the mailbox, among them a rather thick one, and while he walked up to the door, he idly opened them one by one while he was humming to himself. Upon tearing open the thick one, he discovered six colored pictures of Sarah in several different positions, dating from her first year of college. A short letter was attached, which was signed, "As always with love, Roy."

He stopped short. He felt like he had just been punched in the gut. He just stared at the faithful signature, and the world around him grew dim and he heard a faint rushing sound in his ears, while new waves of realization continually crashed over him. The implications of this letter were clear; his life, a moment ago so cozy and secure, with his family at the center, was a sham. These pictures and the love note had in an instant upended his whole existence. He shuddered at the thought of Sarah cheating on him.

For several more minutes, Doug, with all his intelligence, couldn't decide how he should handle this situation. He was certain of only one thing; he wouldn't have a lunch. Instead, he felt like vomiting. Finally, he subdued his paroxysms of anger and despair that came in cycles, and he forced onto himself a calmer demeanor. He went into the house, and without looking into Sarah's eyes, he handed her the open envelope. Sarah looked at it, apparently recognizing the sender, and simply said, "Let's talk about it tonight." They didn't say anything to each other. Doug went back to the clinic without eating, and Sarah made no attempt to stop him.

Before the meeting, Doug had a little time in his office to think about this onerous turn in his life. After the first shock, his optimistic side prevailed. He still hung on to the hope that this tension caused by the letter would prove to be ephemeral. "It cannot be otherwise," he tried to reassure himself over and over again. "During all these years of our marriage I have never experienced anything that contradicted that Sarah and I are inextricably linked together on a rock-solid foundation. This will be cleared up, and we'll be back to normal." Sarah couldn't possibly have anything to do with this scoundrel who had dared to invade their family.

The meeting turned out to be a lecture, with the promising title "Why Don't We All Die at a Very Young Age?" The lecturer, a distinguished oncologist from the Anderson Cancer Center in Texas, was simply brilliant, and he really knew how to hold an audience's attention. He provided many examples to demonstrate how difficult it is for a mutated cell to become a proliferating cancer cell in a hostile environment, like in the sea of healthy cells in the surrounding tissue. It takes many attempts until a small conglomerate of cancer cells can take hold and equip themselves, through several gene mutations, with all the evasive mechanisms required to defend against the attacking immune cells and develop the blood vessels to provide nourishment. For Doug this lecture came at the best possible time. It gave him some time to cool down and divert his brain's attention from Sarah and this mysterious Roy.

But as he drove home, his thoughts drifted back to Sarah. This time he couldn't help it; his pessimistic side began to prevail, and his mood turned foul no matter how hard he tried to remain calm. His instinct, supported by his logical mind, now told him with certainty that at some point in the past, Sarah indeed had had a romantic relationship with Roy, unbeknownst to him, which they had apparently rekindled lately. He grew sick when he realized that Sarah apparently named their eldest son after some past lover. He clearly remembered that they had disagreed over the name, because he would have preferred Brian, his father's name. He had eventually compromised to avoid a melee and agreed to name their second son Brian, but he had never fully forgotten the vehement insistence of Sarah on naming their firstborn Roy. Now it all became clear; Sarah didn't want to tie her life to Roy, but in remembrance of him, she had decided to give his name to her son.

"So how should I react when Sarah will tell me her story?" He tried to anticipate what Sarah would say, but he couldn't. "In any case, I will let her explain herself, and hopefully both of us can refrain from yelling. I would rather divorce her than fight with her."

During dinner their sons did all the talking about the kindergarten and preschool where they had spent their morning hours. This afternoon Sarah stayed at home with them, and they were working on projects that they would present in their classes on the last day before summer break. For Roy, this was the last day in preschool; from September he would attend elementary school, which he wasn't too excited about. After dinner, Doug withdrew to his study to deal with budgeting issues. It was nice to have the grant, but it turned out that if he wanted to employ all the people he had in mind, then $12,000 was still missing from his budget. He decided to compose a short internal grant application to the Mayo Clinic asking for help.

He was so buried in his thoughts that he missed his evening ritual of kissing the boys before they went to bed and raised his head only when Sarah entered the room.

"Are you ready to talk?" Sarah asked him.

"Yes, anytime," he answered. "Just please close the door."

Sarah looked at him nervously and then said in a disquieted voice, "I don't even know where to start."

"Well, I suggest you start from the beginning. For example, it would be a good starting point to tell me how you got to know Roy."

"I don't think there would be any point to go into details, because I want to divorce you." Sarah said this with so much determination and intrepidity that even she was surprised. "There were two men in my life who really, I mean *really*, loved me," she continued. "You aren't one of them."

"I assume they were Peter and this Roy?"

"Yes, Peter and Roy, you are correct. But my problem with both was that at the time, neither of them was mature enough for me, while your intelligence and outlook on life really captivated me. For a young girl, looking for answers, being close to you, who seemed to know the answers for almost everything, was enchanting."

Sarah looked down at her feet and thought for a while about how to make her disclosure to Doug as painless as possible. "I always respected you very much, and I even loved you, but yet, something has always been missing. And that something was that somehow you couldn't make me feel that I was the most important person for you in this universe. Both Peter and Roy made me feel that whenever I was with them. Your altruism is boundless, I know it, it can't be gainsaid, and for that, too, I respect you very much. But compassionate love is different, and at the end of the day, that is what I missed. With each passing day, I understand the importance of it more and more. I know that I should be more tactful and perhaps shouldn't tell you all this, but it's the truth. I can't help it. And I have come to the point that nothing works for me anymore. I'm not strong enough to carry on like this. I can no longer act like your fealty wife. I've wanted to say this for a while, and when you found the package today, I realized it was an opportune time to reveal the truth, my truth, to you."

A short silence fell on them. Then Sarah continued, "Doug, I don't blame you for anything. You are a wonderful individual, and I really should be in love with you, and for quite a while I had been. But lately I just haven't been able to bring myself to truly love you again. And I don't want to waste any more time trying. As for Roy, I had met him in the college, and only recently we've started exchanging emails. It seems that he is willing to divorce his wife, leave his two children behind, and start a new life with me in Madison or Rochester, or anywhere in the world for that matter. He has already been searching for jobs, and he has been offered a computer science teaching position at our college right here in Rochester. But I don't think I should go on. Anything else I would say would be pointless."

Although the pessimistic side of Doug had prepared him for almost anything Sarah might say, he had great difficulties in confronting this new reality. He wished that the earth would open under his feet so that he could escape the predicament he was in. In just a few seconds, he went through a whole range of emotions. He felt terribly betrayed. But what really hurt him was Sarah's coolheaded statement that two other men had loved her more than he did. Sarah had been the center of his universe, and he would have done anything in his power for her. And now she seemed hell-bent to leave him, willing to abandon the family they had built together. Although he was exceptionally skilled at debate, he was so befuddled that he couldn't express his thoughts. He tried to say something to Sarah, but he was stuttering, and his syntax was so snarled and confusing that he just gave up talking. The only thing he could utter with clarity was "Good night." With that, he went down to the bedroom in the basement that was kept for visitors, and it was there where he spent the night with tears in his eyes, alternatively drinking and then dozing off in cycles. When he was awake, he thought about whether he should still try to save his marriage somehow or just give up on Sarah completely; he implored God to help to think with clarity.

The next day he kept confronting this question and slowly settled on an answer to it; he wouldn't attempt to stop Sarah from leaving him, and he wouldn't inflict any retribution on her. Slowly but surely, he began the painful process of preparing his mind for a new phase of life: living alone. He had never believed in the saying that if you are lucky in your business, you will be unlucky in your love life, but now he began to give credence to it.

Fortunately, the next few days were workdays, and being in his office dealing with the myriad of tasks that accompany starting a new project with new people helped to reduce his anxiety, or at least turn it into positive anxiety. After being euphoric for weeks, thanks to the grant, Sarah had certainly brought him back down to earth with a devastating crash. But several days after Sarah's announcement, he once again had the same uncluttered mind he had had before. Now, in the lab he was focused only on his work, and at home he worked on the details of their divorce settlement. By Saturday Sarah had printed the divorce forms from a website, and after a long negotiation, they agreed and completed all pages by Sunday evening. Sarah would get the house, fully paid for by then, with most of the furniture. In exchange, Sarah would receive only a relatively modest $2,500 per month, alimony from Doug until the children turned eighteen. This ensured that Sarah could pay for most of everything even if she chose not to work; however, her intention was to find a job as soon as possible.

After filing the divorce papers, within two months they had the trial and received the divorce decree. Both were free, and Sarah could begin to build her own new life with the two boys. After several more weeks of still living together, Doug found a smaller house in the same area, and from that point, either he visited his sons, or they came to his house for the weekends. Doug never had to do house work before, and it was a strange new reality that now he had to do everything alone, like going regularly to the grocery store and cooking for himself. These extra tasks initially drained a lot of his mental energy, but eventually he developed a workable schedule and a routine that he could live with. There was only one positive outcome of this stressful period: he lost another ten pounds, itching closer and closer to his goal of 180 pounds. He defied the experience of many people who eat more than necessary to alleviate the pressures of stress, but as a result they gain a lot of weight.

One weekend, his eldest son, Roy, told him that on Wednesday a man, also called Roy, had visited them and slept in the same room as their mother. The boys had overheard words like *divorce* and *marriage*, but they didn't know what to make of them. Doug thought he knew; Roy probably had divorced or was planning to divorce his wife, and now he and Sarah were talking about marrying. But as it turned out, this wasn't meant to be the case. Roy kept postponing his divorce, citing his difficulty to make his wife cooperate. Sarah never really figured out the true reason, but after a while, she wasn't even interested to know. When three months later Roy visited her again, she just pointed to the door without saying anything. Roy understood what Sarah meant and disappeared without looking back. That was bad luck for Sarah; she attributed qualities like undying love to Roy that he either never had or had lost during the ensuing years. He was a heroic lover, but only from a distance; when he really had to make a choice, he stayed with his family, because this caused him the least trouble. Sarah, despite being the victim of a mirage that had made her wiser, never considered going back to Doug. She couldn't explain it, but she simply couldn't live with Doug ever again, not that Doug really would have wanted to either. He was hurt beyond the point of return.

CHAPTER 21

AGNES AND DOUG DISCOVER EACH OTHER

Almost a year had passed since Agnes returned to Boston, when one day she received a phone call from Carl, like all the others who had known Peter closely, and she was happy to hear that Peter's case was closed. From that point on, no one would bother her with questions about Peter. Although she felt a little guilty that during the last few days and hours of Peter's life she hadn't been around him, she knew that she really wasn't responsible for anything that had happened. For a while she had enjoyed being in bustling Boston again, rich in culture and sport events, but after a few months, she began to miss the quieter Rochester. In addition, her renewed relationship with Robert had foundered. He was too dull for her, intellectually certainly far less exciting than Peter was. So, except for her parents, nothing really tied her to Boston anymore.

Then fate threw the gauntlet at her. On an early September evening, as she was standing at her kitchen table trying to decide what to make for dinner, she unexpectedly received a phone call from Gary Townsend, an IBM executive in Rochester whom she knew very well. After some explorative small talk, he explained that one of his friends, the head of a prestigious accounting office in Rochester, was urgently looking for an MBA. The friend's firm had won a contract to do all the accounting work for IBM from next January 1, which would require hiring three more people. Agnes appeared to be a good choice for one of the new posts because she not only had an MBA degree, but she already knew the IBM offices and their operations very well. Gary also recommended her for the job based on his assessment that Agnes had a knack for solving difficult problems. Without any hesitation, Agnes accepted the offer. As for the date of her first day, after some thought she said that she could start as early as November 1. In less than two weeks, she had received the job offer in writing, which she accepted immediately. She began to direct her sails toward Rochester again.

The remaining two months were very exciting for Agnes. She could prepare her move back to Rochester unfettered. This time she had been wiser and hadn't bought a house in Boston; all she had to do was to write a sixty-day termination notice to her landlord. Not having a boyfriend or very close friends in Boston, her parents were really the only people she felt sorry for leaving behind. She might miss her tennis partners, but she hoped to soon find new ones. Tennis was something she was very good at, and she planned to continue playing in Rochester.

When people asked her why she was so anxious to return to Rochester besides the job offer, she really couldn't think of a good reason other than it was a relatively calm and easy-to-navigate town; she could get almost anywhere in fifteen to twenty minutes without risking a nervous breakdown. But she knew that none of these reasons were the main one. There was something mystic about her desire to live in Rochester again, which she couldn't quite put into words. Unconsciously, she felt that something important was waiting for her there, but she couldn't put her finger on what that might be.

She bought the airplane ticket for the morning of October 31, but the flight was delayed by two hours. After this unexpected hurdle, the flight was mostly smooth, and when she stepped out of the airplane at the small Rochester airport, she immediately knew that she had come back home, even though for two weeks that home was a cheap hotel room. The weather was still nice; these were the last days of Indian summer, although the mornings were a crisp forewarning that winter wasn't too far away. The housing market's meager recovery worked in her favor this time; it took her very little effort to find a nice apartment in a beautiful green environment yet close to the downtown area.

Agnes's new boss, Jim Cork, a jovial man in his fifties, made a very good impression on her. Unlike some supercilious bosses, he was a down-to-earth, friendly man. On her first day, Jim spent the entire morning explaining her duties from every angle, and then he left a sizable dossier on her desk about IBM's various business transactions and profits. He had already mailed the rest of the documents to Agnes a week ago. Essentially, Agnes's job was to work with three colleagues to prepare IBM's tax return for 2012. Being a big international corporation with many potential tax loopholes difficult to navigate, she would have to make sure that IBM wouldn't overpay its taxes so that it could maximize profits for its shareholders. Although Agnes's real work would start only several months later, she needed that time to familiarize herself with IBM's international business practices, her area of work, and the insanely complex corporate tax code. Also, the large number of transactions and accounting practices could get jumbled together and unentangling them took time. Sometimes she wondered from where she drew the willpower to take on such a demanding number-crunching job, even though she had previous experience with IBM. But as she became entrenched deeper and deeper in the project, she began to get the hang of it and even started enjoying it. From time to time Agnes also conferred with her boss, and their congenial relationship grew every time. He would warn Agnes that complexity can potentially lead to corruption, and since IBM was an extremely complex corporation, she had to be especially careful not to make mistakes. Agnes reminded him that she had worked at IBM before, and she had some idea about its internal matters. Of course, Jim knew that; this was the main reason she had been hired in the first place.

Christmas was approaching, and to get some gift ideas, one day during a lunch break Agnes walked to the recently refurbished downtown Gallery Mall. As she was crossing the road in front of the mall, she casually glanced at a car on her right waiting for the light to turn green. She couldn't believe her eyes; the car's registration number read BGC 178, the plate number that had been burned into her memory during the investigation of Peter's death. And she instantly recognized the woman behind the wheel as well; this was the wife of Doug Lowry, Peter's friend. "My god," she whispered to herself, "that's the woman who visited Peter. That number on the registration plate proves that she was the one." Once Peter had shown her the pictures of his poker buddies and their wives, and she had a very good memory for faces. Peter had also mentioned that Doug's wife had been his first girlfriend. Now she even remembered her name, Sarah. When she had seen that woman in Peter's house, the lights were dim and from the distance she couldn't recognize her. But suddenly, the identity of the mystery woman, who had puzzled and troubled her so much, was solved. She had come back to Rochester to figure this out. "Unbelievable," she murmured.

Once in the mall, she forgot the reason she had come there at all; she sat down and tried to put the pieces of the puzzle together. "Maybe Peter and Sarah met the first time for an innocuous reason, but later it might have led to something else. Peter might have given Sarah some reason to worry about him. Perhaps this led Sarah to occasionally return to Peter's house just to check if he was home. Or was there something more going on? And how does Doug fit into this? What did he know, and what does he know now?" Peter had told Agnes stories about Doug's cosmological theories, which she had found fascinating. "Doug must be a highly intelligent man," she thought. "Sarah doesn't deserve him, if she indeed double-crossed him." She couldn't proceed with her thoughts; she had to go back to work. By the time she got back, she had made up her mind. "I need to meet with Doug. I'll tell him that I knew Peter and Peter told me about him. Maybe I'll ask him to tell me more about his cosmological theories sometime."

It was already January, several days after the year 2013 new-year celebrations, when Agnes worked up the nerve to call Doug at the Mayo Clinic. She introduced herself, and among other things, she mentioned that she had known Peter and they had had some philosophical discussions about God. Once Peter had mentioned that he had a friend who had been developing some interesting theories about Creation. If Doug didn't mind, she would like to discuss his theories with him anytime after work. Doug was about to leave for a meeting, and after thinking for a few seconds, he asked Agnes to call him back after 5:00 p.m. He added that he had heard about her and would be happy to talk to her. Agnes said that she hoped that she hadn't caused him any consternation. "Absolutely not," he said, but after putting the phone down, he felt a little flurry of discomfort at the thought of talking about his budding theory of Creation to someone he didn't know personally.

Doug of course knew that Agnes had worked for Peter, and Carl had kept her on his list of suspects for a while. He had also heard some rumors that Peter and Agnes had been rather close for a boss and his secretary, but he wasn't sure about the extent of their relationship. Once, a young man who attended Doug's church told him that he knew both Peter and Agnes, and he said he wouldn't be surprised to learn that their relationship was closer than people think. The man praised Agnes exuberantly, citing her slender beauty, her raven hair, and her lustrous blue eyes. Overall, Agnes appeared to be an interesting woman, and Doug was really looking forward to meeting her. He already liked her gentle voice over the phone.

A few minutes after 5:00, Agnes indeed called Doug, and they agreed to meet in half an hour in the Starbucks just across the central Mayo building. Agnes wondered if this was really a good time for him, or if he was perhaps too tired. "I don't want to take family time away from you," she said.

"Don't worry," Doug assured her. "I'm a divorced man, and I live alone. My time is fully mine. Look for a tall man with a brownish full hair. I'll have a computer bag in my right hand."

"Don't worry, I'll recognize you. I saw you in an album Peter showed me, and I also saw you at his funeral," Agnes said.

Doug underestimated the time he needed to get to the Starbucks and got there about five minutes late. When he looked around, he immediately noticed a black-haired woman sitting alone at a corner table, in apparent total quietude. He stepped over to the table. "Pardon me, are you Agnes?" he asked.

"Yes, I am," she responded with a small but noticeable blush. "First of all, forgive me if I appeared to be too assertive about this meeting, but I've heard good things about you, and I wanted to see you eye to eye and discuss things that I deeply care about. I hope you have no qualms about it."

"No, not at all," said Doug with a healthy laugh. "I just hope our meeting won't include accounting. I heard that you worked for Peter and that you are an MBA, but accounting is far from my favorite subject. Although, I recently received an NIH grant, and this involves a lot of number-crunching too. But before we go any further, would you like a coffee or something else?" asked Doug.

"Oh sure," answered Agnes with delight; she was an avid coffee drinker, and it was time for one last double espresso for the day.

While Doug waited at the counter to order their coffee, he looked back at the table, and he couldn't help but think, "This woman is a rivetingly beautiful creature. Maybe God sent her my way." His initial discomfort about their meeting completely disappeared.

"Well, let me start introducing myself once again," said Agnes after Doug arrived with their coffee. "You're correct that I'm an MBA and that I worked for Peter. You may not know it, but for a while, shortly before his death, we were dating as well, although never as seriously as to contemplate marriage or live together."

"Yes, I heard that you had worked for him, but I wasn't aware that you had also been in a relationship," acknowledged Doug.

"In any case, as I already mentioned to you," Agnes continued, "with Peter we used to talk about cosmology, religion, and the mind, and once he mentioned that you had rather interesting theories about how the universe was created and that your views about God, Jesus, and afterlife differed from mainstream Christianity. I'm reading books and scientific papers on specific topics of neurophysiology because I would like to have a sense of what the mind and the soul are, or at least how scientists approach these terms."

"And is this the subject you wanted to talk about?" asked Doug.

"Essentially yes, but as I'm starting to see it is almost impossible to talk about the mind and the soul without invoking the other large issues like Creation, how we came about, if we need a god at all, and so on. I'm a semiatheist, or rather an agnostic, with an open mind to everything religious thinkers have to say about the quintessence of our being."

"Well, I'm afraid that discussing all this will take much more time than I would care to spend in a coffee house. For example, we could move our conversation to a restaurant. I must admit that I am getting hungry."

"Don't take my question amiss, but was this an invitation for dinner?" Agnes asked this so breezily, as if they had known each other for a long time.

"Yes, of course it was. I can suggest an excellent Italian restaurant only a few minutes from here."

Fortunately, it was a weekday, and they found a table and could order their meal with practically no delay. It was an interesting coincidence that they ordered the same walleye plate without any prior consultation with each other. Doug suggested some white wine, and they settled with a kind of German Riesling he was familiar with.

After the first few sips, Agnes, momentarily forgetting why they were there, asked somewhat diffidently; "When did you divorce you wife?"

For a few seconds Doug seemed perplexed, because he was prepared to talk about his theories, not family matters, which still hurt him although would not admit to it. Agnes realized that she might have gotten him to emote and was almost ready to withdraw her question as inappropriate. But then he looked at her in acquiescence and said in a plaintive voice without quibbling, "We divorced just a couple of months ago. And perhaps I should add that she basically jilted me. In a way she wanted to run away with another man but eventually stayed in the house we owned. Then I moved out, and now I live alone. That's why I have time to spend with you instead of playing with my sons."

"It's none of my business, and I apologize for my question, but you must be feeling awful, or have you already gotten over it?" Agnes said all this with a very compassionate voice. She really felt solicitous for Doug's personal life, and he sensed it.

"Yes, I like to believe I did come over it, although I'm still not sure." Up until that point, he had been slightly indifferent to

Agnes, although he had instantly noticed her beauty. But Agnes's little remark and particularly the way she said it made all the difference. As they looked straight into each other's eyes, she conquered him, although probably unintentionally. Doug wasn't at all sure that Agnes had any intention to ever take their relationship beyond an intellectual one. Even though he began to feel an impetuous desire toward her, he made all effort to conceal it.

"Well, back to my divorce. Even to this day, I really don't know why she left me. I thought I had done everything right, but apparently this wasn't true, and she had had enough of something I had done or hadn't done to her. It is also true that when I asked her what I did wrong, she gave me a straight answer. I didn't spend enough time with her. I must admit there was some truth to it, but we could solve it without divorce."

"You must have loved her very much," Agnes remarked.

"That is another strange thing. I loved her very much at the time, but by now I have become neutral toward her. Sometimes I think about how odd this is. If it was true love, shouldn't I love her forever, no matter what happens? But at least I never think of her with hostility, and I'm sure that I never will."

Fortunately, in the meantime the meal arrived, which gave him an excuse to stop talking about Sarah. Doug wiped away the sweat that was pouring from his brow before starting to eat, which signaled to Agnes that for the time being they had talked enough about his family. Perhaps she might have gone a little too far by questioning Doug about his personal life. Doug also thought that he might have been too open with his answers; after all, he hadn't even met Agnes before. But both were intelligent enough to avert a calamitous end to their first meeting.

"So, Agnes, what brought you back to Rochester?" he inquired after they both agreed that the meal was very well prepared. "I'm guessing it was a puzzle to some people there why you exchanged Boston for Rochester."

"In part it was about a job that I came back for, although I didn't have to accept it. So it's a little bit of a mystery for myself as well why I chose to come back."

Doug was surprised to hear that level of incertitude in such an important decision, which had brought her halfway across the country. Agnes saw that surprise on Doug's face and added, "Several times in my life I have made decisions which at first appeared counterintuitive, yet something propelled me forward to do them. It's true that when I first came to Rochester it was a hundred percent about a job. I also had opportunities in Boston, yet I chose to come here. Now it is the same thing all over again, except that at least this time I knew what to expect. I received a phone call from an IBM executive about this accounting job, and something told me that I should accept it. Sometimes I wonder that maybe something, some higher power, directed me to come here for a reason beyond job. It's funny that I say this because I'm not religious, but like I said, I'm more of an agnostic. But the more I think about our brain, our mind, and the mysterious thing we call the soul, the more open I become to consider the existence of a higher power, although only at the level of an investigative mind. In other words, I'm floundering on a spiritual road which I don't know where it will lead to. I hope you'll not look at me with disdain because of that."

"Not at all. The way you describe yourself is fascinating and reminds me of my own spiritual journey," mused Doug. "You know, in my youth I was very critical of anyone who believed in a higher being. But with time, I went through the same changes that you just described for yourself, although in my case it was more decisive. It seems to me that I am much further down this spiritual road than you are. What I mean by that is that now I fully accept the existence of God, although not the way traditional churches imagine Him."

"Yes, Peter told me that much, although he didn't elaborate more. All he said was that the Catholics Church would be livid if they heard your ideas," Agnes remarked.

"That is true," said Doug. "This is one of the reasons I don't want to make my thoughts public, at least not for quite a while. The other reason is that my theories change every day, and at this point they are only at the level of science fiction at best. You see, it will not eclipse the astronomers and theorists for a very long time."

"I hope you don't consider me 'the public,'" snickered Agnes. "Honestly, I only wanted to know more about your ideas. That was the main reason I wished to meet you, even if just once, and now here we are doing just that. Isn't this strange?"

"It is strange indeed. But I'm afraid that one meeting won't do it. It is already 8:20, and we really haven't even touched on any of the issues in any detail you're interested in."

They thought of what to do next, and Doug came up with an idea, which was a little unorthodox considering they have known each other for only a couple of hours. "I have a suggestion," he said. "I live alone, and I can afford and would be happy to put aside some time every week for regular meetings and discussions with you. It would be like a regular seminar, with two equal participants. How do you like the idea?"

"I also live alone," Agnes admitted. "I think your idea is excellent. I don't know how confident you are talking about your theories, but I feel like I'm groping my way through a fog, not knowing my surroundings and only hoping that I'm going in the right direction. But I want to stop feeling so blind, and something tells me you can help."

"I can assure you that most people feel like you, myself included. Both of us will profit from our discussions, and I really look forward to them. Now, I believe it is better if I pay and we leave, because the waiter keeps looking at our table. Where did you park?"

"Not far from here, in the parking lot of the public library."

"If you don't mind, I'll accompany you. We can take the skyway. It is quite cold outside." Before they split, Doug promised to call her early next week to set up a time for their next meeting. Amazingly, neither of them hesitated even for a second about their arrangement; it seemed so natural. And that was how their first meeting went. That evening, neither of them could guess where their first hesitant steps would lead to.

On Sunday night snow had started to fall, and the next week remained quite snowy, which gave Doug the pleasure of shoveling snow off his driveway every evening of the following four days. His driveway was relatively short, so in principle, he should have been done in less than half an hour. But on Monday an ice storm hit Rochester, which was followed by almost a foot of snow, making his task exceedingly difficult. As he had been struggling with the ice and snow with a dour face, a car pulled over and Agnes stepped out of it.

"How can I help you?" she chaffed Doug.

"What brings you here?" he asked in surprise.

"It's a funny thing. I live only three blocks away. I didn't know that we live so close."

"What a coincidence. Well, as you can see, I'm almost done. A minute ago, I was quite angry about this weather, but a good hot tea will help. Will you join me?"

"Sure," agreed Agnes happily. "I also had a hard day, and I'm beat. Tea sounds perfect right now."

"But how about your driveway? Do you need help with that?" Doug asked.

"No, fortunately I live in an apartment, and the manager takes care of the snow," answered Agnes warmly.

They went into the house, and Doug instantly started to prepare the tea while telling Agnes what he had done during the day. They were at ease, almost as old friends, and they just talked whatever came to their minds. Agnes led the conversation, mostly talking about her work. She liked her new workplace, except for one fellow who worked in the same office room area. "He's lousy at his job, and he always has a sullen look on his face. We never know whether this is his normal expression or if he's angry for some reason. Anyway, he won't be sitting there for too long, because today I saw the boss hand a pink slip over to him. But the other people are all cheerful and friendly, which is really important for me."

"And how's your work coming along?" asked Doug.

"It's actually going quite well. I've already developed a useful model for dealing with IBM's tax return for international earnings."

"Is it also your task to find as many loopholes as possible?" Now it was Doug's turn to gently tease Agnes.

"Well, I'm just doing what is allowed by tax law, and believe me, a lot is allowed. I'll be the first to admit that. I think people generally think, probably including you, that these allowances equal loopholes. But like every other company, IBM certainly doesn't want to pay more in taxes than what is required by the law. Instead of filling up the US Treasury, they want to fill up the pockets of their shareholders. I would also argue that the less a company pays in taxes, the more money will remain at its disposal to expand and hire more people. And at an official unemployment rate of nearly 7 percent, which in reality is probably higher, don't you think that's more important?"

"Do you really think that large corporations generally pay their fair share in taxes?" asked Doug.

"No, I don't think so. But why should they pay more than what the law requires? I don't know other corporations, but I can tell you that we working for IBM aren't unabashed when putting together the tax return claims."

They further discussed the tax issue and many other things that evening. They had finished their tea long ago, but neither of them wished to rise from the table. Then Doug asked suddenly, "Have you had dinner yet?"

"No, I haven't. I was going to make something later."

"Well, yesterday I made some fish, and I believe that there is enough left for the two us, unless you are extremely hungry," Doug suggested.

"In that case I accept your invitation, if that's what it was," said Agnes smilingly. "First, I had waylaid you outside, and now I let you invite me for an elegant dinner in this quiet ambience. I surmise you probably don't have a very high opinion about me."

Doug liked to see in Agnes a lovely cheerful woman, who could be serious but also so breezy and carefree. "I am very glad to have you here for dinner," he said softly, "even if my dinner is more shabby than elegant as you soon will testify to it. Let me start preparing it right away." He removed a saucepan from the refrigerator, and in minutes he warmed up the fish and rice. "I warn you that the fish could be a little bland, and you may need to add some salt to make it more palatable, but this is how I like it. Too much salt is bad for the heart. Red wine, white wine, or beer?"

"Not much of any alcohol because I'll still have to drive. Maybe a half bottle of beer?"

After dinner it was Doug's turn to tell Agnes about his day, when his phone rang and a voice said in a slightly irritated tone, "Doug, haven't you forgotten about something tonight?"

"Oh my Lord." Doug tapped his forehead and turned to Agnes. "I forgot about one of my few sacred rituals. I was supposed to play tennis tonight at the YMCA." Then he spoke into the phone. "John, I truly apologize, but something urgent came up tonight. Since you're already there, could you make a reservation for tomorrow? Thanks, and again, I am very sorry." Then he turned back to Agnes. "That was one my postdoctoral fellows. We play tennis once a week, most of the time singles, but sometimes we also play doubles. Today I completely forgot about it. First this icy snow and then you. But it's not a big deal. Tomorrow we'll make up for this."

"But this is fantastic!" Agnes exclaimed. "I have been looking for a tennis partner for a while! Occasionally would you also play with me?"

"Yes, and we'll start tomorrow. I'll find a fourth person, and we'll play doubles. How does that sound?" asked Doug, beaming.

"I couldn't have expected a better outcome for this evening. I'm so glad that I stopped my car when I saw you."

"Agnes, I have to admit that you made me very happy that you got out of your car." By then Agnes completely infatuated him.

"I am glad to hear that," said Agnes in an exceptionally soft voice.

They were sitting at the table within arm's length, sipping their beer and looking into each other's eyes, trying to make sense of this evening's events. None of them said a word, but their eyes said everything; there was no need to talk. Then Doug slowly touched her hand. "Ever since we met, Agnes, you have always been in my thoughts. I wasn't brave enough to call you about our next meeting because I was afraid that you would reject me if I told you that since our meeting you have meant a lot to me, and now I feel I'm in love with you." He was surprised that he had somehow found the bravery to tell Agnes how he felt about her. He was also quite anxious about Agnes's reaction.

Agnes made no attempt to withdraw her hand; instead, she gently squeezed his hand and said, "You don't need to worry, Doug. I feel the same way." That was all she said, yet these simple words carried them to the giddiest heights that they never expected to reach. They stared at each other for a while with amazement, again trying to digest what just happened. Then, they both slowly rose and embraced each other. Doug had never experienced such a kiss. He thought that he had truly loved Sarah, and probably that was true to an extent. But kissing Agnes was completely different; the moment seemed to be imbibed with an elementary force that he had never experienced before. Agnes's beauty and intelligence conquered him completely, and he knew it would last without any fickleness until he lived.

After they temporarily separated from the embrace, Agnes said, "I believe that now I know what was the tipping point which brought me back to Rochester. It has been a mystery even for me, but something inside me kept telling me that 'an unexpected pleasant experience waits for you there.' I begin to recognize that this is it. I'm baffled, but at the same time it feels good to be with you, even though I barely know you. I still haven't even heard any of your theories," she added jokingly with her eyes radiating love. She could tell that Doug wanted to go further, but she wasn't ready for that yet. "Now before we go any further, I must leave you. I really think we both need a little more time to figure each other out. So tomorrow evening we'll play some tennis," she said, trying to keep the conversation as neutral as possible. "And how about afterward you come to my apartment for dinner? I'll try to make a meal with very little salt in it, as you like. We live so close you won't even have to drive, so we can have some wine."

"I'll bring the wine," Doug volunteered. Agnes started to put on her coat, and while he helped her, they kissed again. It was difficult for them to spend the night alone, but tacitly they both agreed that all this had come too suddenly, and it was a good thing to cool off for a while and see if they still felt the same way tomorrow.

After Agnes had left, Doug could hardly control his impetuous desire to be with her. He reproached himself, thinking that perhaps he should have tried harder to get her to stay. But the next moment he was glad he hadn't pressed the issue, which wasn't his style in any case. In fact, he had been right to let her go without protest. Agnes was a very independent young woman who didn't like to be pushed into any situation. However, once she made up her mind and latched on to an idea with both her heart and brain, she stuck to her decision. After a long night she mostly spent with thinking long and hard, she concluded that the primary reason she had come back to Rochester wasn't the job but was indeed Doug. "Is nevertheless a higher power out there who arranged this meeting?" She was surprised that she went as far as asking such question, which then she asked many times afterward. From that point on she did not waver, and in the morning when she woke up, she knew that she was about to start a new life, with Doug by her side.

Doug easily found a fourth tennis partner for the next day; it probably helped that he mentioned that one of the players would be a beautiful young woman. The men initially attempted to exhibit some level of chivalry, but after the first few strokes, it was already quite clear that they were completely mistaken; Agnes outmatched them all. She started out partnering with Doug, and they easily defeated the other team in a one-set match. After that they kept switching partners, but this didn't help the men; no matter what the combination was, Agnes's side won. After the match, Doug, flabbergasted, asked Agnes if she had played at a professional level in Boston. She laughed and revealed that during her college years, she had won several state trophies with her team, and after her return to Boston, she again found some challenging partners to practice with. She was very happy to have had the opportunity to play with Doug and his colleagues, and each promised that next week they would play again.

After showering, they first drove to Doug's house to pick up the wine, and then to Agnes's apartment. She had already prepared two big slices of beefsteak before tennis, and she only had to warm them up. Agnes once again led the discussion that evening, talking at length about many different subjects, barely letting Doug get a word in. He didn't mind it; he was simply in rapture listening to her voice. That afternoon, Agnes took a brief break and went to the

Barnes & Noble bookstore to buy Brian Green's book about string theory and the theory of everything. She recently took great interest in hearing about what other people thought about the possible existence of multiple universes, what Greene and others simply call the multiverse. "After I read it, I would really like to discuss this book with you," she said. "May I suggest that this be the first topic on our seminar list?"

"Of course," Doug replied. "I actually read it not long ago, and I still remember Green's major arguments for why string theory could reconcile quantum mechanics and classical physics, although perhaps it would be better if I would go through the book again."

As Agnes served the food, Doug poured the red wine into her most elegant glasses he could find and declared, "It is time for a toast. For me nothing has changed since yesterday evening, and in fact since I first met you. I am very much in love with you, Agnes, and I believe that we could make a great couple. What do you say? Have you thought about it?"

"Yes, I have, and I also feel the same way as I told you yesterday, except that now I'm saying it without hesitation and without asking for more time. Yes, regardless that we don't have a long history together, I love you too, and I believe that we would indeed make a great couple."

Doug was close to tears of joy. "Agnes, you have made me so very happy. I promise you that I'll do my very best to keep you happy too. I have learned from my past mistakes, and I'll avoid them." For some women these words might have sounded too theatrical, but for Agnes they meant a lot; she knew that they came from his heart and that he meant them. She believed wholeheartedly that now they had a strong covenant. They forgot about the food and wrapped themselves in each other's arms, staying like this for quite a while.

They finished the dinner in a short time, and after Agnes did the dishes, she snuggled up to him on the coach and embraced him. As they kissed, the blood went palpably bounding along their veins. "Please don't leave me alone tonight," she whispered.

"How could I?"

That night she gave everything she had to Doug. Perhaps for the first time in his life, Doug felt like he was in paradise. "If paradise isn't like that, then I don't want to go there," he said to himself. He had never felt such harmony with Sarah or anyone else. As Agnes slept by his side, he gently caressed her hair. "What if it was God's hand that arranged all this? What are the chances that otherwise we would ever have met?" With that thought he too fell asleep. Agnes, embracing Doug, fell asleep with the same thought in mind. It was 2:00 a.m.

Agnes was the first to wake up, around 6:00, still feeling quite groggy. It was still murky outside, and the bed was so inviting. But she had to be in the office by 8:00, and before that she still wanted time for a first romantic breakfast together, scrambled egg with ham, in bed. Agnes was afraid that Doug might be annoyed at being woken so early, so she chose to act a little silly. "Please, mister, don't take my question amiss, but I wonder if you would mind if I were to obtrude myself upon you." But she didn't have to worry. Doug just laughed and laughed. He couldn't believe he was about to have breakfast in bed with the most beautiful woman on earth. This was another first in his life. The previous night and the breakfast in the morning became a defining trope of their relationship; there wasn't even a hint of ambiguity that from now on their lives would be tied together. After they had eaten, Doug again felt the same impetuous desire for Agnes that he had had last evening, but she simply flashed him an amused smile and slid out of bed; she still needed some time to prepare for the day. Doug joined her. "You, dissembler woman, how well you can pretend innocence," he said jokingly while dressed. Agnes only smiled again; she had already got the knack of Doug and could tell when he was joking. They left the apartment together, and Agnes dropped Doug off at his house.

During the next few weeks they met almost daily, either in Doug's house or in Agnes's apartment. That May, Doug asked for her hand in marriage. She said yes instantly. She was just surprised it had taken him that long to ask her. Neither had a shadow of doubt that they were meant to spend eternity with the other. They set the date for the last Saturday of August, which would hopefully give them enough time for the preparations. Early in June Agnes moved in with Doug, which greatly simplified their everyday life as well as the preparations for their wedding and honeymoon. It was the honeymoon that both really looked forward to, but for the time being, they had other important projects to deal with.

Agnes liked Doug's house. It didn't have many rooms, but the windows were large, allowing plenty of light in. One of the first things she noticed was that Doug seemed to have a great affinity for cacti. She counted some eighty smaller cactus plants and two large agaves. "What I really like about them is that they aren't only beautiful, but they are also survivors," Doug explained. "Look at this one, for example. It's called sundancer. It's simply beautiful to me. Whenever I look at the arrangement of its leaves, I see a whole, multidimensional universe. Every leaf and flower come with their own space, independent from and yet related to the spaces of the others. If you could make this sundancer more transparent and enhance its size by many billions, you would have an entire universe with many dimensions. But if you went the other direction and reduced it to the size of the particle the universe started from in the big bang, then you could understand how our multidimensional universe was created."

The small cactus had clearly captured Doug's imagination and intellect. "I often wonder how such a beautiful plant can arise from a small seed, which is nothing more to me than an amorphous mass. Of course, we know that all the information needed for the creation of the plant is in the seed just as all the information needed to develop a human being is present in the fertilized ovum. But the fact that all this complex, highly organized information is held in such tiny structures is truly mind-boggling. It's obvious to me that evolution led to the development of the plant seed and ovum and sperms, but how could evolution have led to the construction of the little speck, tiny but containing an enormous amount of information, which developed into the universe? I wish I could ever understand this. What I do think I understand is that evolution wasn't only the driving force behind the development of the universe and of life after the big bang, but it also played a key role in the occurrence of big bang itself."

"You just mentioned something very important," said Agnes, momentarily forgetting about cacti. "Are you saying that all the information present today in our universe was present in that speck from which the big bang started? Wasn't the information born together with the universe from the big bang?"

"Recently I have come to think that information may be the only thing that has existed forever, and its quantity never changes. Since time is eternal, information hasn't been lost and has not been formed. It only takes on different forms through some intermediate steps. Just as there is the 'law of the conservation of energy and matter,' I believe that there also has to be a 'law of the conservation of information.'"

"Oh dear, this is something we will need to discuss later in more detail. I have to acknowledge that today I'm having difficulty conceptualizing this new idea of information," said Agnes.

"I'm not surprised," said Doug. "I can see that the concept of information would be next to impossible to comprehend for anybody. I should add that information might have a dual nature. On the one hand, we usually consider information as knowledge, which doesn't have any physical attributes. For example, a book contains a certain amount of information that we can quantify in any way we want, such as the number of words and so on. Then there may be a different kind of information, a new independent physical or semiphysical entity, from which all known energy fields and matter particles are derived. I want to stress this last point. Very recently it occurred to me that perhaps the big bang started from highly concentrated information content in a preexisting vacuum."

This little conversation reminded them that however much they were in love, they shouldn't forget about the original goal Agnes had in mind when she first contacted Doug to find out more about his apparently ever-changing ideas about the big bang and what came before, God's method of creating and 'running' the universe, His method of communicating with us, and what happens after death. In a nutshell, these were the major topics Agnes wanted to discuss, but she knew that neither of them was likely to come up with anything more than a conjecture. Thousands of failed attempts before them foreshadowed such outcome, but they didn't fret. They made a point not to be intimidated by these previous failures. Success in advancing a theory often comes from unexpected sources. As one evening Doug told Agnes, "Robert Kirschner, an excellent astrophysicist who made landmark discoveries and who wrote the book *Extravagant Universe*, made the following observation, which I can quote only loosely: 'You do not always have to understand the details of the mathematics to contribute to the advancement of science. You just have to face the right direction and go forward with the things that you know how to do.' I believe the same truth applies to those who want to understand the true nature of God. Creativity, our brain's most ineffable quality, doesn't always depend on our level of education."

Considering that neither of them had serious training in astronomy, quantum mechanics, or mathematics, they were really an unexpected team to be grappling with these questions. But, they agreed, they did have one advantage. Unlike theorists and theologians who are often hamstrung by the limits of their own science and faith belief systems, they had completely open minds and therefore the liberty to raise questions and search for truth without any limitation. They could roam free. Of course, this didn't mean that they could be ignorant of the latest developments in science. Quite the opposite; they wanted to assimilate and harness all the latest scientific breakthroughs to advance their own thought. This is where Agnes hoped to get help from Doug, who followed very closely all new discoveries relating to both cosmological theories and quantum mechanics. Once Doug told her, "It's very important to be up-to-date with new science, but it's equally important not to become a prisoner of it. Theories change to accommodate new discoveries, so what today is accepted as fact, tomorrow may be replaced by a new truth. Even the best scientists acknowledge that their theories about the universe might not last longer than a few years. Indeed, this is a perfect example of evolution. But we also need keep in mind the quote that has been variably attributed to Patrick Moynihan, James Schlesinger, and Bernard Baruch that 'everyone is entitled to their opinions, but they aren't entitled to their own facts.' The lesson is that we have to be very careful not to disregard facts."

"Yes, I've started to notice from my readings that scientists can promote their ideas with a lot of elocution often built on nothing but opinion and conjecture. I try to take the ideas of those who build theories without facts with a grain of salt, although this is easier said than done. For example—and I do not mean to hurt your feelings—but it's already pretty clear to me that you also have the tendency to go beyond known facts," Agnes said.

"Well, maybe that is because I am also a scientist," countered Doug with a big smile. "Besides, for the very good reason that you just mentioned, I always refer to my theories as science fiction, which is an entirely different category. However, there is a third category when based on observations and mathematical calculations a scientist can predict the existence of a new phenomenon, which at the time couldn't be proven but later was confirmed. My theories may never reach this third level."

Receiving the grant was a huge deal for Doug, but now he had to put a lot more effort into running his laboratory; almost every day presented a conundrum that he had to solve rapidly. Thus, for months he had no time to fine-tune his theories. In fact, it was more than not having enough time; he was not satisfied with the idea he had concocted before, and he was afraid that he wouldn't be able to come up with better ones. However, now that they lived together, there was no escape from again thinking deeply about these 'ultimate' questions that so much preoccupied him before. Armed with several recently published books on the topic, Agnes proved to be a good cheerleader, selecting the topics and asking tough questions.

They didn't know where those discussions would lead, but they were having a lot of fun. Besides, they also agreed that it was better to settle some fundamental issues before marriage than to be sorry afterward. Once Doug, when he was in a particularly philosophical mood, told Agnes, "I believe that there is nothing more rewarding in a couple's life than having a common interest in developing answers for the reason for our existence. At the very least, looking at the big picture, the circumstances of Creation and what followed, helps us rise above our small, annoying everyday problems. At least for me, it has been a very potent calming anodyne effect. Somehow, when

I realize how lucky we are to have given the opportunity to live on this wonderful planet, all my earthly problems vanish. Of course, for someone who was born into poverty and must be working hard every day just to make ends meet, the last thing that person can afford is to meditate about how it all started. We are very privileged to be able to do that."

"We are very lucky," agreed Agnes. "And I believe that for the advancement of humanity, people like you have always been and always will be needed."

"Agnes, please don't exaggerate. I'm still very far from making a dent in the 'advancement of humanity,' as you put it."

"It doesn't matter to me, because you're at least trying, and, I love you for that, regardless of whether or not you will succeed."

One evening they went to bed a little earlier than usual, and Agnes initiated their first 'scientific meeting' by asking Doug, "Do you believe that the universe sprang forth from an absolutely dimensionless nothing, basically from an enormously energetic singular event, as some scientists suggest, or at that moment of Creation something very small was already there? Hawking and many others say with confidence that the whole universe could have arisen from nothing due to some quantum jigger. I really can't follow his thinking. I guess this theory implies that at the beginning the universe was infinitely compressed into a zero-size starting point, with infinite temperature and energy. But how can something with zero dimensions be infinitely hot with infinite energy with some undefined quantum jiggers going on eventually resulting in an explosion? I also have another problem with this quantum jigger theory. Why after the big bang there has never been another big bang somewhere inside the universe? If the advocates of multiple universes theory are correct, then I believe there should have been many of them. I don't buy the argument that new universes can arise only outside our universe because probably we already would have bumped into some of them with catastrophic outcomes."

Not getting an immediate answer from Doug, Agnes continued. "I also read what string theorists say. They think that Creation started from a very small and very compact aggregation of matter, about 10^{-33} cm long. In this speck, many spatial dimensions were curled up, containing enormous, but not infinite, energy and temperature. When expansion began, within the first 10^{-43} seconds after the big bang, the presently known three spatial dimensions were singled out for expansion, while all the others retained their initial size. String theory suggests that we may be living in a universe that macroscopically has a three-dimensional space, but microscopically we may be surrounded by these multidimensional microscopic spaces. Frankly, this is also something that I can't imagine as real, but I think it is more likely than Hawking's idea. What do you think, Doug?"

Doug raised his eyebrows. "Oh my, oh my, that was a very long question, my dear. I don't even know where to start." He was playing with Agnes's hair, and he had other thoughts in his mind, completely unrelated to deep space. But Agnes looked at him seriously, and he realized that he had to keep to their agreements.

"The speck you referred to as being the very hot, high-energy, tightly curled-up space as the starting point of Creation makes some sense to me, although it is based on string theory, which in this particular context isn't my favorite. More exactly, I can't see how a tiny one-dimensional filament, resembling a tiny vibrating rubber band, could be the basic component of fundamental particles like electrons and quarks. According to prevalent modern thought, these fundamental particles are supposed to be structureless. In contrast, string theory holds that they have a one-dimensional loop structure with a preferred vibration pattern that determines their properties, such as mass and force charges, and I instinctively don't like that part. But maybe with time, string theorists can convince me of the value of their theory explaining how fundamental particles come with mass and structure.

"However, I definitely agree with the idea that right before the big bang, there could have been a very small speck with an unimaginably large and well-organized information content that had already generated an extremely high level of energy. As the energy continued to accumulate, there was a 'breaking point' and the speck exploded. But I can also identify with the part of string theory postulating that the speck had many spatial dimensions, tightly curled up to the smallest possible extent. I don't believe that our universe suddenly arose from *nothing*, whatever *nothing* might be, without some evolutionary intermediate steps before that. So at some point, the matterless *nothing*, which still had certain physical or semiphysical nature including high energy content, gave rise to form the speck. I think there is an important point, which nobody has raised so far. The tiny speck had to exist in space, which we might call the *nothing*, which was different from the space created by the big bang and with which we are familiar with. That space could be very big or small, and it could have two or a large number of dimensions. We will never know how that space looked like because we can never go back to zero time when the explosion was just about to happen."

"What happened to that space, which I think maybe called the primary space, after the big bang?" asked Agnes.

"I see at least two possibilities. It still might exist connected to our universe, or it might be the space that exploded first followed by the explosion of the speck, creating our universe, which is expanding into that primary space. We'll never know."

Then Doug continued. "The most important question for me is really how this *nothing*, or as you call primary space, which I believe wasn't actually completely empty, looked like. I know this might sound like a ridiculous question, because common sense would dictate that we shouldn't be able to attach any attributes to *nothing*. Yet we can't avoid asking what *nothing* means in the first place. If before the big bang there was *nothing*, does that mean that the space's dimension was zero, so there was no space at all? Or does *nothing* imply a vacuum space, with no matter, only information with some energy content? If so, could information be an entirely new entity, which might be the 'mother' of all matter forms, known and unknown, and energy forms we are familiar with? Does the totality of information equal the spiritual God and, in a highly organized form, the material God? These are crucial questions, which we need to find an answer for before we can turn to the subsequent questions of how and from what the speck was formed, what its structure might have looked like or if it even had any, for how long it had existed before it exploded, and what made it explode. Closely related to these issues is the question, were there actually two specks that exploded in two separate big bangs, and if the answer is yes, how do these two events relate to each other? The first time I started to deal with these issues, they made me aghast, but by now I realized I can't be coy by confronting them if I want to progress my theory."

"What makes you to believe that there were two specks and two big bangs?" For Agnes, these ideas came out of the blue, and she instantly wanted to know more about this hypothesis.

"Let us deal with the speck or specks later. Instead, I think we should start discussing the *nothing* first, because my hunch is that this is where we have to start looking for God. One way to characterize God is that He is an all-encompassing information field that isn't comparable to other fields, like the Higgs field."

"The Higgs field," Agnes murmured. "It is interesting that you brought that up. There has been a lot of news about the Higgs boson lately. Is this something produced by the Higgs field? Why is the Higgs boson so important?"

"Well, almost forty years ago, a British scientist named Peter Higgs and several others put forward a hypothesis about the existence of a certain kind of elementary particle, now called the Higgs boson, that would be the manifestation of an invisible field—the Higgs field—that is thought to permeate the entire universe. Higgs and others theorized that the Higgs boson particles gave mass to all the other fundamental particles of the universe after the big bang. Without the Higgs field or something similar, even today all particles would have no mass, and they would just keep flying at the speed of light across the universe, which wouldn't have any mass either."

"Do you mean that without the Higgs field nothing really would exist except photons?" asked Agnes.

"Yes, you are correct. According to this theory, the Higgs boson was required to make order in the inchoate early universe and it made the formation of stars, planets, and life possible. Because of its importance, the Higgs boson is also often called the God particle. Thousands of scientists have searched for it with multimillion-dollar budgets. You have probably heard of the large hadron collider, which was built mainly for that reason. Just to illustrate the size of this facility, it is basically a seventeen-mile-long tunnel built under the Swiss-French border, where high-energy beams of protons are generated and sent crashing into each other at incredible speeds. Collisions between two protons then produce high energy, which leads to the creation of other particles. On very rare occasions, this process could produce Higgs bosons, which are immediately built into these other particles. Now scientists reported that they have identified the Higgs boson, and with a few exceptions, most physicists believe in the report's findings. Isn't it interesting that we deal with mass every day, yet we really have no idea what it is? Something must give mass to the fundamental particles, and through them to everything else, and now it seems that the Higgs boson is the best candidate to do that."

"That's incredible. I never would have thought that fundamental particles may not automatically come with mass," Agnes exclaimed.

"You are not alone. For example, unlike the present form of standard theory which employs sixty-one elementary particles and four forces of nature, string theory, as I partly already explained not long ago, doesn't need the Higgs boson. Let me repeat, string theorists say that the pattern of string vibration determines all properties of all the particles, including mass and charge. So according to them, fundamental particles indeed automatically come with mass. Finding the Higgs boson might in fact kill the string theory. In any case, it will be interesting to see how this will play out. For many scientists involved, the stakes are high, but eventually there will have to be losers and winners unless a miracle happens and it turns out that the string theory and the standard model can be united. This would be a providential solution.

"But back to the bosons, string theory is not the only theory which doesn't require the Higgs boson. For example, Scott Tyson, in his book *The Unobservable Universe*, argues that the curvature of space gives mass to matter. His theory is quite unorthodox, and we shouldn't even try to analyze it now. I would only comment on Tyson's main point, which is, mass is not the property of matter—it is the property of space."

They had gone off track quite a bit, and Doug suggested going back to the concept of God. "I can imagine that God may also permanently exist as a physical being, inseparable from His spiritual counterpart, just like two faces of the same coin. Here is why I think that this may be the case. But before I go any further, I need a little wine. Would you also like some?"

"Sure, why not. We have to loosen up a little bit before we further discuss such a heavy subject." After drinking and kissing, repeating the cycle several times, Doug continued with his theory. "I think it is possible that the physical side of God is the highly organized form of information that not only contains all knowledge but is able to act on it with total freedom. He does everything in complete unity with the spiritual God, who may be structureless yet can provide the motivation for all the actions of His material counterpart. My main point is that since in my present view everything derives from information, there is nothing in the universe which isn't part of God. Literally, God gave part of Himself to produce the universe and everything in it. How this could happen? I believe He organized an enormous amount of information into a very small area of the vacuum that I earlier referred to as speck. This incredible concentration of information somehow must have created, through a kind of phase transition, a very high-energy state that then at some point led to the explosion that we call the big bang and subsequently to the formation of energy and matter particle forms present today.

"I believe that from the point of the speck's explosion, scientists are correct in describing the evolution of the visible components of the universe. However, while practically all physicists think that space as well as dark matter and dark energy were created by the big bang, I beg to differ on that point. I think that the universe expanded into an existing space already containing basic constituents like dark energy, dark matter, and possibly some other similarly invisible components. Again, as one possibility, in a preceding big bang, God might have had created the preexisting space in which the invisible components began to develop a structure. I assume that in this preexisting space the only force was gravitation, which helped with building of the structure. Now I believe that the newborn universe extended into this structured preexisting space, which I call the 'parallel universe.' Furthermore, I think that the parallel universe might have served as a template for the evolution of the universe. Such template could have been composed of dark matter, which with its gravitational influence on matter could have aided in the formation of galaxies according to a certain pattern based on the small differences in the heat and mass densities in our baby universe. Scientists are sure that dark matter plays a critical role in holding the galaxies together, and it's not too far-fetched to think that it was also necessary for the development of galaxies as we know them. It's also known that dark energy plays a role in the expansion of the universe with an ever-increasing speed. Dark energy suffuses space, and its strength is apparently increasing, which accounts for the acceleration of the universe's expansion. The subject of dark energy story is very fascinating to me. Some assume that it also caused the much-faster-than-light inflationary period soon after the big bang, but how that would have been possible if space didn't exist before? In fact, dark energy may be none other than the pulling vacuum force of the larger parallel universe.

"Agnes, think about a smaller circle, representing the space formed in the second big bang, which is within a larger circle, which is the space preexisted or formed in the first big bang. It is inevitable that some of the original components of the larger circle are mixed with the components of the smaller circle. However, the larger circle kept its own components intact *outside* the smaller circle. Let us assume that the larger circle initially contained dark matter and dark energy, which then mixed with the visible components of the universe. We obviously view our universe—the smaller circle—as one space. However, if we were able to look beyond our universe, we would recognize the larger circle, the parallel universe, into which we are expanding. I think that dark energy, present in the parallel universe beyond our universe, is the driving vacuum force which moves each galaxy with increasing speed further and further away from the others."

Agnes was digesting for a while what Doug just said and then came up with a new idea. "You know, something just occurred to me. Do you think that the vacuum force of the dark energy in the parallel space initiated the big bang?"

"Yes, that's one possibility among several others," answered Doug approvingly. "Let's assume that the parallel space was the size of a basketball, and then expanded. At some point it exerted so much vacuum force on the remaining concentrated information content that it exploded and, within a tiny fraction of a second, grew into the size of a tennis ball with speed much greater than the speed of light. However, because of the increasing gravitational forces inside the tennis-ball-sized universe, the vacuum pressure applied by the outer parallel space decreased, and the expansion of the universe slowed. For billions of years the vacuum force of the growing parallel space and the expansion rate of the also-growing universe were nearly in equilibrium. But then, about seven and a half billion years ago, the growth of the parallel universe became disproportionally larger, thereby gradually exerting disproportionally more and more vacuum force on our universe. At the same time gravitation within the universe is reduced due to the increased distances among galaxies. These two opposite effects result in accelerating expansion of the universe in the foreseeable future."

"And what is the parallel universe expanding into?" asked Agnes.

"The first option is that the parallel universe will expand only up to the point where the energy of the initial blast still lasts, then eventually the expansion will stop. Consequently, its vacuum force will stop increasing, and with that, the expansion of our universe will stop too. This scenario may lead to the shrinking of the universe eventually resulting in a big crunch. When the concentration of matter and energy reaches a certain point and forms an incredibly hot speck with incredibly high energy content and gravitation, a big bang may occur in the reverse—that is, the high energy would become information again. The second option is that for some reason, the expansion of both universes will keep accelerating, and at the end, when even the quarks are destroyed, the space will contain only pure information again. I prefer the first option because it is more logical to me."

Another evening they had a slightly different discussion, which this time Doug started. "I already told you that my dream is to see faith and science on the same side of the equation. In real life, this could only happen if people of faith start approaching science with an open mind and try to accept as much of the science's teachings and methods as possible, without of course compromising on the existence of God. I don't know if I can ever make a meaningful contribution to that, but at the very least, I hope that if I find a reasonable and honest solution to bring faith and science together within me, which I need to be at peace with myself for the rest of my life, will you remain with me on this journey, wherever it might lead to?"

"I intend to, every step of the way, and if you will sound convincing, I may give up my agnostic approach to God. But even if you don't convince me, I won't forsake you."

"Agnes, thank you. This is fair and means a lot to me."

"Okay, let's get back to the issue of reconciliation. How do you expect it to work, for example, on the subject of Creation?" Agnes asked.

"You obviously don't want to get me off the hook tonight," said Doug, beginning another round of brainstorming. "I think an important step forward would be the acceptance that it was the material side of God, and not an entirely spiritual God endowed with no other attributes than being spiritual, that had the ability and the means to plan and execute the creation event. Also, to be in line with science, the faithful should accept that after setting the basic parameters for the visible universe, encoded in the speck's structure, He chose evolution as the driving force of the universe. Just like our own extremely complicated, well-tuned development which follows the instructions of DNA present in the fertilized egg, the enormous amount of information present in the speck—the DNA for the universe—also prescribed the main course of evolution for the universe. It follows that He hasn't intervened in the evolution of our universe by any means that would have required any alteration, however small, of any of the physical laws He established. In other words, after the moment of Creation, He didn't and doesn't perform miracles. So in principle, science can explain and understand everything from the moment of the speck's explosion, because the natural laws don't change, so they will remain predictable while the universe exists."

"Let alone the fact that most scientists will not accept God in any shape and form, your theory about the preexisting space also goes against the beliefs of all physicists who believe that the space in which our universe is now embedded was created by the big bang, which was a singular event," countered Agnes.

"Our space, as we know it, was indeed created by the big bang, in agreement with all theories describing this event. What I am saying is that while our universe, with its own fabric of space, was generated by the second big bang, it also contains components created in an earlier big bang. In other words, our universe not only contains the matter and energy formed in the second big bang but also contains components formed during the first big bang. It's like mixing clean water and milk in a cup. According to our naked eyes, milk and water mix very well. But if you looked at it with a microscope, you would see small droplets formed by milk fat and other structures. If you would look even further, you would see thousands of molecules that were present in the milk and absent from water. It means that a mixture made up from water and milk represents a new entity that is neither milk nor water. Yet, in the true sense, the milk isn't lost. The point I'm making is that if there was a preexisting space, then its structure was almost certainly very different from the three-dimensional space of our universe, so macroscopically the mixing might seem complete, but microscopically the two spaces might have retained some important and distinct characteristics. For example, dark matter could have initially been a component of only the preexisting space, but now it is part of our universe too, although it is quite distinguishable from visible matter."

"What other characteristics the two spaces might have?" asked Agnes.

"I obviously can't say for sure, but there are various possibilities. One of them is that the preexisting space is composed of very tiny extra dimensions, now filling each point of our universe as well as the space outside it. This might not be too far-fetched an idea. For example, string theorists say that our universe is filled with microscopic six-dimensional curled-up spaces they call 'Calabi-Yau' shapes that come in many possible forms. Strings would be attached to various Calabi-Yau shapes that would direct their vibration and therefore all their properties. As another scenario, what if we assume that God provided a preexisting space with Calabi-Yau shapes, while the second big bang provided the strings, perhaps from information units? Thus, our three-dimensional space created by the second big bang would extend into the preexisting space filled with Calabi-Yau shapes and would direct their interactions with strings and formation of elementary particles. Thus, while the speck would have had the information to produce certain types of strings, it may be equally important that the information in the Calabi-Yau shapes would have provided the compass for what kind of elementary particles should have been formed, which in turn determined the evolution of our universe."

"This is fascinating!" exclaimed Agnes. "I can't wait to see what you will achieve with this theory."

"Well, what is especially fascinating about this possibility is that just like DNA is double stranded, the initial condition for the creation of the universe also might be visualized as requiring two strands. One strand is the speck that in the event of the second big bang created strings from information units, while the other strand is the preexisting space with its extra dimensions. Both strands may have been needed for the information to be translated into particle formation and the evolution of the universe, just like both strands of the DNA are needed to translate the information into a kind of message that then directs protein synthesis and the development, or evolution, of the organism into an adult."

"So, if I understood you correctly, God might have chosen the same principle for the evolution of the universe and the living, am I right?" asked Agnes.

"Yes, you're correct. That's what I meant," responded Doug. "But before I go overboard with this theory, I need to stress that this is only one possible scenario. But the reason I keep it in mind is that it can be easily understood, because it is based on string theory, probably the most popular theory of everything. How all this is related to the 'cosmic signposts,' concentrates of dark matter or some other source of gravitation, and dark energy that together directed the evolution of visible matter into stars, galaxies, and planets is a whole different story. A lot more discoveries are still needed until cosmologists can build a coherent picture of these events, which would be required to advance my theory as well and perhaps contribute a small piece to this huge puzzle."

"If you like to continue our conversation, then let's first talk about dark energy," Agnes suggested. "I'm not sure that I understood your earlier explanation of dark energy. Sometimes it seemed to influence the visible universe, and sometimes it didn't. What explains this irregularity in its action?"

"Okay, I'll try again, but this is only my theory. I think dark energy might simply be the vacuum force of the parallel universe. This vacuum force was behind the inflation of the visible universe, which lasted for a very small fraction of a second after the big bang. The reason the inflation period was so short could be that the gravitational forces within the baby universe soon became strong enough to decrease the vacuum force of the parallel universe. This was probably the period when the Higgs bosons appeared to endow the elementary particles with mass. As the space of the universe had begun to get filled up with a large mass of matter and associated gravity, the pulling effect of the dark energy decreased. However, about seven and a half billion years ago, the mass of the visible matter had become more dispersed, and the gravitational forces decreased to a point where dark energy began to have the upper hand again, resulting in the ever-greater acceleration of our universe's expansion. The probable cause of dark energy gaining strength at that point was that by then the space of the parallel universe grew disproportionately compared to the space of our universe, while the expansion of our universe decreased the concentration of matter and with that the gravitational force."

"Now I think I understand your line of thinking. And what do you think about dark matter?"

"Well, nobody knows what dark matter is, but I do have a wild theory, although I wouldn't say that I'm dead earnest about it. I think it might represent multidimensional tiny spaces, perhaps leftovers from the first big bang, filled with something other than strings, something that has gravity and attracts ordinary matter but has no charge. This would keep the galaxies together. Without dark matter in and around the galaxies, stars would simply fly away. I believe it is well established by cosmologists that dark matter was also crucial for galaxy formation, but to explain how this could happen, again the two-stranded DNA comes to my mind."

"It seems to me that you always find a reason to come back to DNA. It shows that you are in love with molecular biology," said Agnes, smiling.

"You are correct, but let me explain a common thread here. It is known that in the early universe, there were very small differences in the distribution of matter, creating somewhat denser spots with somewhat greater gravity compared to areas in space where distribution of material was homogenous. A dense spot can be viewed as one strand, which then could have attracted to dark matter, which I call the second, or complementary, strand. You see, just as I believe that God Himself has a dual nature, I think He followed upon this idea of dualism when He made evolution, full of dualistic characteristics, the engine of the universe's development. And having two separate but interwoven spaces, derived from two separate big bangs, would also be testaments to the powerful role of dualism. We might never know the true nature of God or how Creation really happened. But if we can build up a rational approach, then we might be able to bring science and, as I like to call it recently, 'scientific faith' together."

"It is interesting that this time you called your theory 'scientific faith' and not 'science fiction,'" noted Agnes.

"Well, scientific faith is basically science fiction with an element of faith," explained Doug. "Occasionally, I have a penchant for using scientific faith instead of science fiction to describe what I begin to believe in."

Agnes, still trying to wrap her head around what she had heard, felt the need to give a short summary of it. "Well, if I understand you correctly, your idea is that in the universe everything is predetermined by this 'two-stranded DNA-like' nature of the universe. In other words, the multidimensional preexisting space and the speck contained all the complementary information that was sufficient to precisely determine every aspect of how the universe should evolve."

"Not really. It would be nice to believe that, but the situation is probably far more complex. I think that the speck and the primordial space, just like the DNA of a fertilized egg, had the information content necessary to prescribe the main course of events, but there is a certain level of uncertainty to what exactly the evolutionary process will yield. For example, the DNA of our ancestors could not have predicted how we humans would look like today, since there were many other events that impacted our development. If, for example, the Neanderthals had managed to win the war against Cro-Magnons, then today we would look quite different and our brains would probably be less advanced, although I'm not sure about the last point. Heisenberg's uncertainty principle is true not only in the quantum world but also in the evolution of everything in the universe. Just like an electron, evolution seeks out and tries many options at all levels, and there is a certain level of uncertainty to which option will prevail and when. For example, large meteorites coming out of the blue and violent volcanic activities greatly affected the evolution of life here on Earth. However, despite the uncertainty concerning the location and time, somewhere, sometime, a kind of human race would have developed in some form. Evolution is an obdurate but also an opportunistic incessant process, which has been going on since the big bang."

"You mentioned Heisenberg's uncertainty principle. What it is?" asked Agnes.

"Simply put, it means that certain pairs of physical properties of a quantum object, for example, position and momentum of an electron, cannot be determined simultaneously with the same precision. For example, the more precisely a particle's momentum is determined, the less precisely its position can be known, and vice versa."

"I get it," said Agnes. Then she searched for the correct words to phrase her next question. "What you said about evolution implies that God just set the initial conditions and then let things go, without any plan for how things would develop, including life and humanity. Was it uncertain even for Him when and how, for example, the first cells would form or if humans will develop at all?"

"If I want to stay consistent with my logic, then I am afraid to say that after the initial conditions in the speck and preexisting space had been set, God let evolution take its course, wherever it might lead to. So, in other words, my guess is that all that He knew for sure was that the initial information He provided in the speck and surrounding space was sufficient for the eventual rise of intelligent beings, with whom He could communicate. Remember what I said before. Evolution seeks out all possible directions. In my view, this law of evolution is firmly related to the initial set of information and, thus, was established by God. Let me give you an earthly example. If the coach of a soccer team selects good players and gives them good instructions, even if the team's play will undergo many different turns depending on split-second decisions by both team members, the team will eventually win the match and perhaps even the championship. What I'm saying is that while the team wants to execute the coach's plan, many different factors will determine how it will turn out during a long season."

"So you mean that He isn't involved in our lives at all?" asked Agnes with a good dose of incredulity in her voice. "This would mean a lot of debunking for the church to accept."

"I didn't mean that, not at all. I strongly believe that He is involved in our lives, particularly because in a way we are part of Him, just like anything else in the universe. In addition to that, since I assume that God has a material side, I believe that He can also communicate with us. But before we go further with this subject, let's start at the beginning, which is for me information.

"Lately I have been giving more thought how the information would look like, and an idea occurred to me. What if information come in elementary, distinguishable units with no real information content when they remain isolated. However, if three, four, five, or more elementary information units or blocks associate, this will result in an information coding unit, each with a deterministic value. Untold numbers of such information coding units with characteristic information content could exist in the pre–big bang period and in the universe floating around us. These coding units could form various individual super information complexes, some of which, immediately after the big bang, transformed into corresponding elementary particles. These super information complexes, made up from many coding units in variable sequences, resemble the structure of DNA wherein three bases make up a code, each different code corresponding to one amino acid. Just like the sequence of three-letter codes in DNA determines what kind of proteins will be formed in the translation process, the sequence of distinct coding units in the super information complexes would determine which elementary particles or perhaps, as the primary steps, strings and Calabi-Yau shapes would form from them.

"There are an awful lot of questions that I still have to find explanations for before I can make my hypothesis as tight as an anchor bend knot. One of them is how material God, probably Himself a super-super information complex at the highest level, could mass produce the specific super information complexes and how He could make it sure that each type of these complexes will yield a specific elementary particle or a string.

"Another question I need to deal with is the following: Once material God put together the speck from enormous number of super information complexes, how did He make sure that the explosion, i.e., the big bang, occurred just at the right moment, not before the speck was ready, and how did He trigger the explosion? What I can think of is that large concentrations of super information complexes were relatively unstable and could explode if achieved a critical amount due to jiggers of their high-energy content. Thus, God might have built the speck in two parts, each having subcritical amounts of super information complexes. When God decided that it was time for Creation, He pushed the two parts of the speck close to each other to achieve critical mass for the initiation of the big bang. In fact, this is how fission weapons, like atomic bombs, work except that in that case, the fissile materials, like uranium or plutonium, are known.

"Yet another question is, what happened during the first 10^{-43} seconds after the big bang, which is also called the Planck epoch, when the information-loaded speck transformed into something that provided the basis for the universe. Physicists do not really know much about this epoch except that the temperature and average energies were inconceivably high and the presently known fundamental forces, the gravitation, electromagnetism, weak and strong nuclear interactions were together and formed one fundamental force. This is the period, I believe, when the high-energy explosion of the speck led to the creation of the united fundamental force containing the seeds of the four forces, which became separated, as programmed by the super information complexes, soon after the universe started to cool down to a certain temperature. According to this scheme, the super information complexes were eventually also the originators of the elementary particles."

"Do you think that super information complexes form elementary particles even today or if they play any roles in the universe?" Agnes asked.

"Perhaps they do. You know in the quantum world, there are many crazy things happening, which will look crazy only until we recognize the basis for them. One of the mind-blowing observations is that in empty space that contains no energy, no charge, no matter, is filled with active populations of virtual particles, called quantum foam, that individually exist for only small fractions of a millisecond. These particles must borrow energy from somewhere, and one possibility is that empty space, just like any space in the universe, is populated by super information complexes that are able to lend energy for the transient formation of particles. Thus, the newly formed particle will not break the energy and momentum conservation laws for any length of time. But for this phenomenon to happen would require two things: First, because the super information complex is relatively unstable, its energy level will fluctuate, which must be able occasionally to form very short-lived particles. Second, the newly formed particle must be able to give the energy back to the information complex and with that cease to exist. Understanding these energy transitions would require a new physics, which we won't be able to develop until the nature of information becomes known. A similar scenario could explain other weird-looking phenomena, such as electrons emitting virtual photons that are either very quickly reabsorbed or form new short-lived electron/positron pairs. Just think about it, one electron can yield a photon and an electron/positron pair. To explain this and related phenomenon, Heisenberg and other physicists assume that in the quantum realm, energy and momentum need not be conserved if such nonconservation doesn't persist for long, or in other words, it persists only until 'the universe doesn't notice it.' I believe the jury on Heisenberg's theory is still out there. If you ask me which theory I like better, well, of course I prefer mine. I don't like the idea of breaking the conservation law even momentarily."

For the last few minutes Agnes could barely keep her eyes open, which was a clear indication for Doug to stop. He still would have liked to explain a possible connection between soul and super information complexes, but he had no time to do so.

The next evening, a reenergized Agnes rolled out a seemingly different idea. "Have you ever considered that quantum entanglement could somehow play a role in forming our soul and perhaps be the vehicle for information exchange between the brain and soul perhaps somehow involving super information complexes?"

"No, I never thought of that," admitted Doug, widening his eyes in admiration.

"I read that the very perspicacious physicist Erwin Schrödinger was the first to come up with the term *quantum entanglement*, and Brian Clegg, John Bell, and many other notable scientists have dealt with this fascinating quantum phenomenon. Have you at least heard about this?"

"Yes, I did hear about it, and I think it's real," answered Doug. "I read that Einstein hadn't believed in entanglement or, as he called it, 'spooky interactions.'"

That was another of those moments when the student was made a teacher. She was compelled to introduce the idea to Doug. "Incredibly, when two quantum particles, such as two electrons, are momentarily linked together when formed during a quantum event, this in a special way makes them effectively two parts of the same entity, even after their separation. In fact, the two entangled but now separated electrons might reside in different galaxies, yet a change in one is instantly reflected in the other, thereby disobeying the speed limit of light. Nobody knows how this instant communication occurs. And I think it is quite possible that from the time of our conception until our death, such entanglement may continuously occur at the quantum level and somehow might add to our soul, registering the essentials of the quantum world in our body."

"Before we talk about the soul, let's talk about the basis of entanglement between two electrons as I imagine it. To start with, an electron is not a bulletlike structure as it was once considered, but it is both a wave and a particle with indeterminate position surrounded by virtual particles like photons. Again, photons need to borrow energy to come into existence, and I think the super information complexes attached to the electrons can provide it. When a change in one of electrons' properties in the entangled pair occurs, for example when an experimenter changes its spin momentum, the information about the change is transmitted to the super information complex attached to the other electron, which somehow induces an opposite change in its spin. My imagination doesn't extend beyond this point.

"But what you really wanted to know is if entanglement has anything to do with our soul. I can't conjure up any possible mechanism to explain such possibility. First, I believe that soul, conscience, and hologram are the same thing based on how the brain builds and stores memory. This is a many billions of dollars question. The quest of neuroscientists to understand memory formation, consolidation, and retrieval so far resulted in many proposals based on the role of various neurotransmitters, activation of calcium-dependent proteins, signal transducers like cAMP, protein synthesis, and others. I do not follow this field closely, but I believe that according to recent research, synapses in the hippocampus, frontal lobe, and amygdala are the primary locations for memory storage with the frontal lobe gaining more significance as we age. The major question for me is that once the memory of a lifetime is gradually built up in whatever form, how is it stored in a way that survives us as our soul? In a very rudimental approach, I could imagine that soon after the child starts to form memories, the associated biochemical changes and characteristic changes in the frequency and height of membrane potential in the neurons and axons start adding to an initial super information complex we are born with from information coding units or blocks one block at a time. This would require many things, and crucial among them is that memory is not continuous but has a quantal property, one unit of memory corresponding to one specific information coding unit. Of course, another important requirement is that the brain, just like the whole universe, is suffused by information coding units that come with unimaginable variability. Also, the information in the individual super information complex or soul must be retrievable so we can remember our past activities and experiences. In other words, the information transfer between the brain's memory-forming mechanism and the super information complex must be bidirectional. While I think this might serve as a framework for further studies, until the biochemical and biophysical mechanism involved in memory formation and the nature of information are not clarified, I can only believe that my super information complex will survive me. To summarize, I believe that our soul, or you can say hologram, equals our super information complex, which constantly undergoes changes during our lifetime. Conscience on the other hand reflects the fact that every human carries an individual distinguishable super information complex, which serves as the base for the self."

"Just to clarify some things, in heaven we won't have our physical body?" asked Agnes.

"I assume we won't, but we won't even need it. I think all senses that we have now, and probably more, will remain imprinted in our super information complex. This will enable us to recognize others, although at this moment I have no idea how we will be able to seek out those whom we want to meet."

"According to your theory, memory formation from the childhood goes hand in hand with the building of super information complex. This would kill the idea of reincarnation, which would require incorporation of ready-made super information complex into the baby's brain," commented Agnes again.

"Yes, my theory would kill the idea of reincarnation. But I need to emphasize that my theory also includes that we are born with an embryonic form of super information complex, which primarily contains God's image, a moral code, and certain other imprinted information. Units of memory as they are being formed are being added to this information complex."

"I would like to know more about the imprinting process. I find it fascinating what you told me about God's image babies are born with."

"Actually, imprinting is widespread in all living creatures, and they can be inherited. For example, in *The Coming of Age of Quantum Biology* book, McFadden and Al-Khalili describes the migration of monarchs, my favorite butterflies. Monarchs migrate several thousands of kilometers from North America to a special place in Mexico. One of the many interesting things about this migration is that it is the third generation that arrives back to the place in North America where their predecessors started the migration. Moreover, generations after generations, monarchs know where to go in Mexico. The migration route somehow must be imprinted in a special part of their bodies, which is inherited. If such imprinting can occur in butterflies and birds, why should human brain be an exception?"

"But do you have an idea how God's message could have been imprinted in our brain?" asked Agnes.

"I believe that we are all born with an intuitive grasp of God's image embedded in the embryonic form of super information complex God provided. But, in my view, it needs reinforcement. Without reinforcement, the concept of God may fade away because His initial image may be overwhelmed by the vast amount of new information the super information complex receives during a lifetime. This is where Jesus, and before Him, Moses and all the prophets, played enormously important roles. They reinforced in the brains of their contemporaries the existence of God, together with a kind of moral code prescribing how we should behave in society as God-loving creatures and live a purposeful life. Over the centuries, this information, which by now got deeply imprinted in the brains of Christians, has been carried over to every new generation, at first mostly by word of mouth and then by the various gospels written by Matthew, John, and others. Non-Christians are also born with a sense of God's existence, but formation of His image depends on the reinforcing persons how this imprint appears in the brains of people in the corresponding geographical area. One kind of reinforced God's image and associated moral code is imprinted in practically everybody's brain regardless of whether you like it or not, whether the person is an atheist, a Christian, or a Muslim. That imprint may not be in the forefront of your brain's everyday activity, but it only takes the right stimulus to evoke these images, and God occasionally may do just that. After all, what God really wants us to do is very simple: He wants us to remember Him and live our lives according to a moral code He provided, but which would make a lot of sense even without His involvement. Killing someone or taking someone else's property are actions that are against our moral code and deprecated by others whether we are atheists or believers."

As Agnes stared at him, he realized that she still had a problem with accepting the idea of God's image imprinted in human's brain, so he approached this issue from yet another angle. "You know, around two years ago, a neurologist at Stanford University, Josef Parvizi, implanted electrodes into a patient's brain who suffered from occasional seizures. The original idea was to find out the epicenter of the seizure activity. One day, Parvizi decided to stimulate a certain area of the patient's brain through the electrodes, and to the surprise of both, the patient lost his ability to specifically recognize the doctor's face, although he still recognized everything else in the room. This research, and many other examples since then, has revealed that a specific cortical region, more exactly three chunks of the temporal lobe, is engaged in processing faces. Now, if a specific site exists in the brain for face recognition, it shouldn't be too difficult to imagine that there may also be a site reserved for the specific perception of God."

"All right, I guess I got the idea," said Agnes.

The next few days both had a lot of work to do and got too tired by the evening to talk about any serious matter. But the following weekend, they were ready to continue their discussions.

Agnes was again the first to raise questions. "I begin to understand how you conceptualize soul. How could your theory explain God's communication with us?"

"Material God is very highly organized information and omnipresent. Because He is the one who put together the information content of the speck, He also knows how to manipulate the super information complexes in our brains, which, when retrieved by us, provides specific spiritual knowledge and instructions. Some of us get more of these instructions, and some of us receive less. Most likely Jesus received a lot more spiritual knowledge and instructions from God than the rest of us."

Agnes agreed with this conclusion, but then she brought up a new concern regarding God's method of communication. "Another big problem for me is that I don't see how He can specifically direct His message to the right people."

"Well, this is somewhat easier for me to ponder, at least conceptually, based on what we know about directing specific components to specific sites in a cell," said Doug. "A cell contains tens of thousands of proteins that after their synthesis must go to specific sites like the mitochondrion, cell nucleus, cell membrane, and so on to perform specific functions. How do the proteins know where to go? It turns out that there are several ways for the proteins to figure that out. First, most proteins' ends are composed of a stretch of amino acids in specific order that encodes which cell organelle they should move to. But in addition to these so-called sorting signals, there are also signposts in the cell that help direct the proteins to the right places. These include messenger RNA molecules, each corresponding to specific proteins, located in the respective cell organelles. And then there are short RNA molecules that act like zip codes to direct the messenger RNA molecules to the right place in the cell. While there is a lot to be learned about these processes even at the cellular level, the idea is clear: both the proteins, which need to go to specific sites in the cell to perform specific functions, and the sites themselves contain specific signals that allow each of them to seek the other out and conform to it for the proper localization to occur.

"I think the situation when God is sending signals to us might be similar. His signal is essentially an information code, specifically arranged units of elementary information units, which we already talked about, and our brain has a reception site that can only accept such specific signals. Thus, the two signals must fit for the translation of God's message. Again, the idea is to have two strands. In this case, one is God's coded message and the other is the coded reception site in our brain. The two must fit. Otherwise, the message is not taken. You can't open a locked door with just any key. The key must precisely fit into the lock of that door. This principle is an essential one for the seamless operation of our bodily functions. For example, a hormone, like insulin, can't just interact with any structural elements of a cell to produce its effects. It fits only one or two cell membrane components, its own receptor in the cell membrane, and another one more specific for insulin-like growth factor, through which it produces all its metabolic and growth-regulatory effects. It is also possible that God simply incorporates His information code into the developing super information complex, which would serve for Him as the signpost. An important point is that God's message will become part of our memories and will be retrieved like any other memory.

"There is yet another way I can imagine our relationship to God. I can think of a huge circle that contains an almost infinite number of information units that represent God. Then there are billions of small circles with much smaller information content, representing all the lives that have ever existed, that for a finite time have or had some sort of physical appearance on earth but which otherwise come from God. After death, these small circles may become attached to the big one. Thus, we come from God and go back to God, meaning that even during our short existence as physical beings, God is with us. The huge difference between God and us is that while He commands an enormous, in fact close to infinite, amount of information, we can only access a very small part of it, even after we unite with Him in a way. However, it is exactly our inability to ever know everything that allows us to keep our own individuality in the form of individual super information complexes, even after our death."

The next evening, Agnes asked Doug to share his thoughts about Jesus.

"This is a very difficult question to deal with because it goes into the very heart of Christianity. Jesus certainly was a man of exceptional insights into all facets of life's meaning and our relationship to God. To remain consistent with my information theory, I believe that God bestowed on Him an unusually large super information complex with all knowledge He needed for His teachings."

"Doug, do you believe that Jesus was at least part God on earth and could perform miracles?" Agnes asked.

"Once I assume that after creating the universe God performed no miracles, it is hard for me to believe that He transformed Himself into a man and walked among us. On the other hand, I fully accept that Jesus and God were very close and that eventually Jesus merged into God. The question for me is, how could this have happened? My best answer is that since in my opinion everything in the universe, including us, is part of God while keeping individuality, naturally Jesus also was part God and part man. This is just another form of dualism. What distinguished Jesus from us, then, is that for some reason, He was chosen to deliver God's messages to mankind. But the act of God bestowing certain extra knowledge on Jesus doesn't necessarily made him God on earth, but after His death, He became united with God.

"In that context, the way I imagine Jesus's life is that to be able to fully carry out His task, he was born like anyone else. But He came into His physical existence with a far larger super information complex than common people did, meaning he had access to much more of God's infinite knowledge than any of us. This enabled Him to be very receptive to God's messages and instructions and to have the ability to adequately convey them to the people of His time. So, through His knowledge of God's nature, He was very different from the people around Him, and He was much closer to God than anybody else. But I agree with Paul's view of Jesus as he described in some of his letters he wrote to early Christians, namely, that despite being very close to God, during His existence on Earth, Jesus was no God. Only after His death was He reunited with God."

"Do you think that after His death Jesus merged into God fully, or He also retained His individuality, just like we would?"

"I still have to think about this a little longer. But at this moment, I'm leaning toward the idea that Jesus fully merged into God and became one with Him. Then if Jesus comes again, maybe

He will come as God Himself, with full glory and power. What do you think?"

"I think I'm still confused about the differences between spiritual and material Gods and how they relate to the Holy Spirit."

"Let us start with Jesus being born as a man but being very different from common people. God, despite all of us having His image in our brains, gave us free will to manage our lives including how to relate to Him. We can sin, or we can be good, all this being our decision. In contrast, I believe that God had full control of Jesus's brain and soul, thereby blurring the line between Jesus being only a man with messianic messages and being the de facto extension of God. I would stress the word *extension*, meaning that while He didn't possess all the power and knowledge of God, He was very close to Him. As far as we don't insist that Jesus performed miracles, I would readily accept such a quasi-godly nature of Jesus. In my mind He could indeed have been the son of God as presented in the Bible, until the moment He died. After His physical body died, His godly soul, not His body, left earth and fully reunited with God.

"As for the relationship between spiritual God, material God, and Jesus, perhaps the best, albeit still quite rudimentary, analogy is a computer. Let's say that all the information stored in the computer's memory represents the spiritual God. Using that information to perform specific tasks requires its selective organization by a task manager program. In that sense, the material God is the task manager of the general information represented by the spiritual God. Jesus, as a man on earth, had access to the computer through His very developed super information complex and could read parts of the program under specific filenames required by Him to act upon."

"I'm just thinking, could we call God's communication method the Holy Spirit?" Agnes asked.

"I can agree with you on that." With that, Doug wanted to finish the discussion about Jesus, but Agnes raised more questions while on the subject.

"Doug, if you want your theory to be congruent and believe that Jesus didn't perform miracles, then you can't assume that Jesus's body went up to heaven. Am I right?"

"Yes, I want to be congruent," Doug responded, "and therefore I repeat that in my view, Jesus didn't perform miracles, except that He was able to influence people's minds when He chose to. Since no miracles happened, it wasn't Jesus's physical body that went up to heaven."

"But it's so ingrained in Christian belief that after His death He was resurrected and after a while He went into heaven," argued Agnes. "No true Christian will accept what you're saying about this."

"Your argument may be correct, although you must see that I view the Bible mostly as a spiritual, not literal, text that can give us incredible insights into our relationship with God. I don't attempt to approach the Bible with a rational mind. Spiritually speaking, do I believe in Jesus's resurrection and Him leaving the earth? Yes, I do, but I believe that it was Jesus's soul, not His body, that was resurrected and went to haven. God's message then could evoke Jesus's image and His voice in the disciples' minds, who claimed to see and hear Him after His resurrection. The same thing could have happened when people also claimed to see Jesus going to heaven. If this is how it happened, then it is consistent with my theory that in fact no miracles were performed. Of course, if I want to follow my own logic, all these sightings would have required that Jesus's soul was always there, which the disciples saw as His physical body. Just like schizophrenia can cause someone to visualize imagined events and persons as real, somehow the disciples' souls also visualized Jesus's soul as if He would have had physical body. I'm not saying that the disciples were schizophrenic, but their brains performed similar tricks when they saw Jesus after His death. Again, if this is how the sightings occurred, then according to my theory, no miracles were involved."

"However, in the Gospel of Luke, it is written that after Jesus appeared to the disciples, He ate with them. Also, in the Gospel of John, Jesus let the doubting Thomas touch His wounds. Both acts served to demonstrate that He was flesh and bones. These do not seem to agree with your information theory at all," argued Agnes.

"Well, it's also written in the Gospel of John that He moved through the closed door when He appeared to the disciples. This would clearly indicate that He wasn't flesh and bone, unless, of course, a miracle happened, which I don't believe in. Let's face it, these Gospels contain contradictory statements. That's why I said that I view the Bible as a spiritual and not as a literal text. But if I must choose between these contradictory statements, I would choose the one that is consistent with my information theory. In other words, I can believe that Jesus, or rather His soul, could move through any physical barrier, but I doubt that He could perform bodily functions like eating. But speaking to each other is different. Jesus's ability to communicate with the disciples might have been possible as far as the super information complex representing Him could communicate to the similar complexes of the disciples without pronouncing the actual words."

"All this sounds reasonable within the boundaries of your theory," acknowledged Agnes grudgingly, still feeling a little down because the entanglement theory may not work out. "Although many in both the atheist and Christian camps would claim that your boundaries are rather wide," she added. "But going back to the subject of why miracles didn't occur, what would have prevented Jesus from performing miracles?"

"The physical laws of the universe, which God established. We discussed earlier that God wouldn't intervene in the lives of people in a way that would require changing the universal laws of physics."

"You mentioned several times the physical laws that God had established. What exactly are these laws?"

"You know, it's funny," began Doug, "that of all the books I read about this subject, most of them by theoretical physicists and astrophysicists, the best summary of these constants for people like me who aren't formally trained in physics and mathematics was given by the prominent biologist Francis Collins in his book *The Language of God*. So, to answer your question, I need to borrow from his book, although frankly I remember only some of his points with some accuracy.

"For me, the most fascinating fact is that one millisecond after the big bang—what I would refer to as the second big bang—for every billion or so pairs of quarks and antiquarks, the basic units of matter and antimatter on that time scale as you certainly know, one single extra quark survived. The others were annihilated into radiation. This very small portion of quarks made it possible to form the whole visible universe. In the absence of this infinitesimal asymmetry, the whole universe would have instantly and completely transformed into pure radiation with no possibility to form celestial objects and eventually us.

"The other important constants were the total mass and energy of the early universe and the strength of the gravitational constant. For example, if one second after the big bang the expansion had been smaller by even one part in 100 thousand million-million, the visible universe would have already recollapsed. In contrast, if the expansion rate had been greater by even one part in a million, no planets and stars could be formed. These facts are simply mind-boggling. They indicate a truly incredible level of fine-tuning.

"The last thing I remember from Collins's excellent summary is that if the strong nuclear force that holds protons and neutrons together in an atom's nucleus had even been slightly weaker, hydrogen would be the only element in the universe. In the opposite case, all hydrogen would have been converted into helium, and no heavier elements could be formed, for example carbon, which is essential for life. But while these constants and how the visible matter and universe evolved are known, in the books I read about the big bang and subsequent events, I couldn't find a good description of how and when dark matter and dark energy were formed except that everyone assumes that they were formed during the big bang. The few things we know about them is that dark matter played an important role in galaxy formation, and perhaps dark energy provided some vacuum force for expansion. But I have to admit that I have not read original papers on dark matter or dark energy in astrophysical journals."

"How is it that the universe contains only quarks and no antiquarks, although both started out in equal numbers? Where did this extra number of quarks come from that survived annihilation through interaction with antiquarks?"

"As I understand it, this isn't an easy question for even the best astrophysicists and theoretical physicists to answer," Doug responded, deep in thought. "So far, I have found the best explanation for this strange phenomenon in the book *A Brief History of Time* by Stephen Hawking. I'm not sure that I fully understand it, but the explanation Hawking provides is based on the so-called grand unifying theory, and it goes something like this: Immediately after the big bang, the particles had such high energies that quarks changed into antielectrons, while antiquarks changed into electrons. These changes can occur in the reverse as well. Now, as the thinking goes, as the universe expanded, the weak nuclear forces didn't obey a certain type of symmetry between particles and antiparticles, and the number of antielectrons that turned back into quarks exceeded the number of electrons that turned back into antiquarks. As the quarks and antiquarks dutifully annihilated each other, this process resulted in a small excess of quarks from which our entire visible universe has now built. I know that this sounds extremely complicated, and in fact not all scientists accept this mechanism, but so far this is the only viable hypothesis I'm aware of."

"On the surface this seems like an acceptable explanation, although I'm not sure that I really buy it," Agnes concluded. "But if Hawking says it, there must be some truth in it. I truly admire that man."

Once, after one of those long discussions, Agnes commented, "You know, Doug, it occurred to me that with your theories, you simultaneously contradict both scientists and Christian doctrines. Shouldn't you pick only one of them to fight with?"

"Listen, Agnes, I probably will never go public with this or any of my other theories, unless I perfect them to a very high level, because I'm aware that first I have to acquire necessary expertise in physics, quantum mechanics, and even mathematics to contend with the other theories in those fields. Worse, right now I don't even have the necessary confidence to believe in my own hypotheses, or perhaps I should say our hypotheses. At this point, all this is no more than brainstorming sessions with you and previously a few others, and I still haven't decided if I want to continue down this road or if I should just fall back behind the church without questioning its doctrines. Finally, even if my theory would be theoretically unassailable, I don't think I would be ready to receive the rancorous feedback that I would undoubtedly receive from both scientists and believers, as you just pointed out. When I'll become more confident, if ever, then first I'll talk to religious and atheist astrophysicists alike to ask if they see any point in continuing this effort, in particular pursuing the meaning of information, and if they could help me with their scientific tools."

One night, Agnes asked Doug about the multiverse, the existence of many, perhaps billions and billions of other universes. Earlier Doug had already discussed this issue with others without finding good answers but not liking the idea, but by now he was closer to rejecting the idea, although still leaving some loopholes. After a short silence, he began. "Proponents of the multiverse theory wish to avoid acknowledging the existence of God. They deny that our universe was created by Him in an extremely well-controlled process where all physical constants as we know them were fine-tuned to allow the formation of atoms, galaxies, and eventually life. From a philosophical standpoint, it is difficult for me to believe that the event of Creation that eventually led to life was a random event. I simply can't accept that all this could have been an accident. And if Creation was a planned event, it is difficult for me not to conclude that our universe is the only one.

"However, to be fair, and considering the dearth of testable hypotheses, let's also consider the multiverse version. If our universe had indeed been born by a random process, then physicists calculated that on a statistical basis there would have to be 10^{123} other universes for one like ours to be likely to form. This is because it would have taken that many universes to have one where the values of all physical constants were the same that were required for the formation of our universe. Do you have enough imagination to picture that vast number? It would look like this." Doug took a sheet of paper and began writing a 1 followed by a long stream of zeros. As he wrote, he continued. "Most proponents of the multiverse theory say that in this case, one doesn't have to look for a god who carefully designed the little speck that exploded in the big bang, leading to the formation and evolution of a universe in a highly ordered manner. Instead, by random events, after a sufficient number of trials, nature created our universe." He put down his pencil. The paper read, "1000000000 0000000000 0000000000 0000000000 0000000000 0000000000 0000000000 0000000000 0000000000 0000000000 0000000000 000000000000."

"In fact, much to the consternation of many scientists, some theorists go even further. They say that if the number of universes is even larger, say 10^{130}, then it's possible that in the cosmos, there is another universe that is identical to ours in every respect, including having exact duplicates of us."

"It is frightening to even consider that such travesty might be possible," Agnes said. "And that the other Doug would be in bed with the other Agnes, talking about us?" she asked.

"I think this is sheer nonsense, disgusting even. To think that right now replicates of us making love on a distant planet isn't at all an appealing scenario to me," replied Doug. "But you shouldn't be daunted by such silly theories. The physical constants, however fine-tuned, could never fully describe our beliefs, our minds and souls, or our actions. No combination of physical constants could have predicted or determined that the two of us would meet at a certain time and location and fall in love. The whole idea of the development of mankind is that we are unique and have only one copy of each of us. But I'm sure that scientists pursuing the idea of another universe with identical copies of us will not be deterred by my conclusion, particularly because they and us will never know about it."

"Just to confirm, your conclusion is that I shouldn't worry that in the afterlife I might end up with the other Doug?"

"No, I'm certain that it is us who will stay together forever, except that our souls, not our bodies, will meet, although I still have no idea how we'll find each other. Perhaps our souls will be entangled and find a way to seek each other out." That was not a fully reassuring outcome for Agnes, but she had no more questions.

The next evening, Doug returned to the multiverse theme. "You know, Agnes, I've been thinking more about the multiverse hypothesis, this time about the possibility of there being a god who created many universes, which I didn't tell you about last night. It poses many dilemmas for me. One of them is that if God had existed for eternity, why did He decide only 13.7 billion years ago on our time scale to create a universe? Why not earlier? What is so special about the 13.7 billion years? Or is it possible that our universe was preceded by many other universes that, after an expansion, underwent a big crunch from which again a new universe was born? Is it possible that such cycle, always creating and then annihilating the universe, has been going on forever? If so, we might meet an untold number of people in heaven from previous universes. This would make heaven an extremely fascinating place, a meeting point of many different civilizations."

"That could certainly be fascinating to a point," Agnes agreed, "but such a heaven would also be extremely crowded, making it hard to meet each other again even if we are entangled, don't you think so?"

"You have a great point," Doug acknowledged.

"Let's then analyze again the reality of the multiverse. Let's assume that God indeed created a multitude of universes and that such creation is still ongoing and will perhaps continue forever. Now, it is a fact that our universe sets a barrier to the movement of photons so that they couldn't reach another universe. This means that we might never be able to communicate with any of the other universes."

After a short silence, Agnes asked, "How about gravitation as the means of communication?"

"You are correct in bringing gravitation into the picture. Since gravitation is the only force that could transcend the distances between two neighboring universes, it could indeed serve as an inter-universe communicator. But I don't think that today anyone has seriously considered that gravitation could ever be harnessed to such an end. I suggest, let's just assume that gravitation is out of the picture. Now, if there is a material God who resides in our universe, He would also have difficulties communicating with another universe, because He must obey His own natural laws. On the other hand, it doesn't make any sense to have a single universe with a god while having 10^{123} other universes or more with no God at all."

"But what if all 10^{123} universes have their own gods? Or how about if there is only one God in the entire cosmos and He knows how to use gravitation or some other instant means to communicate with any of the universes?" asked Agnes.

"I think the idea of countless universes having countless gods is really going too far. But just staying with the multiverse hypothesis, I have another major objection to it. I can't imagine that having all these universes presumably randomly moving in all directions, our universe has never collided with any of them. One would expect that such a cosmic accident with another universe would have either seriously disrupted our universe or would have distorted it so much that it wouldn't have the observed regularity and homogeneity. With our sensitive equipment, even relatively small ripple effects that happened in the distant past would have already been detected. At the very least, the temperature at different locations in our universe would show some significant heterogeneity, but in contrast, it shows only very small differences, for which there are reasonable explanations."

"I read once that our universe might be the product of two colliding universes. I also read that the inflation period has never really stopped and keeps spawning new bubble universes. What do you think about these theories, Doug?" asked Agnes.

"These theories have no scientific basis." That was Doug's short answer.

"Should we then declare that the idea of a multiverse is dead for us?"

"On my part, definitely, but the existence of the parallel universe doesn't count in that regard," Doug declared. "This is what I call teamwork."

One night after one of these discussions, they suddenly realized they had only two weeks left before their wedding, and they should perhaps be focusing more on that event, including finalizing their honeymoon plans. Yet they felt it was necessary to go through once again what they had discussed before to make sure that they agreed on key points. Thinking similarly about God, the role of the church, and cosmology was important for both, not only because they both knew that it would add a lot to their marriage, but because it would also affect family decisions. Among other questions, they would have to decide if they would go to church and if they should provide religious education for their children.

After another long night, it turned out that they agreed on almost everything. Agnes liked Doug's idea that information had always existed and that there is a material side of God who created a parallel universe to guide the formation and development of our universe. They agreed that information was an independent substance-like entity with certain amount of energy, that information was the mother of all energy fields and matter particles, and that the totality of information equaled spiritual God while the extremely highly organized total information with unlimited knowledge and free act is the material God. She particularly liked the idea that the specks from which the two big bangs originated may have been composed of nothing other than enormous amount of information organized into coding units, which was the source for all forms of matter and energy created during and after the explosions. At this point they acknowledged that the idea of two big bangs will require a lot of theoretical work until it can be taken seriously. She applauded the idea that God has been using evolution as the main driving force for the development of the universe and life and that after constructing the speck or specks, or whatever the big bangs started from, the rest of the events did not require His further direct involvement.

There was however one thing God had been directly involved with, via purely physical means, but without performing miracles: His method for communicating with humankind. They agreed that if even a single cell could figure out how to solve the multiple levels of communication among its constituents, God could surely develop the means to communicate with humans. Since God is omnipresent and we are part of Him, this makes His communication via organized units of substance-like information feasible. In principle, they agreed that everyone is born with a soul or hologram, composed of coded units of information, which also serves as the basis for conscience and God's image. As we carry on with our lives, all experiences become stored in the super information complex in a retrievable manner. It is the soul, defined as super information complex storing all life experiences and memories, that survives us. They were less enthusiastic about the role of entanglement as the mechanism to move information from the brain to the soul, but they weren't ready to fully discard this idea either. However, entanglement may play a role in how we find our loved ones in the afterlife. Previously Doug had thought that Jesus had been "only" a prophet, although the one closest to God. After more discussions, both converted to the idea that it was the event of Jesus's death on the cross that transformed Him from God's chosen man to God, meaning that He became fully united with God at that moment. On her part, Agnes accepted the idea that during resurrection, Jesus left His body behind, and it was His soul that made contacts with the disciples and others during the following forty days. Also, it was Jesus's soul that went to heaven. Finally, they also developed a better understanding of the meaning and role of the Holy Spirit as the permanent link between God and humans.

They recognized that more unanswered questions remained than they could answer with their hypothesis, and of which, one stood out: What is information? Does it have any measurable, therefore physical, properties? Would humans ever be able to construct any instrument to study it, in the absence of any proof for its existence and any idea what could it be? Would the idea of information always remain in the realm of philosophy and represent a dead end for investigative sciences?

Doug was very surprised, and at the same time pleased, that this intellectual exercise transformed Agnes from a semiatheist to a believer, although not in the traditional sense, and that they were able to develop a common understanding. Agnes was happy too; this was the first time in her life she had met a man she really could identify with and whom she was ready to marry.

The brainstorming they went through leading to their common acceptance of a god was enough for them to agree to marry in a church. By birthright Agnes was a Roman Catholic, which greatly simplified the process of signing up for a kind of advisory class and the marriage ceremony. After all, said Agnes to Doug one evening, "Every churchgoer I know is different in their beliefs and level of acceptance of church doctrines, and our distinct views shouldn't pose a problem either. Besides, we owe an explanation only to our God, and as we have agreed, His and the church's standards to evaluate us might be quite different. This actually reminds me that we have never talked about how God will judge us before we enter His empire."

"Oh, not now," Doug begged her. "We'll have plenty of time to discuss that during our honeymoon. We cannot make love all the time. We need to set aside some quality time for these kinds of discussions as well. We just make a list of the topics that have yet to be covered."

"I'll start making the list right away, and I already have the first topic. Will God let you join me in heaven if you cheat on me?" She laughed at the thought of discussing this during their honeymoon.

Finally, it was time to deal with their honeymoon. Agnes proposed, and Doug agreed, that they spend their honeymoon on a three-week trip in Europe. Doug trusted Agnes to put the final itinerary together, and she did, right away. The next day she went to the travel office to begin the reservation process for airplane tickets, car rentals, and hotel rooms. They would start the trip soon after Labor Day, when fewer tourists visit Europe, and as an added benefit, all prices would be lower. According to the plan, they would fly to Berlin and spend four days there. Agnes heard that after the reunification, the city had been going through an architectural renewal, which she wanted to experience. Then she added Munich, Vienna, Venice, Rome, and Paris to the program as well.

Both Agnes and Doug were fully prepared to have children; they talked about parenting issues a lot. It was another pleasant surprise to Doug that Agnes wanted to fully incorporate his sons into the new family without hesitation. He remembered the struggle Peter had had with Helga, and he was a little afraid of going through a similar experience. But from the outset, Agnes was so friendly with the boys that occasionally he came close to crying in happiness. On some of the weekends, when the boys came over and Doug had something to do in his Mayo office, she would play for hours with them. Once she brought them junior tennis rackets and took them to the public tennis courts. They liked the game so much that they asked Agnes to take them there again and again. The two boys would occupy one half of the court on one side, and Agnes would deliver the balls while explaining the concepts of how to hold their right hands for forehand and backhand returns and how they should prepare their feet and body motions respectively. On the fifth occasion, Doug joined them, and it was even more fun when they attempted to play doubles. Tennis was something that Sarah would have never done with the boys; she had a different personality and no interest in sports. In contrast, to the boys Agnes was a sport's authority; tennis broke the remaining barriers between them and her.

There were no topics Agnes and Doug wouldn't discuss with each other, except for two: Sarah and Peter. Doug saw no point in potentially injecting any poison into his beautiful relationship with Agnes by talking about Sarah, who did them a favor by never coming to the house. Similarly, Agnes was hesitant to bring Peter into any discussion. Since she had broken up with him before his death in any case, she didn't see a reason to stir up emotions before her and Doug's major commitment. And she was particularly careful not to mention what she knew about Sarah; this would have required too many explanations. "Maybe I'll tell him after our wedding," she concluded whenever she was close to telling the story to Doug. "Or maybe I'll never tell him. After all, what would either of us gain if he knew about Sarah's visits to Peter? We would both lose."

Sarah's behavior, after her fiasco with Roy and other troubles that came in cascading torrents, had changed from bad to worse. Anxiety about her future personal life infused her very being. She was angry at herself because she realized that what she had done to Doug was, for the most part, uncalled for, and she had hurt herself in the process. One side of her wished to go back to Doug, but her other side realized that it was far too late. Her sons' accounts made it clear that Agnes and Doug were in a very serious relationship.

Sarah knew about Agnes from Peter, and she didn't forget that Agnes had appeared to watch them from the window that night. "She had interfered in my life the second time"; that's how she thought of Agnes. She was also furious to see that after leaving Doug, he had risen from the proverbial ashes and eventually got the better end of the deal. But she also was afraid of Agnes because she didn't know how much Agnes had seen and how much she knew about her visits to Peter's house. Had Agnes somehow identified her? Sarah became obsessed with Agnes, an interesting turnaround; not too long ago, Agnes had been similarly obsessed with her. In her desperation and anger, Sarah stooped to a level she would have never believed she would; she started spewing vitriol about Agnes to their sons, telling them all sorts of awful things about her character. Her most frequent accusation was that Agnes liked to ensnare men, let them fall in love with her, and then abandon them. Then she usually further elaborated, saying that this would be the fate of their father as well. This sounded scary enough to her sons, particularly to Brian, who was only six years old. Besides, their mom's warning was contrary to their experience, which put them in a difficult situation. How could they explain to their mother that her statements belied the facts they experienced?

One weekend, about one week before the wedding, the eldest boy, Roy, found an occasion when he was alone with his father, and he confessed in tears what his mother had kept telling him and Brian about Agnes. Despite his young age, he had the ability to express himself almost like an adult, and he was able to accurately relay his mother's "worries" and how she excoriated Agnes. Doug was shocked; he hadn't been aware of how Sarah had felt about Agnes. "Is this how you know Agnes?" he asked Roy. "Please tell me how you and Brian feel about her. All I can see is that Agnes is making her best effort to fully include both of you in our lives. And I know for sure that she is not doing this for show. I know it is coming from her heart."

"That's the way I see it too," said Roy with a deep sigh. And with that, the matter was over for Roy; he went back to the room where Agnes and Brian were playing chess. But this discussion left a deep scar in Doug's heart. "How can the woman I loved so much just a year ago be capable of using our children to poison my relationship with Agnes?" he thought in shock. She must be anguished. For hours he sat there going over what Roy just said, although he should have been focused on the wedding and packing for the trip. But when he took his sons back home, he reassured them that Agnes was a much better person than their mother had depicted her, and they remained vitally important members of a loving family. This was the most important thing for Roy and Brian that their father could say; they would always remain a family, no matter how their father's marriage turned out.

Two days before the wedding, Doug and Agnes didn't go to work; they made a last-ditch effort to make everything right for the wedding. Agnes's parents were to arrive soon, and she was about to go to the airport to pick them up when the phone rang. She picked it up and almost fainted when the person simply introduced himself as Carl; he wanted to know if Doug was available. Agnes didn't introduce herself, and with her hands shaking, she gave the phone to Doug while telling him in a low voice that the caller was Carl. "Not again… not now…," Doug whispered, but he took the call nevertheless.

"Hi, Carl, I haven't heard from you for a long time. I thought that you were enjoying your retirement."

"Well, maybe next year," answered Carl. "I hope I'm not disturbing you."

"Except that I am getting married two days from now, you aren't disturbing me, but please keep it short."

"Congratulations, and I'll try to keep this as short as I can. Have you ever heard the name James Fletcher?"

"If we are talking about the same person, then he is Sarah's brother who lives in Little Falls."

"Yes, that's the one. He is an artist as you probably know very well, but for some colliding events, the last few years weren't too good to him. For one, in a recession, paintings are the last things that people buy. To cut the story short as I promised, he couldn't pay the mortgage for his house for several months, so the bank decided to start the foreclosure procedure. As they listed his valuable items, a laptop and a cell phone turned up in a paper bag in one corner of his studio, behind a large cabinet filled with papers, brushes, and all kinds of paints. Jim had knowledge of neither the laptop nor the cell phone. He claimed that almost every weekend he used to have visitors, and he really had no control over who was doing what. That was suspicious to the people who did the listings. After a period of indecision, somebody at the bank decided to hand over both items to the Minneapolis Police Department. They looked at them and determined that they belonged to Peter Cartwright in Rochester, Minnesota. Yesterday they transferred both items to me, and now I have to decide what to do with them and how to deal with Sarah's potential involvement."

A long silence fell upon them; Doug had no idea what to say. "Are you still there?" asked Carl, not knowing what to make of Doug's silence.

"Yes, I'm here, and I have a suggestion. I don't know what you really expect from me, but if there is anything I can help you with, could we do it after returning from our honeymoon in Europe?"

"Yes, of course, and I am truly sorry that I had to call you at such an inconvenient time. I had no intention to roil your preparations for the marriage. But you see, since the Minneapolis police are also involved, I may have no other choice than to reopen the case and follow this line. I can prolong the preparations, but after your return, I would like to have a talk with you. You see, last time we talked, I thought that closing this case would be as easy as flipping a switch. Unfortunately, things didn't turn out that way."

"I understand," said Doug, still wondering if he was daydreaming due to his fatigue.

Reopen the case? This part of Carl's sentence kept reverberating in his brain. Agnes looked at him with concern. "What that call was all about?"

Doug had his first very real test of honesty with Agnes. He was an honest man to start with, and he particularly wanted to be honest with Agnes. But on the other hand, this was suddenly such an awful situation. Yet he decided not to parry Agnes's question. "Carl wants to reopen the case concerning Peter's death because he got some new evidence that he must follow upon."

"What new evidence?"

"Well, as far as Carl could tell me, they found Peter's missing laptop and one of his cell phones in the house of Sarah's brother in Little Falls."

"That sounds unbelievable," said Agnes, but then she hesitantly continued. "Doug, there's something about Sarah that I decided to never talk to you about, but under the circumstances, perhaps I should."

"Tell me later, my dear. I still have to do something in the lab, and you'll have to go to the airport soon anyway. Where will your parents stay?"

"I booked a hotel room for them just two blocks away from us. All right then, see you soon."

Agnes and Doug both thought the other looked fabulous at the wedding ceremony, and everything went exactly as they planned. It was a very moving experience for their parents and the 150 other guests gathered in the Catholic church to listen to their vows. Almost all their close relatives and friends from both sides were there; it was quite a crowd. The dinner afterward was in the Olive Garden in a large special room, and most guests stayed until well after midnight. Both Doug and particularly Agnes were good dancers, and the guests left little time for them to rest. During one of the breaks, Mike and Steven sat down with Doug and Agnes, and somehow the conversation shifted to poker. The friends all missed it; the last time they had played together was that fateful night when Peter didn't show up.

"We should start playing again," said Mike hopefully. "I'm sure we can find a fourth person if we try."

"Maybe I could be your fourth," Agnes suggested. They were all taken aback.

"You never told me you played poker," Doug remarked in surprise. "Do you have any other skills I'm not aware of, perhaps like playing for the NFL?"

Agnes laughed. "Working as an accountant is almost the same thing as playing poker. But I used to play with my friends too."

Without much additional discussion they agreed that after the new couple's return from their honeymoon they would once again play every Monday. "If for nothing else, it was worth coming here tonight," teased Mike. Doug grinned and slapped him on his back.

Then George Carrick came up to them jokingly, asking if Doug needed any more PDC. "No, I don't need more, but you be careful because overeating tonight will swamp the benefit of your exercise tomorrow," answered Doug teasingly while looking at George's outsized stomach.

That evening Doug was in such a good mood he could take any joke and then respond in the kind. When everybody finally left, except their parents, Agnes nestled close to Doug and said, "I'm finally home. There must be a god who arranged that I came back to Rochester, and I promise you again that I won't disappoint either you or Him. We have somebody looking after us. It just can't be otherwise."

"This was the nicest thing you have ever told me," Doug told her, embracing and kissing her with all the love he had. In the background their parents could hardly disguise their weeping as they watched their adult children embarking on the greatest experiment on earth: starting a new family with God's name on their lips.

And finally, their honeymoon trip arrived, which was what they had really been looking forward to during the weeks leading up to the wedding. Their first destination was Berlin, where they stayed for three days. Their hotel was in downtown, within walking distance from all the major tourist attractions. The first agenda on their list was visiting the Pergamon museum, which didn't disappoint them and to which they devoted almost a full day. They had dinner in a restaurant atop a revolving tower in the eastern part of the city that was known before 1990 as East Berlin. The tower was built during the era of communist East Germany, and the guests had the pleasure to view Berlin, including parts of West Berlin, always from a slightly different angle as the tower slowly turned. The second and third days they sped up their program and were able to get acquainted with most of Berlin's cultural, architectural, and political inheritance. They saw remnants of the Berlin Wall that, during the Cold War era, had separated the west and east sides of Berlin for decades. They also visited Charlie's Point where once Soviet and American soldiers looked each other steadily in the eyes, never knowing what the next moment might bring for themselves and the world.

Early in the morning of the fourth day they flew to Munich, where they spent two full days. Hand in hand, they listened to the music of the great clock on the top of the city hall and then spent some time in a big beer garden where beer was served in one-liter mugs. Munich is a very lively and romantic city; they both wished they had more time for it, but they agreed that they would come back someday. Early in the morning of the seventh day, they rented a car and first visited Ludwig's unfinished castle at Chiemsee, a beautiful lake between Munich and Salzburg. Legend says that Ludwig II was a crazy German grand duke suffering from delusions of grandeur who had no concept how much building a castle would cost and run out of money well before it could be completed. But at least the few rooms that were finished and furnished looked beautiful. Agnes and Doug counted around twenty-five of them, although they couldn't agree on the exact number. Then for the sake of curiosity, they drove to Hitler's bunker in the Bayer Alps, and by late afternoon they had gotten to their hotel in Salzburg, where they still had the chance to enjoy the evening with people strolling on the beautiful streets.

Their goal was then to relax for two days in Bad Aussee, a small town not too far from Salzburg and Hallstatt. Both Bad Aussee and Hallstatt are jewels of the Austrian Alps, and they spent the best days of their trip there. Their next destination was Vienna, where on the tenth day they arrived just before dinner. Fortunately, they could park the car in the hotel, although they used it only once to visit Schönbrunn where the emperor of the Austrian-Hungarian Empire had lived. They were awed at everything they saw in Vienna; the Hofburg, the St. Stephan church and the vibrant square around it, the opera, the streets, the cafés. After they left Vienna, they knew that this magnificent city would remain in their memories forever. Their next destination was Rome. They spent three days there and then another three days in Paris, and finally, they couldn't believe how quickly their three-week honeymoon had gone. On the plane back, Agnes made Doug promise her that they would come back to Europe. "I still want to go to Venice, Florence, Greece, London, Barcelona, and many other places if you would accompany me."

"Well, if we'll do it in moderation, nothing would make me happier," answered Doug with a big smile.

The airplane hadn't yet touched the tarmac, but Doug's mind had already focused on the next trip. For some people, traveling is like a drug; once hooked, it is difficult to control the desire for it. But as soon as they cleared customs, they were back to reality. Doug was anxious to go to his lab the next morning. His coworkers had come back from vacation a week ago, and by now they all had started new experiments, so there were a lot to catch up with. And soon he would have to meet with Carl and talk about what was found at the house of Sarah's brother. Although before the trip Agnes had hinted that she had to tell Doug something about Sarah, they had never gotten a chance to discuss whatever it was. But one night in Munich, a memory that he had never thought was important surged forward in his brain. Before Peter's funeral, Sarah had unexpectedly gone to visit her brother. Strangely, she took a big computer bag and another smaller bag with her that from the distance he identified as something destined for delivery to Goodwill. What had really been in the computer bag and the other bag? Could Sarah somehow be behind Peter's computer and cell phone turning up in her brother's house? Who else could that person be, considering she both knew Peter and had access to that house? The whole situation seemed rather suspicious, but he put aside his anxiety on the trip. Now that he was back, he had to deal with it. Should or should he not tell Carl about his observation?

The first night after their arrival, they were too tired to do anything other than curl up together and sleep. But on the second evening, when they were already in bed, Doug reminded Agnes that she still owed him the story about Sarah.

"Oh yes, I promised you, didn't I? Frankly, I was afraid of this moment and the thought I would have to reveal something that might hurt you. But since you asked, I'll tell you everything. Just please promise me not to fall into a fib."

"I promise I won't," said Doug with a deep sigh. He already did not like what he was going to hear.

"From my kitchen window," Agnes continued her story, "I had a direct view of Peter's living room, which had large windows with the curtains never drawn together. You know that I love to watch birds. Seeing those beautiful alpine birds in the forest between Bad Aussee and Hallstatt added so much joy to the trip, you wouldn't believe it. But I love all animals, and in that area between my house and Peter's, there were always many birds and rabbits, squirrels, and sometimes even deer. Whenever I had a little time, I would stand at my window to watch the birds and whatever else was moving. The birds and squirrels seemed to like the trees that were around the two houses."

Doug began to circle his eyes. "My dear, how does all this relate to Sarah?"

"I am getting there, but it isn't easy. A few days before Peter's death, just around dinnertime, I saw a white car parked in his driveway and your wife sitting in his living room, perhaps for about an hour or so. I didn't keep track of the exact time. Then, the day before his death—as I recall it was a Monday—I again saw Sarah in his house. It was late afternoon, and I don't know if I should be telling you this or not, but… they were clearly kissing. I swear I wasn't spying. I was only watching the birds, but I couldn't possibly avoid seeing what was happening in that house."

"How did you know that it was Sarah?"

"At that time, I didn't know that it was her. She was too far to recognize. But with my accountant's brain, I remembered her car's license number. More exactly, I took a picture of a blue jay bird, and the picture also included the white car and its license number. Later I looked at the picture, and this is how my brain imprinted the license number. Then after my return to Rochester, one day I saw her in the same white car that was once parked in front of Peter's house. That time she was close enough, and I recognized her."

"But how did you know her? Have you met her before?"

"No, I have never met her personally. But in his office, Peter kept relatively new pictures of his friends and their wives, and once when I asked him who they were, he told me about all of them, including you. I specially remembered her face because Peter mentioned to me that she had been his first girlfriend. And one more thing, as you know, that Monday Peter didn't go to your poker meeting. That evening, Sarah came back and parked in front of my house. But after some fifteen minutes or so, she left. I didn't see her, so I don't know if she had ever gone into Peter's house or not."

Doug was shocked to hear Agnes's account of Sarah's visits to Peter's house; he couldn't know that certain small details in her story were either not accurate or missing, but he had gotten the big picture, which was accurate. Without his knowledge, Sarah had visited Peter at least three times or perhaps many more times? After all, Agnes wasn't supposed to be aware of all of Sarah's visits.

"Thank you, my dear," he finally said. "You have helped me a lot to develop a strategy for what to say to Carl."

"Well, actually, about Carl. You should also know that after my return to Boston, I felt obliged to report to Carl that a woman visited Peter, and I gave him the license number of her car. However, I obviously couldn't guess at that time that it had been Sarah's car. I thought that Carl would find out the identity of that woman if he wanted to follow this line."

"What made you tell all this to Carl?" asked Doug.

"That was a preemptive action on my part. I knew that because I was one of the neighbors, sooner or later Carl would come to me asking whether I saw anything suspicious. And he might have had other, more personal questions, which I wanted to avoid. I thought that it would be better if I told him that much up front without him asking me first. It wouldn't have looked good if I had kept it to myself because it might have given the wrong idea to Carl that I had something to hide, which I obviously didn't. You see, I simply wanted to spare myself from extensive interrogations and Carl digging into my personal life, and that's really all."

"It's all right, I believe you did the right thing. I love you so much. And, indeed, thank you for wielding the scalpel so delicately."

Doug didn't have to wait for too long for Carl's call. It came the very next morning. To Doug's chagrin, Carl wanted to see him at the local police headquarters. Doug grudgingly agreed, adding that on that day the lunch hours would work best. Carl was almost as nervous as Doug during the first minutes of their meeting. But he slowly unveiled to Doug that from Peter's secretary, Agnes Lambert, he had received the information about Sarah's visits to Peter's house just before the day of his death. He then added that the Minneapolis police figured out that somebody had hid Peter's computer and one of his cell phones in the studio of Sarah's brother. "This is just too much of a coincidence, and I unfortunately have no other choice than to investigate whether Sarah had anything to do with Peter's death. I understand that you and Sarah divorced and that you remarried. Your wife is in fact Ms. Agnes Lambert, correct?"

"Your information is correct, except that Agnes Lambert is now Mrs. Agnes Lowry."

"Thank you for the correction. I sort of assumed that. Now, what do you think about this new situation?"

"First, I have something else I want to clarify. Agnes told you the car's license number without knowing the owner, is that correct?"

"Yes, that's correct too. I figured out the name of the car's owner later," acknowledged Carl.

"Well, then I really don't know anything you already don't," said Doug. "Agnes told me that she saw Sarah in Peter's house twice. But whether she was the last person to see Peter and whether she was the one who took the laptop and cell phone to her brother's house, I really have no idea. What I do know is that her brother's house was a kind of rendezvous point, and many artists and friends visited him there even from Rochester and elsewhere. He started his artistic career in Rochester and moved to a sort of farm near Little Falls about ten years ago. So he had many friends in Rochester who visited him frequently."

"Can you recall when the last time Sarah visited her brother was?" asked Carl.

Even though Doug was still angry at Sarah, he couldn't bring himself to reveal what he had seen on that day when Sarah unexpectedly left with the two bags, so he decided to fib and said, "I can't remember."

"Could she have visited her brother after Peter's death?" pressed Carl.

After a little hesitation, Doug answered, "She could have. She apparently did quite a few things I wasn't aware of." He decided that this was all he had to say and remained reticent for the rest of the interview; nobody could force him to say or remember more. Now it was up to Sarah to decide how much she would tell Carl.

"All right, Doug, thank you for coming, and I hope I won't have to bother you again with this matter."

After leaving Carl's office, Doug thought of his sons; if Sarah got in any real trouble, then he and Agnes would have to take care of Roy and Brian. They would obviously be happy to do it, but this would mean a huge change in his sons' lives. The present arrangement seemed to work best for everybody involved.

When Doug arrived home that evening, he found Agnes and the two boys talking about something with great excitement. "What happened?" he asked. "Are you done with your homework?"

"We are," Roy answered his father with the hesitation of a typical boy who isn't so sure about the accuracy of his answer.

"I sense some excitement here. What is it all about, and how much will it cost me?" Doug asked lightheartedly.

Roy didn't seem to know quite how to answer his dad, but Agnes came to his rescue. "They want to get a dog, but their mother won't allow one in her home."

"Ooh, that is a serious matter. Well, let say we have a dog. My question is then this: Who will take care of it? Do you know that a dog needs to be regularly walked and cleaned? And that's not to mention the medical care like vaccinations. Where would we keep a dog? When I was a young boy like you, my family had one. Whenever anything dared to move outside during the night, he would bolt over to the window and start barking like crazy. I loved that dog during the day, but in the nighttime, I hated him because he regularly woke me up and wasn't docile at all."

But the kids weren't deterred, and Agnes said that she would volunteer to take care of the dog when the boys couldn't. "Listen, Doug, have you ever considered that walking a dog before going to bed would be good exercise for us too? And we can take the dog to a trainer to make it docile."

Seeing that Agnes was on the boys' side, he gave up his resistance. "Well, all right, if that's what you all want, let's get a dog. I know that the Human Society has many rescued dogs. Perhaps Agnes can take you two there and you can pick one out. But try to choose a very young one that we can still train well. And don't get a Labrador.

They are very territorial and can be dangerous to people passing by on the street, although I guess we could get an invisible fence." The two boys were ecstatic, full of zest for life, and Agnes was very pleased too.

It took them less than a week to find a puppy, and from that day on, Roy and Brian came to the house almost every day. Doug and Agnes also changed their daily routines and went on a long walk with the puppy every day after dinner. This wasn't only healthy but also came with the added benefit of once again being able to engage in prolonged discussions about their future and even their cosmological theories. Astronomers now discovered new planets almost daily, some of them having characteristics that might be suitable for life. The list of habitable planets even included our closest companions. Mars was suspected to carry large water reservoirs under its surface, and on one of Jupiter's moons, Europa, water had been detected. "If habitable planets can be found all over the universe, then perhaps we on Earth are not the only children of God," said Agnes on one of those evenings.

"Yes, we might not be, although to tell you the truth, I still doubt it," commented Doug. "Sure, a lot of queerness are going on in the universe, but I believe that all other planets are too inchoate to host life."

The possible existence of alien worlds led to some conflict of opinion between them. Agnes most recently had come around to embrace the thought of the existence of many different intelligent life forms in different galaxies, while Doug still stuck to his view that Earth was unique and the only possible home to life. He cited some statistical calculations, showing how many different factors and conditions must have been met for the first cell to develop, and he concluded that it was already almost a miracle that it had happened at all even once in the universe.

"Maybe on that occasion God provided a helping hand here on earth, and elsewhere too," suggested Agnes.

"Earlier I too considered this possibility, but I ultimately rejected it because I wanted to remain consistent with my idea that immediately after the carefully designed events of the big bang, evolution took over the development of the universe and life without further direct help from Him.

"To better clarify what I think about God's hand, let's go back to the issue of the parallel universe. Imagine someone sprinkling a jar of iron filings onto a sheet of paper. From your vantage point above the sheet, the filings should end up in a random arrangement on the paper. But, unseen to you, there are magnets arranged in a pattern under the paper. So the filings will arrange themselves, seemingly by magic, in the same pattern. Ultimately, whoever set up the magnets also decided the arrangement of the filings. In the same way, I believe that God arranged the outcome of the second universe by controlling the first one. Then, all that He maintained is some form of material-based communication method with us. Future science may belie every facet of my theory, but it never will be able to rule out the role of God in Creation and His communication with humans."

Every evening during walking the dog, they had a conversation like that, always finding a new angle with which to view God, His methods of communication, the universe, evolution, and so on. Doug admitted to Agnes that getting the dog had been a great idea, a kind of double blessing, because besides exercising by walking the dog, it also brought her and his sons even closer to him.

It was Thanksgiving evening and time to celebrate everything that had happened in 2013. Doug and Agnes decided to stay at home and instead invited their parents, along with Roy and Brian, to come over for dinner. They were all busy, the women working in the kitchen and the men laying the table. Suddenly, as Doug reached to place a serving dish in the middle of the table, he felt a kind of dull pain in his chest. For days he had been coughing and had felt a little weak, but this was the first time he felt actual pain. Since the pain was more on his right side, he was reasonably sure that it couldn't have anything to do with his heart. Perhaps it had been there before, but he hadn't noticed it. He couldn't be sure. It is hard to distinguish between chest pain caused by lung disease or inflamed muscle. Whatever it was, it was slightly beyond the level of being annoying.

During the whole dinner, and in fact from that point forward, the pain never went away. Agnes soon noticed that Doug wasn't in his usual cheerful mood. After dinner, when for a moment they were alone in the kitchen, she asked him if anything was wrong.

"No, no, I'm fine," he told her. "I think I just strained a muscle, and it doesn't seem to want to go away."

"Maybe you have the flu. I noticed that you keep coughing too."

"No, it's not the flu," Doug insisted. "Maybe I just lifted something heavy, or last time at tennis I served the ball too hard. In fact, my shoulder hurts a little bit too." For the reminder of the evening and the next few days, he tried not to think about the chest pain and the coughing. But about a week later, he suddenly noticed he was starting to have trouble breathing too. Doug was worried that the symptoms were connected to and all originated in his lungs. That was the tipping point when he decided that he needed to see a doctor. Christmas season wasn't long away, and months ago he had made reservations for a hotel room in Little Marais, a ski resort in northern Minnesota close to Lake Superior. They were spoiling for some good skiing and at the same time celebrate the New Year's Eve there. But Doug was starting to realize that skiing would be next to impossible with the chest pain and shortness of breath.

Being a Mayo doctor, he could get an appointment with a specialist in two days, who was by chance a good friend of Doug's. The doctor's hunch was, in agreement with Doug's suspicion, that the discomfort was originating from his lungs. After a series of examinations, the doctor decided that a PET scan was needed to find out more. Doug's worst fears were realized; the scan revealed densities in several regions of the right lung that could indicate tumors. An accurate diagnosis still required a biopsy, taken from his lung tissue by fine-needle aspiration assisted with computed tomography imaging guidance, for histology. The procedure was performed three days later. Afterward, as Doug sat in his doctor's office waiting to hear the results, the man wasn't sure how to give him the news. Although he had been in the exact same situation countless times before, he still was averse to communicate the results. The histology results confirmed that Doug's right lung contained a non-small cell carcinoma already in stage IIIB, which meant that it was advanced and surgical removal was no longer an option. Doug knew that this type of lung cancer is most often drug resistant and comes in several forms, but before they concluded anything, they had to wait for the opinion of the pathologist, who the next day informed them that Doug's tumor was the adenocarcinoma subtype.

About 80–90 percent of lung cancer cases are due to smoking, but Doug had never smoked in his life. However, the adenocarcinoma subtype isn't as strongly associated with smoking as the other subtypes of lung cancers, so in Doug's case, some other environmental exposure or a spontaneous genetic mutation must have been responsible. Asbestos, often found in old houses, could be the culprit. But finding causation was the last thing in Doug's mind. Now all the focus, his and his doctor's, was on the treatment. Unfortunately, the PET scan also indicated metastasis to the liver, making the development of an effective treatment regimen very difficult, in fact nearly impossible. In these cases, a combination of radio and chemotherapy is used, and Doug was perfectly aware that his life expectancy was anywhere from several months to perhaps one year. It was a distorted grimace of life that one of his jobs at the Mayo had been to give advice about various treatment modalities for cancer patients, and now he had to be involved in devising a treatment schedule for himself.

After an initial period of total despair and incense, he eventually calmed down and accepted his fate. He always looked at cancer as an enemy that needs to be defeated on the battlefield, and now he felt as being a vanquished opponent giving up fight. During his initial treatments, he was confined to his hospital bed. Now his focus was on how to direct his coworkers to finish as much work in the lab as possible and how to make sure that Agnes and his sons would be taken care of. But as he approached the barrier between life on earth and whatever came after, perhaps continued existence in heaven, he also decided to attempt to put down his philosophical thoughts in a more organized manner and in a somewhat finalized form. Perhaps, he hoped, Agnes would be able to do something with it.

Agnes visited Doug whenever she could and encouraged him to write. Converted from an agnostic to a firm believer in God's existence, she promised Doug that she would ask as many people as she could to pray for him. She had found studies suggesting that seriously ill people have greater chance for healing if people pray for them. Doug was aware of these studies, but he had also read other studies that seemed to completely debunk the idea. And he couldn't find any logical reason to explain why God would be persuaded by prayers to intervene on his behalf. Can God interact with our minds? Doug strongly believed that He certainly had the ability to. But it would be against his core belief—God isn't in the business of performing miracles—to expect Him to perform a miracle and cure him. And why would He specifically cure him and not the millions of other people worldwide who die of cancer every year? He didn't mention to Agnes that in his mind the theory of the seriously ill helped by praying to recover from illness is all cart and no horse.

The treatments soon started and took away his strength, along with his will to perform any serious mental work. Within weeks he was weakened to the point that all he wanted to do was stay in bed. He lost several pounds every week, and he knew that the doctors' hesitation to maintain the same level of the treatment grew every day. This fact of life for many cancer patients is exactly what he used to battle when he would give his recommendations for the right dosage for cancer patients. How would he have to take the patients physical characteristics, such as their serious weight loss, into account and still able to supply the dose that afforded the maximum impact but fell short of killing them? Now he was the patient, and if the clinicians would have asked him about his opinion, he would have told them that any treatment at any dose was too much to bear.

However, Agnes tried to convince him that giving up and staying in bed all the time wasn't the answer. The difficult part was to persuade Doug to move without antagonizing him. Occasionally she, Brian, and Roy could still force him to get up and walk with them, and once he crossed that physical and mental barrier, he was happy to do so. But to get out of bed and walk proved to be more and more taxing for him with every passing day. After several weeks of treatment, his brain still worked but couldn't focus on the outside world for more than a few minutes; after that he was once again embroiled in his own thoughts. His cancer lacerated his will to do anything constructive; he always needed a push. Agnes realized this; she regularly brought a notebook with her and decided to register everything that she thought might be important for a book.

Thanks to Agnes's ability to elicit some excitement from Doug, even if only for short time periods, at various times they still had short conversations about God, the big bang, evolution, the Holy Trinity, Jesus, and why God doesn't unify the existing religions. This last question that Doug raised once was particularly interesting to them: "Why does God not communicate to us which religion He prefers so people could do away with the others?"

"Possibly, God wants to hear all the competing ideas about Him," suggested Agnes. "After all, all the monotheistic religions worship only Him, even though most people don't recognize that. And I think that God wants to see competition, which is part of the evolutionary process, everywhere. Competition drives progress, and that is what God wants to see, healthy competition—that is, when the two sides aren't killing each other for their religious views. If athletes would not compete from time to time, they would make much less progress in their quest to achieve newer and newer records." Agnes had been so engrossed in her own thoughts that she didn't notice Doug falling asleep in the meantime. Agnes kissed him and left with the notebook; she realized that Doug was too weak to hold himself up and write into it.

Even though Doug's condition deteriorated rapidly as more weeks passed, in fact more rapidly than even the doctors had anticipated, in his clear moments Agnes still tried to share with him recent developments in cosmology. She explained to him a recent finding that black holes were already present at very early periods after the big bang. These are called primordial black holes, formed by the extreme density of matter present during the universe's early expansion. Normally, black holes are formed by the gravitational collapse of a large star. According to the hypothesis of the scientists who made the discovery, such primordial black holes would be stable and could still exist to this day. Some astronomers went as far as to suggest that the mass of these black holes accounted for dark matter and played important roles in the formation of galaxies. Doug seemingly absorbed most of what Agnes said but couldn't add much. But some other times, when the treatments temporarily helped him regain a little strength, he would quickly seize on a good opportunity to twist the new information a little bit here and there and melt it into his own theory. But these were rare moments; most of the time he didn't seem to possess the mental power that he had drawn so much admiration for, and that was by far the hardest thing for Agnes to witness.

CHAPTER 22

ULRICH TAKES DOUG'S JOB

Ulrich's initial success with his company ended precipitously. Longer-term toxicology studies with his drug revealed that in rat's hearts it induced certain inflammatory events resulting in detectable structural alterations. In humans, such a process could eventually lead to heart failure, a risk that's not acceptable. This finding also effectively and irrevocably killed Ulrich's idea to make his compound an antidiabetic drug. In fact, he also stopped taking the drug to treat his own diabetes, hoping that it wasn't too late. Instead, he started taking metformin, a prescription drug that prevents the liver from making too much glucose. He wasn't particularly enthusiastic about taking a drug that someone else had discovered, but he grudgingly accepted this turn of events. Then, right after he had started taking it, he came across an article claiming that metformin also reduced the risk of heart disease and certain cancers such as prostate cancer, which he was afraid of above all else. This news somewhat reduced his anxiety, and he continued to take the medication twice a day, even though it caused him occasional light bloating and diarrhea when he forgot to take it with meal.

Slightly compensating Ulrich for all his misfortunes, some of his more recent experiments revealed that when his drug was applied directly to the skin after initial treatments for basal carcinoma, a form of skin cancer, it had preventive effects against its recurrence. Such direct application seemed to carry no risk of heart diseases, and he was able to sell his company for a modest sum, which would provide income for him for at least a year or so. However, he once again found himself in need of a new job to avoid an eventual financial debacle.

Ulrich and Helga, after hearing the news from Carl that he finished his investigation of them, got in regular contact again and frequently updated each other on their respective situations. "What if you called Doug or your boss at the Mayo?" Helga suggested during one of those conversations. "They might be able to find a position for you. After all, you spent quite a few years there".

"That's a brilliant idea," Ulrich acknowledged, "but do you think I can believe Carl, and it's safe for me to return to Rochester? I don't want to impugn Carl's promise, but I better to be very careful. On the other hand, I miss you very much, and I still have the wherewithal to buy a ticket and live on it for a year without indigence."

"Oh, I'm absolutely sure you can safely return. Although, I assume that you might not be able to get a license to practice medicine, because that matter would surface during the review of your application. But my guess is that you wouldn't even want that."

"No, I really wouldn't. I would want to do research." Eventually Ulrich gathered up his courage and called Doug to inquire if he knew of any research opportunities at the Mayo. At that time Doug was in the middle of preparations for his wedding but promised that after his return from the honeymoon, he would look around. For months nothing happened, and Ulrich decided to take the job search into his own hands. At the end of October in 2013, he visited Helga in New York, only to learn that the gallery showing her paintings had gone out of business. When she did a frantic search to find another gallery and realized that at least fifty artists were competing for one spot, she gave up hope. She was ready to relocate to Minnesota, where she had the opportunity to reclaim her nursing position at the Stewartville hospital.

Perhaps the fact that their dreams had been shattered around the same time made them ready for a new start in life, which included renewal of their intimate relationship at deeper level than before and with full commitment by both. Early that November they moved back to Rochester, where soon after finding a small house and Helga getting her old job back, they got married. It had been long in the making, and they both felt it had been inevitable. The most interesting thing was, as they discussed and agreed on it, that despite all the baggage that they both brought with them, neither of them carried chips on their shoulders. They had been created for each other, and now they treated this as a fact that they would never again question. Their resilience won out; the past was behind them, something they could now look back on without regret. The past errors they had made hadn't been in vain, and those "lost" years weren't really lost; they had prepared them for a long and happy married life, truly respecting and appreciating each other's values and paying more attention to each other's needs. They didn't regret the failures in their professional lives ether; as Ulrich put it in paraphrasing a famous scientist, "it's not worth doing something if it's not hard." They certainly didn't hold back in making up for lost time; shortly after their marriage, Helga learned that she was expecting a baby.

Now the most important task for Ulrich was to find a job in Rochester or close to it. When Doug became hospitalized, it was clear from the beginning that he would need an immediate replacement at the Mayo Clinic to continue providing essential services. During one of Ulrich's visits to the hospital, Doug suggested to him that he requests an appointment from the department's head and inquire about the possibility of taking over his position. Ulrich's chances were good. His knowledge of scientific equipment and methods was excellent, and at the time being that was all the Mayo Clinic required of him, although later he was also expected to get significant grant support. It also worked in his favor that the clinic couldn't go for too long without replacing Doug. His interview with the chairman of the institute was concluded very positively. Doug already was on the treatment, but he still had a sufficiently clear mind to write a recommendation letter although not without panting. A little later, Ulrich's previous boss sent to the institute's chairman another one. Thanks to all this warm support and his successful interview, the apparition of unemployment for Ulrich was over. He got the position for a twelve-month period, starting on January 1, 2014, with the stipulation that upon Doug's return, he wouldn't be extended. But by then nobody expected Doug to return, although nobody talked about it openly. This reality quieted Ulrich's initial ambivalent feelings about getting the job at Doug's expense. He immediately betook himself to jump into the middle of tasks the new job demanded.

CHAPTER 23

Doug's Last Struggle, a New Beginning for Agnes

Even though Doug didn't show any encouraging signs of improvement—his condition only kept worsening—everybody around him was optimistic that he would live a little while longer, but it wasn't meant to be. During the night of April 21, he grew so sick and weak that the doctors gave up the fight and took him off his treatment. By 6:00 a.m. his vital signs practically reached zero. Agnes received a phone call summoning her to the hospital, though the nurse noted that he was unlikely to regain consciousness again. In fact, she was already about to leave the house with her laptop in hand. She rushed to Doug's bed, but instead of finding him dead, to everybody's amazement, Doug, his head drooped low, opened his eyes. He couldn't talk, only whisper to Agnes's ear, "I was in heaven, not my body, my soul. Agnes, now I know for sure that there is a god. I couldn't see Him, but I felt Him. I felt Him everywhere. Heaven is very different from anything we ever experienced or thought of. Without seeing or hearing Him, my soul could still understand what He said to me. He let me know that He would allow my soul to come back into my body and tell you about Him, that after our deaths, our souls will experience a very different sort of existence. On earth our bodies did the living. In heaven, our souls will continue to live equipped with all the senses we have now and more."

"Oh, my dear, can you tell me what heaven is like?"

"First, I entered a huge glittering space." He stopped and laid his head on his pillow. He closed his eyes, and for a while he appeared to be in the twilight zone, but Agnes had the feeling that he still wanted to say something. About ten minutes passed until Doug could open his eyes again. Agnes moved close to him, trying to make out his whispers. "Inside I saw a giant endless cloud with many leaves, like the sundancer cactus."

Agnes quickly opened her laptop and searched for cacti. She found one that resembled the sundancer cactus. She held the picture in front of Doug. "Was it something like that?"

Doug's answer was barely audible. "Not exactly, but similar. I saw clouds forming giant leaves in all directions partly covering each other and thus creating a huge multidimensional space." These were his last words. The muscles needed for him to talk stopped working, and within a minute, he completely stopped breathing. The room suddenly became a somber place. Outside, a large black cloud covered up the sun. Was it the mourning veil of sun mourning the loss of Doug, who dealt so much with nature's secrets?

Agnes diffidently closed Doug's eyes and called for the nurse. She didn't cry. There are people who weep inward and not outward. Agnes was one of them. She felt simultaneously wretched and hopeful—hopeful that their separation would only be temporary. That after her death, their souls would meet again.

After the funeral, Agnes had many major tasks ahead of her. She was into her fifth month of pregnancy, and she knew that the baby was a girl. Agnes had never had the strength to talk about their child to Doug; she felt that the news would stir up his emotions to a level that would be life-threatening. Had Doug noticed her growing belly, that would have made all the difference. Later, Agnes regretted that she had deprived Doug of the joyful news and the pleasure of finding a name for their daughter together.

Fortunately, the major tax-filing season at the firm was over, and she had a little break with which to organize her life. Doug's sons often visited her; the three of them spent long hours together, sometimes in silence, sometimes talking about Doug with their eyes suffused with tears, sometimes walking the dog together. The weather also got better, and Agnes occasionally took them to the tennis court and coached them, which was a good way to bring back some normalcy into their lives.

Sarah hadn't attended the funeral and never called Agnes to offer her sympathy. She became a completely self-absorbed continuously querulous woman. Initially Agnes thought that Sarah's ignorance was rather rude, but few days after the funeral, one of Doug's old colleagues let Agnes know that for months Sarah had been receiving regular psychiatric and drug treatments at the Mayo Clinic for severe depression. Carl's ongoing investigation of Sarah's possible role in Peter's death almost certainly contributed to her condition. All this didn't bode well for the boys, and Agnes started to prepare herself for the time that may come when she would have to take care of them too, along with her own daughter.

CHAPTER 24

AGNES GOT NEW FRIENDS

Ulrich had met Agnes twice at Doug's bedside in the hospital, but Helga had never met her until Doug's funeral where they had a brief conversation. Agnes heard about Helga from Peter, but Helga didn't know that Agnes was Peter's one-time girlfriend. The fact that Ulrich got Doug's job and that both Helga and Agnes were expecting their children to be born around the same time brought them close. After the funeral they met almost weekly, most of the time in Agnes's house, and these meetings meant a lot for both expectant women. One day they got an invitation from Mike Collins to the St. Paul opera house where his opera piece was scheduled for late June, as a kind of after-the-season performance for emerging artists. As much as they wanted to go, neither Agnes nor Helga wanted to take any risks; their due dates were drawing closer and closer.

So instead of going to the opera that evening, Helga and Ulrich visited Agnes again, ready for some light conversation. Helga mentioned that once they had been the guests of Doug and Sarah and at that time their impression was that she was a suave woman.

"That night Doug talked about some fantastic new ideas he had about how God might have created the universe and the role of evolution in shaping the universe from the early moments of the big bang, eventually giving rise to mankind. He also had an idea for how God might communicate with us, although, to be frank with you, we didn't see it as something feasible. Yet, although we are atheists, we were really shaken by Doug's theories, and I have been thinking ever since, what if he was even just a little bit correct? Agnes, can I assume that the two of you talked about these things a lot?"

"Sure, we did. In fact, his theory about Creation, which was constantly changing, was the reason we met in the first place. I heard from Peter that Doug was developing his own original ideas about Creation, and I wanted to talk to him about them. Shortly after we met for the first time, we fell in love, but this didn't prevent us from having long discussions about God's role in Creation and our lives. In a way we were pandering each other as we discussed and further developed aspects of his theory like evolution. During my intellectual interactions with him, I completely changed my view about God. I was a semiatheist before I met him, and now I'm a strong believer. Don't talk to me about these kinds of subjects if you want to remain atheists," she joked. "Although I can testify that even in atheists, the image of God is there, only slumbers, and then it either bursts to the surface or remains dormant forever."

"I don't mind hearing about them. I'm open to any theories, as long as they are logical," said Helga. "So how did you progress together?"

"As far as I can judge it from the retrospective, most of the basic elements had already been present even in Doug's earliest theories, but during our many discussions, I think we were able to better fine-tune some key points."

"Like what?" Helga asked.

"That would be a long list for one conversation, but one key point was that God put together two key sets of information to initiate two consecutive big bangs. The first one created space as well as some forms of invisible material and energy, like dark matter and dark energy, that directed the evolution of the visible universe that arose from the second explosion, generally known as the big bang. In Doug's view, our universe would be very indigent, if it would exist at all, without the first big bang. According to him, the explosion prior to the second big bang could also have created the Higgs field, producing Higgs bosons that are probably needed to give mass to fundamental particles. These pre–big bang events precisely determined that during the initial fractions of a microsecond after the second big bang, when the energy of information translated into the precursors of our universe, the universe was nothing more than hot gluon-quark plasma. These events then further determined how microscopic structures, like protons, neutrons, and atoms, would look like. The role of dark matter in shaping the formation of our universe particularly interested Doug because these particles, if they can be considered particles at all, have gravity, and this is the only thing we know about them."

"I remember that Doug indeed mentioned something about dark energy and dark matter, but I can't recall whether he told us how they were supposed to shape galaxy formation," commented Helga.

"I think Doug came up with quite an original idea," Agnes replied. "He built on the prevalent view of scientists today that dark energy is responsible for pulling the galaxies away from each other with ever-increasing speed, while dark matter is concentrated in the intergalaxy space and around the galaxies, and its gravitational impact on matter accounts for the stars not flying out of their galaxies. Doug's contribution was that he assumed that large aggregates of dark matter had been dispersed throughout the pre–big bang space, what he called parallel universe, serving as sort of signposts and templates for galaxy formation in our universe. In other words, the size and distribution of galaxies could follow the pattern of the dark matter aggregates. The distribution of matter in our universe was also not completely homogenous, and the small denser spots were sufficient to attract dark matter and slowly build up the galaxies. This is how far Doug proceeded with his theory to explain why galaxies are composed of both visible matter and dark matter and how the latter had a crucial role in the development of our universe based on the precise information that was built into the origins of the two big bangs." Whenever Agnes explained something this way, she took care to emphasize that Doug's wish was to treat his theories solely as science fiction or science faith, and nothing else.

"But I still don't get it," Helga said. "What was his goal? Why produce an elaborate theory without attempting to communicate it to others?"

"He was hoping, he told me, that one day some of his ideas may spur thinking about religion and science in new directions. His plan was that after completing his theory to a reasonable degree, he would meet and seek collaborations with some highly respected religious and atheist mathematicians, theoretical physicists, and astrophysicists willing to use the weapons of modern elementary particle physics and mathematics to advance his theory. As he was a good scientist, he could guess how much more was needed to elevate his idea from science fiction to a more acceptable science theory. He also knew that such an undertaking was a very long shot. I think he wanted his theory, when completed, to serve as an exhortation to both atheist scientists and religions that there is a reasonable ground to work together on the long path of understanding the universe and what or who is behind it."

"You alluded to information from which the universe was created. What is it?" asked Ulrich.

"Doug thought that information is the precursor of all forces as well as energy and matter forms that compose the universe. He imagined that several information units form a coding information unit, just like three-letter bases code for individual amino acids, and then large numbers of such information coding units form super information complexes similarly to the DNA in our cells. God put together the two specks for the two big bangs from a very large number of super information complexes, which provided a blueprint for the formation of the two universes."

During the following weeks they found more time to discuss some other aspects of Doug's theories. These discussions were useful for Agnes because they helped her further hone her own ideas and see where she could add to Doug's theories. She also regularly read relevant articles in the journal *Science* that Doug subscribed to, which was still delivered to the house every week. It was particularly interesting to her to learn that cosmologists ostensibly traced ripples in space from the first moments after the big bang, which, if confirmed, would prove the theory of inflation, an exponential expansion of the universe stretching subatomic distances to cosmic lengths in just 10^{-32} seconds. For days, Agnes's thoughts reverberated around this finding. "Is it really possible, as Doug thought, that the vacuum force of the parallel universe was the force behind this rapid expansion of the newborn visible universe?"

Only two weeks passed, when another article appeared in the same journal echoing concerns by several scientists in connection with the inflation theory, adding that at least two more years were needed to experimentally confirm it. These controversial articles led Agnes to conclude that the dust had not quite settled about this issue. Reading and talking about subjects she and Doug used to talk about was a special pleasure for her; on those occasions she felt his strong presence.

One evening she explained to her guests that she had agreed with Doug on the central role of evolution in the development of the universe, including humans, and that God is unlikely to interfere with our lives, since that would require changing the natural laws. Both Ulrich and Helga applauded Doug's idea of putting evolution into the center of the development of the universe and life. "If the church could pull this off, with this single act it could retain half of its halfhearted churchgoers," predicted Ulrich.

"You may be surprised to hear that the new pope, Francis, recently stated that evolution of nature is real," Agnes informed them.

"If so, this is really an amazing development," Ulrich exclaimed.

"But he also said," Agnes added, "that God created human beings and let them develop according to internal laws that he gave to each one, so they would reach their fulfillment." Invoking God in the creation of humans took Ulrich's enthusiasm back a little bit, but he and Helga were happy to see that the "church has started to move in the right direction," as he phrased it.

All of them had problems with Doug's ideas about the soul or hologram, although none of them could come up with a better model for the soul, or if the soul and the hologram are the same or separate things; in fact, the guests questioned that they existed. Yet, despite being atheists, both Helga and Ulrich held the view that the brain's activity can't explain everything, and certainly not whom we fall in love with or what makes us accept or reject God.

"What if we are holograms as well?" Ulrich jokingly posited. "It would sound uncanny to me."

Agnes raised her eyebrows. "I know you meant it as a joke, but Leonard Susskind, in his book *The Black Hole War*, already introduced the concept of the holographic principle, which means that all the information in the universe, including us as three-dimensional subjects, might be situated at the two-dimensional boundary of space. In other words, we could be viewed as holograms, and the information about us may be found at this boundary. Could that information about us out there be our souls? You see, it wasn't only Doug who dealt with issues that many would consider science fiction. Numerous respected scientists did that too and still do it. But in the end, who knows? None of these ideas may turn out to be real science after all."

At their last meeting they also talked about the absolute *nothing*, the eternal amount of substance-like information, and how such information might be transformed into energy fields and matter particles.

"We already dealt with the information in cursory manner, but after thinking about it more, I still don't get it. What do you mean by information?" Ulrich asked Agnes, with his eyes radiating confusion. "I know that what you mean must be something very complicated, but what is it? I doubt that you meant information as words in a book. But if you meant that information is something that can transform into energy and matter forms, then shouldn't it be also a kind of physical entity?"

"I'm only at the very beginning of understanding what physicists and Doug meant by information," said Agnes. "To cut a very long story short, as far as I can understand, scientists measure information in bits, which are different from the bits used in computers. Some physicists think that a single bit is probably the most fundamental building block of the universe. They estimate it to be 10^{-33} cm long, which is the smallest meaningful length in nature according to quantum mechanics. This single bit is also often referred to as the Planck length. All material objects, including neutrinos, electrons, proton, atoms, and macroscopic objects, are believed to be composed of bits of information. They call this theory digital physics. I know that this definition of information is very difficult to conceptualize, but this is one of those fundamental questions that we somehow must be able to get a grip on if we want to understand how the four forces, space, energy, and matter, probably in this order, could originate from apparently nothing. As we discussed with Doug, information might be a substance-like entity that exists on its own right and that is separate from the known energy and matter forms. However, it can transform into energy fields and eventually matter particles, perhaps through some intermediate entities.

"For Doug, dualism was an important feature of everything, and once he told me that he thought that information exhibits the greatest dualism of all, because it simultaneously represents both the absolute *nothing* and the *something* that may explain the great mystery of how the universe apparently popped out from nothing some 13.7 billion years ago. A few months ago, Doug raised the possibility that perhaps information was the only entity that has existed forever and that the spiritual God equals the totality of information, while the material God is the extremely highly organized form of information that has free will and the unconstrained ability to act on any plan He wants to act on."

After this tiresome conversation, they turned their attention to more earthly things, like the names of their soon-to-be-born babies. Helga and Ulrich had already decided to give their son the name Jonathan, and Agnes had chosen her own name. "Have you had a special reason to name your daughter Agnes, other than it is a beautiful name?" Helga asked.

"You know, I like my name. If I'm correct, Agnes comes from the Greek name Ἁγνή *hagnē*, meaning 'pure' or 'holy.'"

CHAPTER 25

AGNES'S BIG DECISION

Agnes still expected to carry on for at least two more weeks before going to the hospital to give birth to her daughter, and to pass the time, she went through the notes she had recorded in the hospital. Suddenly, an intrepid idea struck her. Cancer prevented Doug from cresting as a brilliant scientist and thinker. His best legacy would be a book, turning his science fiction and science faith ideas into a more realistic and more coherent science theory, backed by mathematics, physics, and the latest results of cosmology. The more she thought about it, the more she became intoxicated by the thought of writing the book in which Doug would be the post-humus coauthor. "Some of these ideas are just too important to be lost, even if right now they are nothing more than conjectures," she thought. "Doug would support my decision to write this book, provided that I can accurately relate to his theories." She figured that if she started working on the book soon after little Agnes's birth, she could proceed very well until going back to work six months later.

As soon as she began to develop a framework for the book, she realized that she wouldn't be able to avoid a discussion of Doug's revelation about heaven. Should she believe Doug's testimony that he had really seen heaven and that he knew for sure that God existed, but he could only feel but not see Him? This was a key issue that in the eyes of many future readers could erode the credibility of Doug's ideas about other important matters. In addition, her own future life was at stake too, if she became a victim of incredulity, even if the source of information about heaven was Doug.

Although at first hearing she had wholeheartedly believed Doug's testimonial, by now she had developed a certain level of incredulity herself that further complicated her decision of how to deal with Doug's testimony. To aver seeing the other world, which some people in the near-death state describe as heaven, is a very touchy issue. Most neuroscientists believe that some specific chemicals trigger certain events in the brain, plain and simple, which can produce these images that the moribund person then accepts as real. What particularly disturbed Agnes was that when Doug described heaven, he compared it to a certain kind of cactus that he had always been fascinated with. For Doug that cactus, the sundancer, even resembled the multidimensional space that just before his illness he had started to deal with in connection to string theory. Perhaps during his last hours this earlier vision somehow suppressed his other brain activities and this is why he pictured heaven as a multidimensional space wherein individual spaces are partly separated by giant leaflike structures.

Agnes didn't know much about the brain chemistry that may lie behind visions and life-after-death experiences, but she suspected that there must have been a lot of research done on the specific neurological events that Doug had probably experienced. On a website and then in a neuroscience book, she found abundant information on the topic. She learned that the two hemispheres of our brain sense the world differently; when the left hemisphere dominates, as can be experimentally established, the person thinks that he is part of the world. In contrast, when the right hemisphere dominates, then the person thinks that the world is part of him or her. These are diametrically opposite views about the mental relationship between a person and the outside world.

She also learned that the lack of oxygen and many other pathological stimuli can cause instability in the right temporal lobe, due to increased discharges of the neurons that may lead to hallucinations as well as mystical and religious experiences. Agnes reasoned that Doug's brain was almost certainly short on oxygen supply, which could have triggered pathological activity of the temporal lobe and the underlying limbic system; this could have then evoked the false experiences that Doug was trying to communicate. And perhaps the sundancer cactus and his vision of multidimensional space had linked his vision to reality, so his left hemisphere wasn't entirely suppressed. Agnes began to lean toward the possibility that Doug, like probably everybody before him who had experienced heaven or something supernatural, had only undergone a complex and vivid hallucination. Although this wouldn't necessarily negate the idea of an afterlife, she decided to deal with this episode in the book very gingerly. If the readers would conclude that she was a superstitious, unscientific person, they wouldn't look at the book as a serious attempt to bring religion and science together. On the other hand, in deference to Doug, she didn't want to negate his experience about heaven. She was in a tough spot.

At her next visit to her doctor, who had known Doug very well, he asked Agnes how she had been dealing with her terrible loss.

"I'm trying to cope with it by building a legacy for him. I'm thinking of writing a book about his cosmological theories and how scientists and people of faith could close the gap between them. I also contributed a little bit here and there, after being converted to most of his ideas. But there's something that especially troubles me. The minutes before he died, he claimed that he had seen heaven and communicated with God, who allowed him to come back and tell me about his experience. I've been reading up on the subject, and I'm starting to think that his vision was just the mischief of chemicals in his brain. I'm afraid that coming out with a book with statements about seeing the heaven included would induce many people rile against it. Such negative publicity would take away the attention from the major points of his theory."

"Well, this is a very complex topic, and I don't think we have enough time to discuss this in depth," the doctor said. "But as I understand from my colleague, your husband came back from a state of clinical death when you entered the room. That happens very rarely, and as a doctor, I can tell you that nobody has found a satisfactory scientific explanation for that. It seems to me that to come back to life, even if only for a very limited period, like few minutes, his brain must have received a very strong stimulus that temporarily reactivated the areas of his brain associated with short-term memory and some motor activity that enabled him to communicate. What was this stimulus and who provided it? Was it you, or a sudden strong flash of light or sound, or could it have been a supernatural being? Until brain researchers find out more about how some signals and stimuli can still generate visions like that in the dying brain, there is room to believe that Doug was correct, and he indeed saw what he described to you. But I also should tell you that there is a debate raging on among neuroscientists about afterlife experiences. Some, like Andrew Newberg and Sam Harris, think that any afterlife experience can be explained by changes in the brain's activity and chemistry. Some other experts, like the highly regarded psychologist Emily Williams Kelly, believe that even when the brain seems to be virtually disabled, people still can have afterlife experiences. You can see that this issue is very murky, and I myself don't know where I stand." They didn't have any more time to discuss the meaning of Doug's vision; the doctor had to concentrate on the baby's growth and health. Everything was all right with her daughter, the most wonderful thing to hear for an expectant mother.

After she got home, she thought long and deep about what her doctor had said. Putting everything in the context of Doug's idea about the soul and the material basis of God's communication with people, she decided that in her book she would most likely provide two alternatives for how the life-after-death experiences may occur. Without prioritizing them, the first mechanism would be the brain tricking the moribund person via substantial changes in brain's chemistry. The fact that certain chemicals, like ketamine or dimethyltryptamine, can evoke hallucinations gave strength to such a scenario. The second mechanism assumed the existence of soul, which equals conscience and hologram, which is a super information complex with the property of being released from the brain upon irreversible inactivity of the corresponding area, which means the person died. However, even almost completely inactive neural cells and synapses can be reactivated for a short time by a strong signal that in theory could in some cases be provided by God, resulting in the reabsorption of the super information complex and in the reactivation of memory and the muscles involved in speech. Then the person may even survive, while in other cases, as happened to Doug, such reactivation lasts only for a short time and then the person eventually dies. It will be important to emphasize that whenever the neural cells are reactivated, the soul returns, reestablishing an information exchange with the brain leading to the reappearance of consciousness. She was satisfied that by providing the two possibilities in her treatment of afterlife, it will not be viewed as inane talk, but the readers will have the option to choose what they want to believe in. This kind of duality gave her book a new purport.

During the few remaining days before giving birth to her daughter, she kept working on developing the book's outline, and she began putting down her thoughts in a thick notebook. First, she again returned to Doug's afterlife experience, and after more thinking, she began to shift to the idea that the soul, after leaving the temporarily dead brain, could indeed witness the other world. Then, reactivation of the neural cells by God or another strong signal would allow the soul to reincorporate into the brain and communicate what it saw. "It appears," she thought once, "as if God applies Heisenberg's uncertainty principle, certainly true in the quantum world, to life and death situations as well. For the observer, he may or may not have seen heaven. The truth is in the eye of the beholder. Only Doug and others who claim to have witnessed the *'twilight zone'* would know the truth about heaven, while the living is entitled only to believe and not to ever know for sure. Whether we admit it or not, we all live with this uncertainty, perhaps the greatest duality on earth, either believing or not believing, but always doubting and wondering if we made the right choice, whichever it was." Had God tested her faith in a crucial way through Doug's testimony? Slowly she concluded that in the book she would probably have to present the option that Doug went through an out-of-body experience, although his brain probably seriously distorted what he saw.

Another serious problem Agnes needed to clarify before starting to write the book was to find a central idea or theme. Then, while reading a book that she planned to use as a major reference, she found the focus, the central issue around which she would develop a more scientific idea of Creation and the role of God. She marveled at the fact that even 13.7 billion years after the big bang, the universe was still relatively well organized and far from a state of complete entropy that in layman's terms is an expression of the level of disorder or randomness in a system. "This must mean that in the original two halves-speck, the information had to be extremely well organized, containing the lowest level of entropy possible, which allowed its explosion. In other words, the speck had very little disorder or entropy in the form of information jiggers just enough to bring together the two halves and initiate the big bang. Even if we, highly organized creatures, die, we rapidly disintegrate and along with it rapidly increase entropy. Such a highly organized speck with a very low level of entropy can't be formed by itself. And even after the big bang, could the gravitational force of dark matter, helping to keep the universe organized, play a role in slowing the rate of entropy?" After thinking all this through, she decided that she would attempt to construct mathematical models of the big bang event, with two separate assumptions: i.e., it started out either as a singular event or was the result of consecutive events resulting from two separate explosions of two specks. Based on astrophysicists' data, she will relate today's level of entropy in the universe to the entropy level in the speck or specks as well as the influence of dark matter.

"I will definitely consider the possibility that the speck or specks contained only pure extremely highly organized substance-like information with near-zero entropy and extremely high energy content. Then, information transformed into a quantum field in a very violent manner during the big bang, subsequently decaying into the gluon-quark plasma soup within fractions of a second."

One evening, seemingly so long ago before their marriage, she and Doug had also considered the scenario that God had provided both a preexisting space with multidimensional Calabi-Yau shapes and the big bang that created the strings from information during the very short period of explosion. Thus, the three-dimensional space of the universe with its strings would extend into the preexisting space filled with the multidimensional Calabi-Yau shapes. Then interactions between the strings and Calabi-Yau shapes would direct the formation of matter particles and thus the universe.

The next day she continued her brainstorming. "A related central idea should be to understand a deeper physical, or perhaps philosophical, meaning of information and its relationship to spiritual God and material God, as well as the many forms of energy and matter. If the big bang originated from highly organized information, then information must be, at least in part, something objective, perhaps the most elementary physical entity that can yield other physical entities. This is distinct from information as a subjective category, without any physical attribute, which solely functions to help people to organize their lives. I will also deal with Doug's concept that information has the dual nature of representing both the absolute *nothing* and the *something* with important physical attributes like energy. I will also have to deal with the organization of information determining how the universe is evolved and providing the basis for the soul and God's communication with mankind.

"Yet another central idea should be to understand the nature of empty space, or vacuum. Had empty space always existed with some characteristic attributes? Is it possible that the spiritual God is present in the empty space as the totality of information? How to put all this together, with the concept of the material side of God? Has the material God always existed, or did He develop from the spiritual God during an evolutionary process, or is the concept of the material God flawed to begin with? Could a special kind of entanglement exist between the spiritual God and the material God, explaining their unity?"

After preparing a lunch and then walking the dog, she went back to adding more central issues to the outline. "Then the book will have to attempt to deal with the issue of gravity, which so many scientists have been unable to fully understand. Does gravity entirely reflect space curvature caused by mass, as Einstein postulated, or do the hypothetical gravitons, flying with the speed of light, carry it? What is the nature of dark matter's gravity? Is it possible that the gravity of visible matter is different from the gravity of the invisible matter? Just like there are different quarks and bosons, maybe there are different gravitons or different forces carrying the gravity of visible and invisible matters. Does visible matter also exert a gravitational pull on dark matter, or is it only the other way around? This would be crucial information for understanding galaxy formation and maintenance. I am not sure where the science stood on this last point.

"I assume yet another central idea will be the evolution of the universe and the life as well as their functioning requiring some signposts, greatly reducing randomness and providing direction to these processes."

She had no more ideas for the book, but she was happy that she could get at least this far. It might take her a decade to finish the book, but she was on track. This task gave a new meaning to her life, and she also saw the effort as a continuation of Doug's life. To get some fresh air and hoping to get some more information, she decided to walk to the Barnes & Noble bookstore. On her way she thought that constructing mathematical formulations for either completely random or directional events, both at the level of the universe and of cells, should be possible, and for that she would seek the cooperation of leading theoretical physicists and cell biologists. This was something that didn't even require mentioning God.

At the bookstore she saw several very interesting books that she immediately bought. The first book, *To Heaven and Back*, was written by a medical doctor, Mary C. Neal. She was drowned on a river during kayaking and went to heaven, according to her account. Her coming back to life couldn't be described in presently understood medical terms; those familiar with the case described it as a miracle. Then there was another book by Eben Alexander, a neurosurgeon, who was in a coma for seven days. Although in his previous professional life he had never believed in people's accounts coming back from clinical death and claiming that they had afterlife experience, after waking up from the coma, he also claimed to have been in heaven. After reading these two books, Agnes decided that in her own book she would treat Doug's account of heaven as something that probably really happened but which didn't require changes in laws of physics. In principle, at least in her mind, the super information complex, or soul, leaving the brain and then coming back would not break the conservation of energy law.

The author of yet another book, *A Universe from Nothing*, was Lawrence M. Krauss, a highly recognized cosmologist at Arizona State University. He presented a materialistic point of view of how the whole universe could have popped out of *nothing* without any involvement from a superior being. At first, Agnes was both annoyed and happy to see that a materialist can also produce science fiction when it comes to Creation. Even after reading the book twice, she still couldn't see the scientific basis for the author's contentions. She realized that the book was filled with axiomatic statements like "the *nothing* is unstable," implying that such instability, which others might call quantum jiggers, in the vacuum-like empty *nothing* could have led to the creation of the universe. "Oh, really, what is the evidence for that? At least Doug honestly admitted that his theories are in the category of science fiction. I'll definitely point out the baseless statements made by Krauss."

But then, after giving it some more thought while still browsing through the pages, she realized that essentially, Krauss and Doug had said the same thing. She desperately pondered for hours and hours where the difference was and what the implications were. "The difference is that Krauss didn't invoke God in the creation of the universe from nothing, and Doug did. Again, all comes down to the meaning of the *nothing*. Krauss didn't have a good answer to this question, but Doug and I found a possible feature of the *nothing* that preceded the big bang. It contained all the information that was needed for producing the quantum field, which then produced the elementary particles and the whole universe. So the speck or specks, according to Doug, contained highly concentrated and highly organized, almost entropy-free information with high energy content. I wonder whether there is a balance between the amounts of information as well as energy fields and matter particles in the universe, or if the relationship is one directional, with only information producing energy and matter. Or can energy fields and matter particles possibly decay back into information? Could giant black holes be the places where energy and matter revert to information? Could giant black holes then be the birth places of new universes? Is it possible that God has created and keeps creating more and more universes and that ours isn't in fact unique? And on a more personal level, could any one of us carry a soul, which is composed of pure information, highly organized with the capacity to store all lifelong experiences that then survives us forever?"

Close to Krauss' book she discovered yet another interesting one with the title *The 4 Percent Universe* by Richard Panek. The foreword promised to take the reader through the path astronomers and particle physicists had taken to come to the conclusions that about 96 percent of universe, made up of dark matter and dark energy, is invisible, the latter being responsible for the expansion of the universe with ever-increasing speed. As she quickly browsed through the pages, toward the end of the book her eyes were drawn to a paragraph referring to a hypothesis that gravity may reflect the existence of a parallel universe. According to this hypothesis, if two giant brans, representing two universes, get close, then they could influence each other through gravitation. In fact, this hypothesis also raises the possibility that the large influence of the other universe may cause the effects of dark matter and dark energy in our universe. This book discussed so many interesting possibilities about the beginning, development, and end of the universe that she simply couldn't resist to buy it as a future reference to her own book.

She also bought two books from Richard Swinburn with the titles *Was Jesus God?* and *Is There a God?* Despite being converted by Doug, she didn't like these books; in fact, she found them somewhat nettlesome, because in her opinion, the author appeared to approach these big questions with a predetermined mind; he practically prejudged the outcome, giving affirmative answers in both books. Agnes liked Doug's approach much more, because he was led by the curiosity of a materialist, and he accepted the existence of God only after he exhausted all other possibilities to explain the fine-tuning of the universe and the incredible complexity of life. As a result, Doug's God was very different from Swinburn's God. Doug's God was much closer to humankind; in fact, humankind was part of God.

Agnes was about to pay and leave the bookstore when she noticed yet another book, this one by Mark Vernon, with the simple title *God*. She bought that book too, and after reading it, she was very happy that she had seen it. The author, a very colorful personality with a wide-ranging expertise in physics, theology, and philosophy, approached God's image in a more rational manner, and overall the book gave a lot of ideas to her as to how to deal with religion. She particularly liked to see that some people, as referenced by the author, started to bring the concept of evolution into religion. This is exactly what Doug wanted to see; religions should accept evolution as the means to incorporate science, God's science, into their teachings.

Then, later in the year, she read in a journal that in the same large hadron collider, which in the previous year had been used to find the elusive Higgs boson, physicists had created a new form of matter by colliding protons with lead ions. This new kind of matter is called color-glass condensate, a liquid-like wave of gluons that stick quarks together inside the protons and neutrons. What really caught Agnes's eyes was the finding that some pairs of the particles were flying off from the collision point in the same direction, showing that they somehow kept communication with each other. The scientists explained this phenomenon with quantum entanglement, meaning that two particles retained a connection after their separation, and an action on one instantly reverberated on the other. "I have to deal with this quantum entanglement issue very seriously," she concluded. "How can these particles exchange information instantly? Is information really a third objective entity in addition to energy and matter, and if so, can it be linked to quantum events? Since the connection between entangled particles is instantaneous, could that mean that information can spread through the universe instantaneously? Wouldn't this be a good time to find a different word for this kind of information, an entity between the known and presently unknown physical worlds? Is it possible that information represents the fifth force and it provides a link among the four other forces? Could description of information lead to the grand unification theory so many theoretical physicists are working on?" She couldn't express it with words yet, but she felt that a deeper understanding of this strange phenomenon, quantum entanglement, together with information might help to better understand our relationship with God and what was waiting for us in the afterlife.

Now she had plentiful ideas for her and Doug's book, all relating to unsolved mysteries of the universe and life. One evening, thinking about the title of the book deep and hard, she suddenly snapped her fingers; "of course, it's so obvious; after little Agnes is born, I'll continue to write our book with the title *God, the Big Bang, and Other Unsolved Mysteries.*"

Main Literature

Observations and Theories Related to the Big Bang, Subsequent Events, and Multiple Universes

Cho, Adrian. "Hints of Great Matter-Antimatter asymmetry Challenge Theorists." Science 328 (2010): 1087.
Cho, Adrian. "The Morning After: Inflation Result Causes Headaches." Science 341 (2014): 19–20.
Greene, Brian. The Hidden Reality: Parallel Universes and the Deep Laws of the Cosmos. New York, NY: Alfred A. Knopf, 2011.
Kirshner, Robert P. The Extravagant Universe. Princeton, NJ: Princeton University Press, 2002.
Krauss, Lawrence M. A Universe from Nothing. New York, NY: Free Press, 2012.
Penrose, Roger. Cycles of Time. New York, NY: Vintage Books, 2010.
Penrose, Roger. The Emperor's New Mind. Oxford, New York, NY: Oxford University Press, 1989.
Singh, Simon. Big Bang: The Origin of the Universe. New York, NY: Harper Perennia, 2005
Tyson, Scott M. The Unobservable Universe: A Paradox-Free Framework for Understanding the Universe. Albuquerque, NM: Galaxia Way, 2011.

Astrophysics, Cosmology, Planetary Science

Bhattacherjee, Yudhijit. "Where Are the Missing Baryons?" Science 336 (2012): 1093–94.

Blanford, Roger D. "A Century of General Relativity: Astrophysics and Cosmology." Science 347 (2015): 1103–08.
Cartlidge, Edwin. "What Reionized the Universe?" Science 336 (2012): 1095–96.
Clery, Daniel. "Supernova Breaks the Mold: A Massive Explosion in a Nearby Galaxy Both Confirms and Confounds Astronomers' Expectations." Science 345 (2014): 993.
Cordes, James M. "Radio Bursts, Origin Unknown." Science 341 (2013): 40-41.
De Grasse Tyson, Neil., and Goldsmith, Donald. Origins: Fourteen Billion Years of Cosmic Evolution. New York, NY: W. W. Norton & Company, Inc., 2004.
Finkbeiner, Ann. "The Mystery of the Dead Galaxies." Science 346 (2014): 905–7.
Kerr, Richard A. "Cassini Plumbs the Depths of the Enceladus Sea." Science 341 (2014): 17.

Laws of the Universe

Hawking, Stephen., and Mlodinow, Leonard. The Grand Design. New York, NY: Bantam Books, 2010.
Penrose, Roger. The Road to Reality. New York, NY: Alfred A. Knopf, 2005.

Black Holes

Clery, Daniel. "The Dark Lab." Science 347 (2015): 1089–93.
Cowen, Ron. "Decade of the Monster: An Infalling Gas Cloud and Other New Probes Herald a Revealing Period of the Milky Way's Supermassive Black Hole." Science 339 (2013): 1514–16.
Fender, Rob., and Belloni, Tomaso. "Stellar-Mass Black Holes an Ultraluminous X-ray Sources." Science 337 (2012): 540–44.
Hawking, Stephen W. A Brief History of Time. New York, NY: Bantam Books, 1988.
Maldacena, Juan. "Testing Gauge/Gravity Duality on a Quantum Black Hole: A Numerical Test Shows that String Theory Can

Provide a Self-Consistent Quantization of Gravity." Science 344 (2014): 806–8.

Susskind, Leonard. The Black Hole War. Boston, MA: Little, Brown and Company, 2008.

Thorne, Kip S. "Classical Black Holes: The Nonlinear Dynamics of Curved Spacetime." Science 337 (2012): 536–38.

Witten, Edward. "Quantum Mechanics of Black Holes." Science 337 (2012): 538–39.

Dark Matter, Dark Energy

Cho, Adrian. "How Hot Is Dark Matter?" Science 236 (2012): 1091–92.

Cho, Adrian. "What Is Dark Energy?" Science 236 (2012): 1090–91.

Harvey, David., Massey, Richard., Kitching, Thomas., Taylor, Andy, and Eric Tittley. "The Nongravitational Interactions of Dark Matter in Colliding Galaxy Clusters." Science 347 (2015): 1462–65.

Panek, Richard. The 4 Percent Universe. New York, NY: Houghton, Mifflin Harcourt Publishing Company, 2011.

Randall, Lisa. Dark Matter and the Dinosaurs: The Astounding Interconnectedness of the Universe. New York, NY: Harper Collins Publishers, 2015.

Seymour, Percy. Dark Matters: Unifying Matter, Dark Matter, Dark Energy, and the Universal Grid. Franklin Lakes, NJ: New Page Books, the Career Press Inc., 2008.

Schmiedmayer, Jorg., and Abele, Hartmut. "Probing the Dark Side: Fundamental Questions about Dark Matter and Dark Energy Probed in Laboratory Experiments." Science 349 (2015): 786–87.

Spergel, David N. "The Dark Side of Cosmology: Dark Matter and Dark Energy." Science 347 (2015): 1100–102.

The Xenon Collaboration. "Exclusion of Leptophilic Dark Matter Models Using XENON100 Electronic Recoil Data." Science 349 (2015): 851–54.

Quantum and Particle Physics, Entanglement, Quantum Biology

The ATLAS Collaboration. "A Particle Consistent with the Higgs Boson Observed with the ATLAS Detector at the Large Hadron Collider." Science 338 (2012): 1576–82.
Cho, Adrian. "Furtive Approach Rolls Back the Limits of Quantum Uncertainty." Science 333 (2011): 690–93.
Cho, Adrian. "Hints of Greater Mater-Antimatter Asymmetry Challenge Theorists." Science 228 (2010): 1087.
Cho, Adrian. "Primordial Matter Comes into Focus in Many Tiny Big Bangs." Science 338 (2012): 324–25.
Cho, Adrian. "Who Invented the Higgs Boson?" Science 337 (2012): 1286–89.
The CMS Collaboration. "A New Boson with a Mass of 125 GeV Observed with the CMS Experiment at the Large Hadron Collider." Science 338 (2012): 1569–75.
Franson, James D. "Pairs Rule Quantum Interference." Science 329 (2010): 396-397.
McFadden, JohnJoe., and Al-Khalili, Jim. The Coming of Age of Quantum Biology: Life on the Edge. New York, NY: Broadway Books, Bantam Press, 2014.
Merali, Zeeya. "Quantum Mechanics Braces for the Ultimate Test." Science 331 (2011): 1380–82.
Musser, George. Spooky Action at a Distance: The Phenomenon that Reimagines Space and Time—and What It Means for Black Holes, the Big Bang, and Theories of Everything. New York, NY: Scientific American/Farrar, Strauss and Giroux, 2015.
Negra, Della M., Jenni, Peter., and Virdee Tejinder S. "Journey in the Search for the Higgs Boson: The ATLAS and CMS Experiments at the Large Hadron Collider." Science 338 (2012): 1560–68.
Vedral, Vlatko. Decoding Reality: The Universe as Quantum Information. Oxford: Oxford University Press, 2010.
Werner, Eric. "Meaning of a Quantum Universe." Science 329 (2010): 629–30.

String Theory

Greene, Brian. The Elegant Universe. New York, NY: Vintage Books, 1999.

Yau, Shing-Tung., and Nadis, Steve. The Shape of the Inner Space. New York, NY: Basic Books, 2010.

Life Sciences, Evolution

Dawkins, Richard. Selfish Gene. Oxford: Oxford University Press, 1976.

Dawkins, Richard. The Blind Watchmaker. New York, NY: Norton & Company, Inc., 1986.

Dawkins, Richard. The God Delusion. London: Transworld Publishers, 2006.

Dawkins, Richard. The Magic of Reality. New York, NY: Free Press, 2011.

Good, Matthew C., Zalatan, Jesse G., and Wendell, Lim. "Scaffold Proteins: Hubs for Controlling the Flow of Cellular Information." Science 232 (2011): 680–86.

Hordijk, Wim. "The Living Set." The Scientist (June 1, 2015): 30-35.

Lesburgieres, Edith., Gobbo, Oliviero L, Alaux-Cantin, Stephanie et al. "Early Tagging of Cortical Networks Is Rewired for the Formation of Enduring Associative Memory." Science 331 (2011): 924–28.

Service, Robert F. "Origin-of-Life Puzzle Cracked." Science 347 (2015): 1298.

Teinholz, Nina. The Big Fat Surprise: Why Butter, Meat & Cheese Belong in a Healthy Diet. New York, NY: Simon & Schuster, 2014.

Consciousness, Memory, Psychology

Edelman, Gerald M., and Tononi, Giulio. A Universe of Consciousness: How Matter Becomes Imagination. New York, NY: Basic Books, 2000.

Miller, Greg. "Why Loneliness Is Hazardous to Your Health?" Science 331 (2011): 138–40.

Nagel, Thomas. Mind and Cosmos: Neo-Darwinian Conception of the Nature Is Almost Certainly False. Oxford: Oxford University Press, 2012.

Park, Alice. "Love Hurts." Time, April 11, 2011.

Wimmer, Elliott G., and Shohamy, Daphna. "Preference by Association: How Memory Mechanisms in the Hippocampus Bias Decisions." Science 338 (2012): 270–73.

Religion

Aczel, Amir D. Why Science Does Not Disprove God. New York, NY: Harper Collins Publisher, 2014.

Chopra, Deepak. The Third Jesus. New York, NY: Harmony Books, 2008.

Gervais, Will M., and Norenzayan, Ara. "Analytic Thinking Promotes Religious Disbelief." Science 336 (2012): 493–96.

Keller, Timothy. The Reason for God: Belief in an Age of Skepticism. Dutton, NY: Penguin Group, 2008.

Mailer, Norman., and Lennon, Michael. On God: An Uncommon Conversation. New York, NY: Random House, 2007.

Meacham, Jon. "Like Your Faith Depends on It." Time, March 25, 2013, p. 46.

Miller, Lisa. "Visions of Heaven: A Journey Through the Afterlife." Time, 2010.

Scott, Robert A. Miracle Cures: Saints, Pilgrimage, and the Healing Powers of Belief. Berkeley, CA: University of California Press, 2010.

Sullivan, Amy. "The Rise of the Nones." Time, March 12, 2012, p. 68.

Swinburne, Richard. *Was Jesus God?* Oxford: Oxford University Press, 2008.

Swinburne, Richard. *Is There a God?* Oxford: Oxford University Press, 2010.

Warren, Rick. *The Purpose Driven Life.* Grand Rapids, MI: Zondervan, 2002.

Acknowledgments

I am very grateful to my wife, Eszter, for the many discussions, usually over dinner, about the role of the Almighty in our lives, who never abandons us. She also helped me better understand Jesus's message to mankind and the role of meditation in healing both the body and the mind.

I am also deeply indebted to my children, who gave their honest opinions after reading various versions of the manuscript. Gregory helped me throughout the writing process by eliminating grammatical errors and pointing out flaws in certain concepts in the first version. Zoltan Jr. insisted I expand on the role of zinc in human physiology, which I did gladly, to good effect. Kornel, Dora, and Veronika helped me eliminate "overly fancy" words, which detracted from the text's flow.

I am thankful to Elizabeth Varadi and Peter Barati, who helped to develop the title page.

Thanks also to the Editorial staff at Page Publishing for their expertise and patience in correcting the grammar in the original manuscript and dealing with my several revisions.

Finally, without knowing him personally, I thank Dr. Francis Collins, the director of the National Institutes of Health, for the inspiration he gave me with his book *The Language of God*. His idea that science and faith can be mutually enriching and complementary resonates with me greatly. This is one of the main ideas that some of the characters also discuss in the book. I hope this important message will come through, despite being embedded in various storylines, to give the reader a break from some of the "difficult to digest," unorthodox concepts presented in the book.

About the Author

Zoltan Kiss, PhD, is a biochemist and cell biologist by training, mostly interested in a better understanding of altered signaling inside and outside of cells that underlie chronic diseases, such as diabetes and its complications, as well as cancer. After thirty years of basic research, he left the academia and founded three biotechnological startup companies to develop better and fully tolerated treatments for these ailments.

Millions of highly specific and precise interactions among tens of thousands of different proteins and many other cell constituents residing in many different cell types occur in the human body every second. An untold number of internal and external regulatory signals contribute to this cavalcade of interactions, ensuring the internal harmony of the human body. Although Kiss is a firm believer in the theory of evolution, the immense complexity of the human body convinced him that there must be an organizing principle provided by a higher authority in preparation of the big bang, which gave some broad directions but which allowed and still allows evolution and natural selection to take their courses. An organizing principle also underlies the evolution of the universe as directed by the very fine-tuned laws of nature also established by a higher authority. As it stands now, neither materialist scientists nor theologists can prove how the creation of the universe happened, but they seem to agree that it popped out from nothing, whatever nothing might be, as thoroughly discussed in the book. The Catholic Church also begins to accept the science behind the role of nature's evolution, so there is some common ground between atheist scientists and Christian churches. This gives some hope that Christians and scientists will not treat each other as adversaries. This is what Kiss is hoping for as well. Thus, after writing numerous scientific papers and patent applications, Kiss decided to write a book, *Dualities in Heaven and Earth*, to expand on these ideas.

CPSIA information can be obtained
at www.ICGtesting.com
Printed in the USA
LVHW011932160222
711307LV00002B/63